Advanced Structured Materials

Volume 236

Series Editors

Common engineering materials are reaching their limits in many applications, and new developments are required to meet the increasing demands on engineering materials. The performance of materials can be improved by combining different materials to achieve better properties than with a single constituent, or by shaping the material or constituents into a specific structure. The interaction between material and structure can occur at different length scales, such as the micro, meso, or macro scale, and offers potential applications in very different fields.

This book series addresses the fundamental relationships between materials and their structure on overall properties (e.g., mechanical, thermal, chemical, electrical, or magnetic properties, etc.). Experimental data and procedures are presented, as well as methods for modeling structures and materials using numerical and analytical approaches. In addition, the series shows how these materials engineering and design processes are implemented and how new technologies can be used to optimize materials and processes.

Advanced Structured Materials is indexed in Google Scholar and Scopus.

Visakh P. M.
Editor

Nanomaterials for Defense, Aeronautical and Aerospace Applications

Springer

Editor
Visakh P. M.
Natural Bioactive Materials Laboratory
Department of Bioengineering
Ege University
Bornova, Türkiye

ISSN 1869-8433 ISSN 1869-8441 (electronic)
Advanced Structured Materials
ISBN 978-3-031-92722-5 ISBN 978-3-031-92723-2 (eBook)
https://doi.org/10.1007/978-3-031-92723-2

This Springer imprint is published by the registered company Springer Nature Switzerland AG
The registered company address is: Gewerbestrasse 11, 6330 Cham, Switzerland

Contents

About the Editor

Dr. Visakh P. M. (M.Sc., M.Phil., Ph.D.) is a prolific editor with more than 50 books already published. Currently he is working in the Natural Bioactive Materials Laboratory, Department of Bioengineering, Ege University, Bornova/Izmir, Türkiye. He was working as Associate Professor in TUSUR University, Tomsk, Russia from 2017 till 2022. He did his Post-doctoral Research in Tomsk Polytechnic University, Tomsk, Russia (2014–2017). He obtained his Ph.D., M.Phil., and M.Sc. degrees from School of Chemical Sciences, Mahatma Gandhi University, Kottayam, Kerala, India. He edited 50 books from Scrivener (Wiley), Springer, Royal Society of Chemistry, Elsevier and more than 25 books in press (from Wiley, Springer, Royal Society of Chemistry, and Elsevier). He has been invited as a visiting researcher in Russia (2014–2022), Portugal (2013, 2014), Czech Republic (2012, 2013), Italy (2009, 2012), Argentina (2010) Sweden (2010–2012), Switzerland (2010), Spain (2011, 2012), Slovenia (2011), France (2011), Belgium (2012), and Austria (2012) for his research work. He visited 13 countries; He visited 15 universities in Europe. He published 23 publications, four reviews, and more than 55 book chapters. He has attended and presented more than 28 conferences, he has 3094 citations, and his h-index is 27. He acts as guest editor for four International Journals.

Chapter 1
Nanomaterials for Defense, Aeronautical, and Aerospace Applications: Challenges and Opportunities

Shiv Kumari Panda, P. M. Visakh, and N. B. Shibin

Abstract In this chapter, some discussions on different topics such as carbon nanotube based nanocomposites of aerospace applications, nanomaterials coatings for aerospace and defense applications: current prospects and future trends, spray coating process of MWCNT for aerospace applications, introduction of fiber reinforced polymers-polymers and composites for aeronautical and aerospace applications, metal nanoparticles for aeronautical engineering applications, impact simulation of polycarbonates in aeronautical and aerospace applications, synthetic nanofillers in the aerospace sector and green nanoparticles and nanocomposites for aerospace applications are provided.

Keywords Aerospace · Aeronautical · Carbon nanotube · Polycarbonates · Green nanoparticles and nanocomposites

1.1 Carbon Nanotube Based Nanocomposites of Aerospace Applications

There are three kinds of nanocarbon materials: carbon nanofibers (CNFs), carbon nanotubes (CNTs) as well as carbon nanospheres, based on the excellent strength-to-weight ratio and versatility of CNTs, which can be widely used in composite materials with high performance structures. In 1993, Iijima and Ichihashi reported that CNTs

S. K. Panda (✉)
BioengineeringAssistant professor, Department of Chemistry, Udayanath Autonomous College Of Science & Technology, Adaspur, Cuttack, Odisha, India
e-mail: shivkumaripanda@gmail.com

P. M. Visakh
Natural Bioactive Materials Laboratory, Department of Bioengineering, Ege University, Izmir, Türkiye

N. B. Shibin
Qualitas Material Testing Laboratory, Dubai, UAE

Visakh P. M. (ed.), *Nanomaterials for Defense, Aeronautical and Aerospace Applications*, Advanced Structured Materials 236,
https://doi.org/10.1007/978-3-031-92723-2_1

with only one atom thickness were observed with high resolution transmission electron microscopy (Iijima and Ichihashi 1993). Carbon nanotubes (CNTs) are allotropic isomers of carbon with a similar structure to fullerenes (also known as buckytubes), which are nanomaterials with excellent properties composed of a two-dimensional hexagonal lattice of carbon atoms (Rao et al. 1995; Chen et al. 2021). CNTs have good capabilities, such as high electroconductibility, thermal conductivity and specific area, as well as excellent mechanical capabilities, and have different latent applications in accumulated energy, nano-sensors, biology, emission devices and mechanical systems. Han et al. (2019) found that CNTs could be filled with epoxy adhesives to develop adhesives with high mechanical and electrical conductivity, studies showed that when the amount of CNTs was 0.25 vol%, the tensile strength was improved and the mechanical properties were very good. SWCNTs have the merits of excellent electrical and thermal conductivity, high mechanical strength, high elasticity, and light weight (Kharlamova and Kramberger 2021). Organic functionalization can change the surface properties and enhance the application potential. Common covalent functionalization methods include amidation and esterification of oxidized SWCNTs and addition of carbene, nitrobenzene and free radicals to SWCNTs (Singh et al. 2009). Preparation of MWCNTs/carbon black nanoparticle coated wood and irradiated evaporation rates. Epoxy resin, as a matrix material in polymer composites, has excellent properties, but epoxy resin alone can't satisfy high performance requirements composite materials, so MWCNTs are added to epoxy resin for the development of high-performance nanocomposites (Singh and Gupta 2019; Zhang et al. 2022). In 2001, Li et al. (2001) synthesized small-diameter carbon nanotubes using silicon carbide rods as electrode anodes. Wang and co-workers (2005), showed a simpler and more convenient way to synthesis SWCNTs using arc discharge in NaCl solution.

CVD is divided into two main forms based upon the presence of a catalyst: matrix and floating methods (Kingston and Simard 2003). The matrix method involves attaching the catalyst at a high heat. Carbon gas is used to grow CNTs on the catalyst. Meanwhile, in the floating method, the catalyst is in a gaseous state and enters the reaction chamber with the carbon-containing gas. Here, both undergo decomposition and deposition to produce carbon nanotubes. Non-covalent modifications are achieved through non-covalent bonding interactions. Through π–π bond interactions, van der Waals interactions and adsorption on the carbon nanotube surface (Polyakov et al. 2020). Non-covalent modifications can introduce not only polymers onto carbon nanotubes but also surface-active catalysts to their surfaces, thus changing their surface activity. Direct addition of carbon fillers, carbon nanotubes, and other similar materials to composites can result in clustering because of van der Waals forces. This can result in undesired electrical conductivity of the composites (Lee et al. 2022; Jang et al. 2022). In order to optimize carbon nanotubes' interaction and reactivity with polymers through hydrogen bonding and enable their dispersion in polymer matrix, the sidewalls of the nanotubes must be chemically modified. Nowadays, carbon nanotubes have been widely used in modern aerospace and other fields, and these fields require lightweight materials with very high strength and stiffness.

Sanchez-Romate et al. (2021) characterized the dispersion of nanoparticles in polymers using different ultrasonic conditions to show the variation of electromechanical properties of GNP/MWCNTs reinforced epoxy-based nanocomposites. Huang et al. (2020) studied the effect of T–ZnO on the dispersion quality, rheological, electric and mechanical characteristics of CNTs/epoxy nanocomposites. Cao et al. (2023) prepared MWCNTs-reinforced nickel-based alloy nanocomposites by means of laser powder bed fusion and investigated its properties, and the potential of CNT-reinforced nickel-based nanocomposites for applications such as aerospace and defense, where materials with excellent mechanical properties are required, was demonstrated.

Chen et al. (2022) et al. used the fused deposition modeling (FDM) technique to 3D print special engineering thermoplastic materials such as polyetherimide (PEI). By adding carbon nanotubes (CNTs), they were able to enhance the mechanical characteristics of the composite, improve the interfacial bonding strength, and increase the tensile strength and Young's ratio of the material. Nguyena et al. (2020) modifies nyl66 nanofibers with CNT. SWCNTs and MWCNTs were separated into nylon 66 polymer solution, and then directly mixed with normal Portland cement by electrospinning with the improved collector. Kumar (2023) investigated the influence of the agglomeration of carbon nanotubes on the elasticity of CNT nanocomposites by means of a multiscale FEM model. Carbon nanotubes and interfacial phases (between CNTs and between CNTs and matrix) were introduced into the representative volume element (RVE) of nanocomposites as discrete structures, and the aggregation of CNTs significantly decreased the modulus of elasticity of nanocomposites with short CNTs. Chen and Han (2020) significantly improved the thermal conductivity of the composite by hydroxy-functionalizing carbon nanotubes. With the increase of the volume concentration from 0.05 to 0.8%, the effective heat conductivity was raised from 0.25 to 0.400 W/(m K). Compared with polymer materials, an increase of 0.8% in volume of hydroxy-functionalized multi-walled CNTs results in an increase of over 80% in thermal conductivity. Carbon nanotube-based composites have become important components of aerospace materials for the reason of their above-average weight to strength proportion, low density, fiber-like structure as well as the mechanical features. And the addition of carbon nanotubes (CNTs) has been commonly used by scientific researchers to improve the mechanical properties of composites and has been studied and investigated in different ways with varying degrees of success (Evers et al. 2023). Karthikeyan et al. have discovered that incorporating carbon nanotube reinforcement to ceramics can enhance some characteristics, like its mechanical and thermal qualities, to provide high strength and low weight, which is of great important for aerospace applications (Ramachandran et al. 2021). Carbon nanotube yarns have been developed in composite applications, but another breakthrough in terms of mechanical attributes is still required to fulfill the requirements of big aircraft constructions for structural design. Raunika et al. (2017) prepared polydimethylsiloxane/SWCNTs using 80 °C curing technique for shielding proton radiation and found that its PDMS/SWCNTs exhibits strong proton blocking performance at 105 MeV and is very lightweight. Buzea et al. (2007) used carbon nanotube polymer composite wiring instead of copper wires, which could reduce their weight. The research results show that there is an increasing demand for structural materials in the aircraft and aerospace sectors.

1.2 Nanomaterials Coatings for Aerospace and Defense Applications: Current Prospects and Future Trends

Accordingly, several publications describe a nanomaterial as a substance that is at least one of the three dimensions (Findik 2021; Gaffet 2011; Pal et al. 2011; Buzea et al. 2007; Afolalu et al. 2019; Afolalu et al. 2019; Nagarajan 2008; Abdalkreem 2018; Mrinmoy et al. 2008; Santos et al. 2014) and has a size between 1 and 100 nm. Nanotechnology has garnered a lot of attention throughout time because of the novel and fascinating uses of nanomaterials for the upcoming industrial generation. Nanomaterials are used in many different industries, including agriculture, biomedicine, electronics, energy, pollution control, food engineering, transportation, communications, cosmetics, coatings, materials, and mechanical engineering (Joudeh and Linke 2022; National Research Council 2009; Shaikh 2022). Compared to micro particles, nanoparticles have a larger surface area and more particles per unit mass. At first, tiny particle measurements were connected to gold. There are several synthetic approaches for creating nanoparticles, including top-down, hydro/solvothermal, and bottom-up. One of the most popular and widely used techniques for creating nanostructured materials, etc., Nano capsules; environment and energy, including fuel cells, photovoltaics, and water and air purification filters; military and energy, including biosensors, weapons, and sensory enhancement; electronics, semiconductors, memory storage, photonics, and optoelectronics; and scientific tools, including atomic force fields, such as a microscopic and scanning tunneling microscope (Baig et al. 2021; Samuel et al. 2022; Khan and Hossain 2022). These days, nanomaterials are increasingly crucial to human progress. Green technology that relies only on nanomaterials, for instance, is the only way to lower the danger of global warming and climate change in the first place.

Nanomaterials are classified as dendrimers, which are materials based on metal. When it comes to structural composition and arrangement, the options include carbon-based, inorganic, organic/dendrimers, and composite (Cho et al. 2019; Khan et al. 2019; Dolez 2015). Examples of organic nanomaterials include ferritin, liposomes, dendrimers, and micelles. However, metal-based or metal oxide-based nanomaterials make up inorganic nanoparticles. For instance, boron nitride is used as a coating material in aerospace due to its high thermal conductivity, chemical and thermal stability (hydrophobic), fiber form, continuity of properties on composites, and good insulation properties (constant wide band gap above 5.2 eV on BN nanotube, independent of diameter, chirality, or number of tubular walls) (Vidales-Herrera 2020; Lopez 2020; Umaira et al. 2021; Barshilia 2021; Kiesel 2021; Vasilev 2021; Abbasi 2020; Liao et al. 2020). Every nation's military forces made an effort to improve the agility of their vehicles and soldiers without sacrificing the capacity of their protective gear to withstand severe impacts. Aluminum, magnesium, titanium, and nickel are a few common lightweight metals and alloys. Because of their monolithic character, these lightweight materials were considered unsuitable for use in military and defense applications in the past, when steel was the sole common metal. Liao and his colleagues developed a novel lightweight Tix (AlCrNb)100−x (x = 45−80) alloy for use in transportation vehicles. The resulting alloys can demonstrate the necessary mechanical characteristics, such as an increased compression plasticity of over 30% at room temperature, a high compression yield strength of around 1500 MPa, and a greater compression fracture strength of approximately 1800 MPa (Singh et al. 2018). Lightweight composites called syntactic foams are composed of hollow particles scattered throughout a matrix. These lighter materials are frequently used in applications where weight is an issue, such as pipeline thermal insulation, aircraft construction, and maritime vessels. Metal-reinforced polymer foams have demonstrated good electromagnetic interference shielding at relatively low densities (Venkataramanan et al. 2021).

Metal matrix composites (MMCs) are far more effective than alloys in the defense sector. Israel's Light Improved Ballistic Armor (LIBA), which can withstand up to 20 strikes from projectiles of 14.5 mm caliber, is another example of metal matrix nanocomposites (Thakur et al. 2023). Similarly, the rapid development of military and aerospace communications systems is largely due to the high device precision needed for telecommunication applications. Polymers are thought to be reliable and secure materials for a variety of military uses, from compass boxes to parts for precision weapons. Graphene nanofiller-reinforced polymer composites were used to create the "Star Wars" high-tech armor for its lightweight, high ballistic capacity, superior relative strength, high hydrophobic ability, and remarkable electrical and thermal conductivity. Additional developments in nanomaterial coatings for defense and aerospace include advanced insulating materials, self-healing coatings, chemical agent detection, and safety enhancement. The worldwide aerospace and military coatings market is expected to develop at a compound annual growth rate (CAGR) of 6.62% between 2023 and 2029. Certain materials, including magnesium alloy with an outstanding strength to weight ratio and Ti–Al-based alloy (for jet engine components), are being investigated for potential future uses. The defense is also improving its defensive technologies to make it more creative and efficient in the face of contemporary threats. Current technical developments include space militarization, directed energy weaponry, and supersonic flight (Bahaskaran et al. 2013; Yadav et al. 2024; Naghdi n.d.).

1.3 Spray Coating Process of MWCNT for Aerospace Applications

With the ability to apply coatings on flat surfaces (like glass or plastic) or larger, non-planar surfaces, spray coating has become a popular deposition technique in the industrial sector as an alternative to spin coating. Applying a liquid coating substance, such as polymer solution or photoresist, on a substrate that rotates quickly produces a thin, consistent layer in the case of spin coating. Uniformity, thickness control, fast processing speed, minimal material consumption, and the use of a variety of materials are some benefits of spin coating. Nonetheless, spin coating finds use in energy storage, biological applications (such as biosensors and implantable devices), semiconductor fabrication, and other fields (Ichiki et al. 2004; Girotto et al. 2009; Steirer et al. 2009; Qadir et al. 2024). Thermal spraying has been successfully used in many fields, including corrosion and oxidation resistance, high-temperature protection, wear and erosion, abradable coatings and dimensional repairs, biomedical applications, and electronics, because of the special combination of a broad range of deposition techniques and coating materials (Ann Gan et al. 2015; Agarwal et al. 2022; Britain 1985; Kumar et al. 2024, 2021, 2019, 2020a; Fauchais and Vardelle 2012; Kumar and Kumar 2021; Mauer et al. 2013; Hardwicke and Lau 2013; Wilson 2013; Bedi et al. 2019; Sharma et al. 2020; Crawmer 2013). Technologies that use the same concept to deposit metallic or nonmetallic coatings are included in thermal spraying methods. The feedstock material is melted by a heat source, and the molten particles are then given kinetic energy by a jet. They constantly accumulate the required thickness after impacting the substrate surface and quickly cooling to create a solid splat (Crawmer 2008; Kohler et al. 2008). According to Song et al. (2007), PU–CNT coatings have improved electrical and thermal mechanical qualities. When compared to conventional films, Liu and Guo (2010) discovered that PU-MWCNT films had a higher coefficient of friction and a higher level of wear resistance. Research has examined the impacts of PU-MWCNT and plain PU's glass transition temperatures and found significant variations between the two (Xia and Song 2005). A flux blockage type hypothesis is the most widely accepted idea, even if the processes underlying changes in thermal stability are still mostly understood (Kukovecz et al. 2013; Li et al. 2017).

Stacking and foaming are steps in the process of creating multilayered composite foams. In order to improve the microwave-absorbing capability of PUG composites, Xu et al. (2008) created multilayered thermoplastic polyurethane/graphene (PUG) composites by stacking single-layered PUG foams together. This technique allows for the unrestricted customization of each layer's thickness, cell shape, and graphene concentration; nevertheless, the application is limited by the irregular interface between neighboring layers. On the other hand, a practical and practical method for creating multilayered foams with a continuous interface is to first stack several sample sheets together using hot melt pressing or multilayer coextrusion, and then foam these laminated sheets (Sumey et al. 2019; Li et al. 2006). Flexible membranes coated with nanocomposite materials have a number of intriguing qualities that make them valuable for aerospace applications, including charging mitigation layers and thermal blankets. Because of their remarkable multifunctional qualities, carbon-based nanocomposite films are the subject of much research (Clausi et al. n.d.; Laurenzi et al. 2018). The main goal of producing nanocomposite films for the aerospace industry is to achieve outstanding tailored material qualities while significantly reducing weight and outperforming conventional materials' electrical, thermal, and mechanical characteristics. Nanocomposite films may be used for electromagnetic interference shields, UV sensors, radar absorption, radiation protection, and thermo-optical coatings (Dinită et al. 2024).

1.4 Introduction of Fiber Reinforced Polymers-Polymers and Composites for Aeronautical and Aerospace Applications

Fiber-reinforced polymers (FRPs) represent a class of composite materials that have gained widespread use and recognition across various industries due to their exceptional mechanical properties and versatility. These materials combine the strength and flexibility of fibers, such as carbon, glass, or aramid with the lightweight and moldability of polymers, resulting in a high-performance material with superior mechanical properties compared to traditional materials like metals and plastics in specific applications. Composites have become increasingly important in the main aircraft structural construction due to advances in material research and manufacturing techniques (Rajak et al. 2019). According to a report published in Marketsandmarkets, the aerospace composites market is expected to grow at a compound annual growth rate (CAGR) of 11.7% from 2022 to 2027, from an estimated USD 29.7 billion in 2022 to USD 51.6 billion (Gürgen et al. 2016). Due to its remarkable performance qualities and ability to endure harsh environments, aerospace composites are experiencing rapid market growth. When compared to one another, each of these polymers has benefits and drawbacks regarding cost, application technique, and durability traits. As a result, linked criteria determine the type of polymer to use in the fabrication of the composite material. The reinforcing fibers might be arranged as particles, discontinuous, or continuous. The fibers exhibit high stiffness and strength, homogeneous cross-sectional properties with minimal material property variation between individual fibers, and high aspect ratios for their geometry (extremely high length to diameter ratio). Polymer-based composites more specifically FRP composites are 25% stronger and 25% lighter than steel. They also offer resistance against deterioration and rust. The car parts are safer and more fuel-efficient since they are stronger, lighter, and more moldable than aluminum car parts. Fiber-reinforced polymer (FRP) composites are widely used in the transportation industry for a variety of applications, including automobiles, boats, ships, airplanes, and helicopters.

The two most frequent issues that most metals face are fatigue and corrosion. Aircraft metallic components are prone to strain or stress, particularly when the aircraft is used for many flights. In addition, these parts corrode more quickly than carbon fibers. Carbon fiber is undoubtedly resistant to fatigue and corrosion (Ertekin 2017). Government laws to reduce vehicle pollution and the growing need for fuel-efficient vehicles, along with rising demand for carbon fiber reinforced plastic (CFRP) from the aerospace and automotive industries, are anticipated to boost the market. Ballistic protection uses vast amounts of aramid fiber, which was first commercialized in the 1960s. Commercial fiber products are derived from six main aramid compositions. The degree of fiber characteristics are determined by the molecular weight, content, and manner of fiber creation of the polymers. Aramid fibers have found widespread application in body armor, vehicle armor, military helmets, protective gloves, and fireproof suits for firefighters (Parveez et al. 2022). Since the mid-1990s, the use of fiber-polymer composites in aircraft and helicopters has grown significantly, and the development of more efficient and affordable manufacturing processes has been a key driver of this rise. According to the Federal Aviation Agency (FAA), composite materials have been extensively utilized in the production of aircraft since World War II (WW2). Typically, an aircraft's structure contains up to 70% composite material. Nonetheless, the aerospace sector necessitates substantial production costs. As a result, it is crucial to ensure that the composite manufacturing process is economical for use in aerospace applications (Chen et al. 2023). The aerospace industry has gradually seen a decline in the use of the hand layup technique since the products frequently contain several voids, wrinkles, and other faults that cause them to operate below the intended level. Computer and machine-aided Automated Layup Technology (ALT) was introduced to expedite the layup process and decrease errors in composite constructions while increasing manufacturing efficiency (Pultrusion Market: Trends, Opportunities and Competitive Analysis [2023–2028] 2023). Pultruded FRPs are being used in a variety of industries more and more as businesses become aware of the advantages of pultruded products. Global demand for Pultrusion products is predicted to reach $3.4 billion by 2024 at a compound annual growth rate (CAGR) of 4.0%, according to a report by market research firm Lucintel (Jayakrishna et al. 2023).

Pultruded FRP composites are inexpensive and have great performance characteristics, making them essential for aerospace innovation. Due to its economical production, wide selection of materials, and adaptable design, aerospace injection molding is one such cutting-edge manufacturing technique that is quite useful. In comparison to die casting and machining processes, the setup costs for this mold are comparatively cheap, and it is less costly than traditional die casting molds (Tao et al. 2023). Aerospace plastic injection molding can be done with a broad range of materials, including reinforced composites and standard thermoplastics. Among other FRP composites, carbon fiber reinforced polymers (CFRP) offer the finest mechanical qualities, but the raw ingredients needed to make them are scarce. Although they are rarely utilized in reality, basalt (BFRP) and aramid (AFRP) bars are in the center of the pack in terms of mechanical qualities and production complexity. Among the various FRP varieties, glass fiber reinforced polymer (GFRP) bars are the most widely used because they combine environmental resilience with a relatively cheap cost for structural fibers. GFRP bars are extremely durable and can withstand stresses up to 5.6 times greater than those of structural steel (Siengchin 2023). The aerospace industry's never-ending quest to increase the efficiency of both military and commercial aircraft is what propels the advancement of better high performance structural materials. Among these materials, composites are used extensively in both present-day and future aeronautical components. Because of their remarkable strength, stiffness-to-density ratios, and superior physical qualities, composite materials are particularly appealing for use in aviation and aerospace applications. The aerospace sector in North America is a major user of FRP composites due to its requirement for lightweight materials that have the potential to decrease carbon emissions and increase fuel efficiency (Zhang et al. 2023). The need for FRP composites is anticipated to increase in the upcoming years as they are utilized in a variety of aircraft components, including the wings, fuselage, and tail sections. The aviation and propulsion industries benefit from additive manufacturing (3D printing) in a number of ways, including its suitability for many material groups, its ability to produce complex geometries without the need for tools or dies, and its ability to produce designs that are lighter and more functionally superior. Particularly useful for low-volume production—which is common in the aviation industry—is additive manufacturing. It shortens machining times, gets rid of precious alloy waste, and decreases buy-to-fly ratios. The 2012 Boeing 787 Dreamliner is one of the aircraft models that employs fiberglass. According to them, fiberglass makes up 50% of airplanes (Dhinakaran et al. 2021). Manufacturers are using composite materials in their aerospace goods more and more. Fiberglass has several benefits for airplanes, the first of which is its reduced weight. Because of the fiber-based matrix architecture, this material is stronger than aluminum. Additionally, fiberglass offers a smoother surface.

1.5 Metal Nanoparticles for Aeronautical Engineering Applications

Metal nanoparticles exhibit tremendous potential for a variety of aerospace applications for having distinct physical, chemical and biological properties at nano-scale level. Aeronautical systems function better, last longer and consume less energy by incorporating metal nanoparticles into a variety of parts and materials. Additionally, components for aviation and spacecraft are made robust, lighter and more resilient by incorporating these metal nano materials in components of aircraft wings, fuselages, and engines. Presently used composites, coatings and electrical devices are other areas of applications (Wu et al. n.d.). They have enhanced features such as light weightness, weatherproofing, thermal stability and different sensing and monitoring systems inside an aircraft (Wenwu et al. 2019). Their application in environmental monitoring systems, gas sensors, and structural health monitoring helps to raise the bar for safety and dependability in aircraft operations (Wenwu et al. 2019). They are appealing for a variety of aeronautical applications because of their high surface area, size-dependent characteristics, and adaptability. Because of their special optical qualities, like surface plasmon resonance they are useful in optical and electronic applications in the aerospace industry. They are employed in the creation of cutting-edge optical sensors, displays and communication systems for aircrafts (Vela-Cano et al. 2021). Door handles, armrests, tray tables, and other interior surfaces of airplanes can all be coated with silver nanoparticles contained materials. To improve cabin air purification, silver nanoparticles can be added to air filtration systems and water purification systems (Stephen et al. 2019). Being catalytic is a well-known characteristic for platinum nanoparticles. They can be utilized in aerospace to improving fuel combustion efficiency in jet engines. Platinum catalysts have the potential to lower emissions and enhance propulsion systems overall efficiency (Gawande et al. 2016).

Aircraft surfaces can be made more resistant to corrosion and erosion by using copper nanoparticles. This is especially important for parts like the leading edges of wings that are subjected to adverse weather conditions while in flight. When flying over bodies of water, coatings for an aircraft's exterior can be made with copper nanoparticles to prevent the growth of marine life. This can lower drag and increase fuel economy (Kannaiyan and Sadr 2017). When aluminum nanoparticles are added to aviation fuels, more energy may be released during combustion. Higher thrust and overall better engine performance may result from this. Emissions may be decreased as a result of aluminum nanoparticles' which enhanced combustion efficiency. Less unburned hydrocarbons and other pollutants by complete combustion, makes aviation sector greener and cleaner (Shi et al. 2013). Titanium dioxide's photocatalytic activity might aid in the breakdown of organic pollutants on the surfaces, thus maintaining the cleanliness of the aircraft. Because titanium dioxide has the capacity to block ultraviolet (UV) light, it is frequently found in sunscreens. Titanium dioxide nanoparticle contained coatings can be used on airplane windows or surfaces to reduce the effects of UV radiation and shield passengers and interior items from UV-induced deterioration (Sauvik Raha and Md 2022).

Zinc oxide nanoparticles possess the ability to be used in the production of photo detectors that sense light. They are appropriate for uses in light-sensing applications, flame detection, UV monitoring and other situations requiring sensitivity to ultraviolet (UV) radiation. Temperature sensors can benefit from the addition of zinc oxide nanoparticles, which will increase their sensitivity in identifying temperature changes. Numerous industries, including aerospace, automotive, and electronics, depend on this application (Parveez et al. 2022). In the aerospace business, protecting aircraft from environmental elements that can cause corrosion, like moisture and salt, is essential. Applications for heat management can make use of iron or copper sulfide. These nanoparticles can improve a material's thermal conductivity, which qualifies it for use in engine and electronic components heat dissipation. Aluminum nanocomposites are anticipated to surpass the weight reduction, attained through their use in aerospace applications since these metal matrices are stronger and more rigid than fiber composites and nanocomposites based on polymers. It was discovered that adding ZrO_2 nanoparticles to polyimide decreased the coefficient of friction and lowered the wear rate. Materials used in aircraft constructions must be extremely strong and stiff to retain mechanical integrity due to the high phase temperatures qualities (Nhiem et al. 2023).

To enhance heat dissipation and thermal stability, metal nanoparticles can be added to thermal coatings. Coatings offer defense against inclement weather, radiation, and corrosive substances in severe environmental conditions. Nanotechnology-related coatings greatly improve aluminum alloy corrosion resistance as compared to traditional coating materials (Aruna Prabha et al. 2023). Titanium alloys are among the lightest materials utilized in contemporary airplanes, especially for producing parts for gas turbine engines (Chaisaenrith and Taksakulvith 2021). Enhancing titanium alloys' resistance to heat and wear will enable the substitution of less costly, lighter substitutes for pricey nickel alloys, greatly lowering the weight of the airplane. A multilayer metal-ceramic nanocoating is intended to shield the turbine and compressor blades of contemporary jet engines constructed with titanium alloys (Nan et al. 2023). Metal nanoparticles can be added to coatings or polymers that are used on aircraft parts. This is especially true with conductive metals like copper, gold or silver. These substances contribute to the formation of a conductive shield that either reflects or absorbs electromagnetic energy (Kausar 2017).

1.6 Impact Simulation of Polycarbonates in Aeronautical and Aerospace Applications

Aerospace engineers face the significant challenge of combining lightweight materials with high impact strength to meet strict safety standards. Polycarbonates have emerged as a leading solution due to their ability to withstand drastic atmospheric changes and resist harsh environmental conditions, making them a common choice in aviation and aerospace applications. Polycarbonate (PC) is an engineering plastic widely utilized in the aviation, automotive, and aerospace industries because of its high impact resistance, thermal stability, fatigue resistance, light weight, and strength (Costa et al. Aug. 2021; Ibrahim et al. 2011, Hamouz et al. 2006). For instance, bird strikes frequently cause windshields and canopies to crack or perforate. Polycarbonate has been employed in various high-impact applications across multiple industries. Aromatic polycarbonates derived from bisphenol-A are particularly important for their rigid molecular structure, optical clarity, exceptional impact resistance, toughness, and chemical stability (Kember et al. 2011). Polycarbonate, derived from Bisphenol-A, is known for being partially crystalline, transparent, and possessing a high impact strength of 9.1 Ncm^{-1} and a high glass transition temperature of 149 °C (Rokaya et al. 2018). For various applications polycarbonates have been used in Dentistry for its high impact resistant and mixing with anti-bacterial agents (Safety glass with improved impact resistance and safety strength and method for manufacturing the same 2020). PC is used in bullet-resistant windows, riot shields, and safety goggles due to its high impact resistance. Advanced glazing technologies combine

PC with reinforced films and bulletproof glass to increase strength while reducing thickness, enhancing compact safety doors and windows (Polycarbonate composition with excellent wear resistance and low-temperature impact resistance and preparation method of polycarbonate composition 2019). Additives like dicarboxyl-terminated nylon 1212 improve wear resistance and low-temperature impact resistance, making PC ideal for machine tool windows (Polycarbonate-based composite material and preparation method and application thereof 2019). Incorporating photosensitive and acid-producing groups enhances photoresist properties, while UV absorbers and flexibilizers improve light yellowing resistance and long-term UV stability (Application of polycarbonate as ultraviolet photoresist material 2020; Maamoun et al. 2018).

In the aerospace sector, impact resistance is essential. Aerospace materials must withstand high-impact forces and temperature variations while preserving performance. The durability and efficacy of aircraft and spacecraft are contingent upon the materials selected. Aluminum is mostly used in the making of an aircraft (Batool et al. 2018; Khan et al. 2021). Impact properties are vital in aerospace structures to mitigate damage from various sources like maintenance incidents, collisions, and environmental factors. Composite materials play a pivotal role in aerospace technologies, focusing on improving thermo-mechanical properties for unmanned aerial vehicles, underscoring the significance of lightweight and high-strength materials in aerospace applications (Zai et al. 2020). The outer skin of the fuselage is susceptible to impact from various sources, such as bird strikes, ground handling, emergency landings, hail, and debris. Ensuring the impact resistance of aircraft skins is crucial for maintaining structural integrity and safety (Artero-Guerrero et al. 2013). Runway debris may impact the underside of wing structures, leading to hydrodynamic ram effects in fuel tanks and significant aircraft vulnerability (Yu et al. 2019). The wing leading edge is identified as one of the most vulnerable zones to bird strikes. Impact on the leading edge can result in structural damage and aerodynamic disturbances, affecting the overall performance of the aircraft (Barbagallo et al. 2024). Accurate constitutive modeling of PC is crucial for predicting its material behavior. Constitutive models were studied to understand the strain rate sensitivity and deformation mechanisms and to enhance simulation accuracy (Mirone et al. 2024; Andriß et al. 2022). The development of a true stress–strain curve through experimental analysis allows for better predictions of PC performance under dynamic loading conditions. Discrete Element Method (DEM) can be utilized to predict the mechanical properties and failure behavior of polycarbonate materials under various loading conditions. By incorporating data from different research papers, DEM can simulate the behavior of polycarbonate composites with nanoparticles (Wang et al. 2017), investigate the rate- and temperature-dependent deformation of polycarbonate analyze the mechanical response of polycarbonate under uniaxial tests at different temperatures and stress levels (Krop et al. 2016), and predict the intrinsic mechanical response of polymers filled with particles. Studies have shown that DEM simulations can effectively capture dynamic fracture instability characteristics in brittle solids like PMMA, reproducing crack propagation and branching under dynamic loading conditions (Zhang et al. 2015). In the aerospace industry, particle-based modeling is essential for predicting the impact behavior of materials such as polycarbonate. Due to polycarbonate's exceptional impact resistance, it is widely used in aerospace structures. Polycarbonate (PC) composites are being studied to use it in aircraft fuselage structures. A constitutive model: ZHU-WANG-TANG was employed to observe mechanical properties under dynamic load and the result was validated with experiment test. Researchers found out the mechanical properties by incorporating composites suitable for withstanding high load optimal composition of Polyethylene Terephthalate (r-PET) and Polycarbonate (PC) ND and organopolysiloxane-polycarbonate block copolymers improves heat aging performance and impact resistance. The yield stress of PC increases with strain rates under dynamic loading conditions (Krishna et al. 2011). Blending PC with other materials, such as Acrylonitrile Butadiene Styrene (ABS) and glass fibers, can further enhance impact performance (Su et al. 2019).

1.7 Synthetic Nanofillers in the Aerospace Sector

Aircraft have been protected from electromagnetic radiation using metal materials because of their excellent electrical conductivity and ability to create a shielding effect, similar to the Faraday cage. However, the materials made from metal have some disadvantages owing to their heavy weight and susceptibility to corrosion (Asif et al. 2022). Because of corrosion as a crucial factor, it becomes challenging to account for it due to the convolution of predicting when corrosion will start and the potential damage it can cause to structures (Garg et al. 2023). Such corrosion-resilient materials can be prepared by nanotechnology using nano-scaled composites. Nanotechnology is an innovative tool that involves modifying bulk materials to nano-scale (Yadav et al. 2020). Damage repair in structural aerospace composites has often necessitated significant quantities of healing agents. Various synthetic nanofillers have been integrated into polymer matrices to successfully obtain self-healing polymer-based nanocomposites for applications in the aerospace industry (Guadagno et al. 2019). The physical and chemical properties including the size and distribution of reinforcing nanomaterials of these novel materials are characterized using sophisticated techniques (Meure et al. 2009). These nanomaterials possess application in industries as catalysis, sensors, and for bioremediation and in aerospace. This analysis highlights the enhancements provided by the composite material in comparison to its unreinforced counterpart (Krishnakumar et al. 2020). Several methods have been used for synthesis of nanocomposite materials that depend on the intended application, shape, size, reinforcement type, and available resources. These requirements have been made possible by the recent advancements made by scientists in selective synthesis designing and handling nanoparticles within polymeric matrices (Zhang et al. 2013). A combination of different materials like polymers, metals, and ceramics make the composite and are responsible to strengthen a matrix material. The matrix fulfills the purpose of securely holding the reinforcement material in place, providing a strong structural foundation. These matrix materials include resins such as polyester, epoxy, and vinyl esters (Kumar et al. 2020b). The mechanical characteristics like resistance and rigidity are determined by reinforced materials which transfer the loads to the matrix and also lead to cost reductions (Sharma et al. 2023).

Iron oxides, such as magnetite and maghemite, have also attracted a lot of attention as nanosized fillers. These materials have been widely used in various fields such as pigments, medical applications, catalysis, photocatalysis, and environmental remediation (Mendoza et al. 2023). The tiny magnetic materials such as ferromagnetic and/or ferrimagnetic particles are entrenched within a matrix to form ferrogel nanocomposites. Ferromagnetic nanoparticles can be successfully incorporated in polymeric composites because they can serve as templates and self-assemble. The nanoparticles can be coated with a surfactant and then self-assembled or grown in situ within a template (Li et al. 2021).

The combination of magnetism with other properties in polymer nanocomposites have resulted in many innovative possibilities like preparation of magnetic conductive materials using magnetic nanoparticles with conductive polymer matrix. The dielectric properties and energy storage capabilities of dielectric polymer nanocomposites have been found dependent upon the polarity of polymer matrix (Li et al. 2018). Zhu et al. (2019) modified the poly(vinylidene fluoride-hexafluoropropylene) matrix by incorporation of $BaTiO_3$ nanofibers. A study by Paolillo et al. (2021) shows successful development of self-healing nanocomposites using UV-responsive nanocapsules. These materials could efficiently protect the aerospace structures from radiation damage and also promoted self-healing of mechanically damaged or scratched aerospace structures. An investigation by Hu et al. (2018) studied appealing self-healing properties of epoxy/polyethylene-co-methacrylic nanomaterials for aerospace applications and exhibited effective repair damages. Xie et al. (2022) conducted tests on self-healing epoxy/graphene oxide nanocomposites obtained by free radical copolymerization of the β-cyclodextrin and azobenzene functional graphene oxide. The self-healing effect was examined by conducting experiments involving ultraviolet curing and thermal treatment. The healing in polyurethanes has also been observed with reversible chemical bond formation in the polymer chains resulting in interlinking (Huynh et al. 2019). Self-healing nanocomposites have proven to be valuable in aerospace applications, such as engineered structures, adhesives, coatings, and more (Wang et al. 2016).

Self-healing in epoxy/carbon nanotube nanocomposites has also been explored. Self-healing properties have been observed through the reversible hydrogen bonding between the nanofiller and matrix. The self-healing progressed with the mobility of the polymer chain under the influence of an external stimulus. Both physical and covalent interactions were considered to facilitate the self-healing of thermoplastic or thermosets nanocomposites. Ramesh et al. (2020) successfully created self-healing epoxy/carbon nanotube nanocomposite laminates with carbon nanotube sheets and carbon fibers arranged in alternating layers. Additionally, copper foam sheets were used to create composite laminates. The laminates containing carbon-nanotube-based nanosheets exhibited excellent dispersion of nanoparticles. Coatings for metallic alloys like magnesium, aluminum, and titanium are crucial for both military and civil aircraft. Their parts require effective corrosion protection to ensure their longevity and reliability. Self-healing coatings and adhesives have lately proved to be highly effective at repairing substrate damage (Gu et al. 2016). A study using silver nanoparticles incorporated composite matrix of cellulose nanofibril functionalized with polydopamine was examined in which the matrix showed highly efficient EMI shielding with specific shielding effectiveness value of 93.8 dB. Vertuccio et al. (2018) developed shape-memory nanocomposites by utilizing polylactide-based polyurethane (Fe_3O_4/PLAUs) and achieved a uniform dispersion of Fe_3O_4 nanoparticles with the assistance of oleic acid. Epoxy-based carbon nanotubes were used by Liang et al. (2019) to prepare intelligent coatings for industrial carbon-fiber-reinforced plastics that carry highest value for the gage factor (Rana and Fangueiro 2016; Parveez et al. 2022) among thermosetting resins that possess improved mechanical performance (Vertuccio et al. Vertuccio et al. 2018). Liang et al. (2019) created a nanocomposite by combining reduced graphene oxide foam, Fe_3O_4 decorated carbon nanotubes, and epoxy in a three-dimensional structure.

1.8 Nanoparticles and Nanocomposites for Aerospace Applications

The aerospace industry is characterized by its demanding material requirements, driven by the need for high performance, safety, and efficiency in airplane and spacecraft (Bhat et al. Apr. 2021; Masrafee et al. n.d.). Aerospace materials must endure extreme conditions, including high temperatures, significant stress, and exposure to harsh environments. Traditional materials like aluminum alloys, titanium, and composite polymers have been the mainstays of aerospace engineering due to their durability, lightweight properties, and strength. However, the industry progressively seeks advanced materials that perform better while meeting rigorous environmental and regulatory standards (Olaniyan et al. 2024; Rashid and Hoque 2022). Nanocomposites, incorporating nanoparticles into a matrix material, further extend the benefits of nanotechnology (Haque et al. 2024). Compared to their traditional counterparts, these composites can be tailored to exhibit enhanced electrical, thermal, and mechanical properties. In aerospace engineering, nanocomposites can contribute to lighter, stronger, and more efficient materials, potentially transforming the design and performance of aircraft and spacecraft (Rashid et al. 2024).

According to research, thermal and mechanical designs that rely only on polymers are considered less trustworthy for aeronautical applications (Liu et al. 2014; Rai et al. 2021). As a result, polymeric nanocomposites have gained significant attention in industrial applications because of their exceptional characteristics (Muhammed et al. 2020). Examples of natural polymers include polyesters, cellulose, starch, chitin, and natural rubber (Koli et al. 2015). In addition, synthetic green polymers such as polycaprolactone, polyurethane, poly(vinyl alcohol), poly(lactic acid) (PLA), and poly(ethylene glycol) (PEG) have been synthesized (Adil et al. 2024). Aluminum-based composites have improved physical and mechanical qualities compared to the base alloy, making them suitable for aerospace uses (Prater 2014; Heidari et al. 2024).

The aluminum-based metal matrix composite (MMC) reinforced with 30% silicon carbide (SiC) has a comparable density to 2219 aluminum-based alloys, but exhibits a specific modulus that is 70% higher, an elastic modulus that is 60% higher, and a tensile strength that is 15% higher (Karchin et al. 2011). The materials used to make aviation industry components must withstand high durability, superior heat resistance, and lightweight, and electromagnetic interference (EMI) resistance. Initially, it was accomplished with composites (CMCs, MMCs, PMCs), which are materials composed of different materials according to anticipated results in application areas, such as an improvement in the material's tribology behavior, an increase in temperature resistance and an intensification in the strength of the base material, alongside other advantages. In some applications within the aerospace industry, achieving high performance is crucial in addition to having a lightweight design (Heidari et al. 2024).

These applications include aircraft interiors, cockpit components, equipment enclosures, coatings, solar array substrates, nozzles, shrouds, housings, space-durable mirrors, heat shrinkage tubing, and crew gear. Composite materials provide chemical stability and fire resistance in addition to the benefit of cheap running costs resulting from their lightweight nature. Three significant disadvantages restrict the use of composites in most aeronautical constructions. Higher electrical resistance imposes limitations on many applications, such as lightning strike protection, antennas, circuits, and electromagnetic shielding. Clay, alumina, CNT, and nanodiamond are essential components used in the production of aircraft coating. Spray painting is a method specifically used for painting airplanes. The use of fillers enhances the viscosity, leading to challenges in spray painting and insufficient wetting of nanoparticles and additives. Alternatively, the electrospinning approach involves propelling a stream of solution via a capillary tube onto a surface (Karchin et al. 2011). Incorporating nanofibers and carbon nanotubes (CNT) into the solution has the potential to create coatings that exhibit excellent area-to-volume ratio, strength, and flexibility.

References

Abbasi S (2020) Functionalized nanomaterials for the aerospace, vehicle, and sports industries. In: Micro and nano technologies, pp 795–825

Abdalkreem TM (2018) Master thesis, Sudan University of Science and Technology

Adil MM et al (2024) Effect of jute stick charcoal powder on characterization of aluminum metal matrix composite. Results Eng 22:101990. https://doi.org/10.1016/j.rineng.2024.101990

Afolalu SA, Soetan SB, Ongbali SO, Abioye AA, Oni AS, Conf IOP (2019) Ser Mater Sci Eng 640:012065. https://doi.org/10.1088/1757-899X/640/1/012065

Agarwal H, Breining WM, Sánchez-Velázquez G, Lynn DM (2022) Reactive multilayers and coatings fabricated by spray assembly: influence of polymer structure and process parameters on multiscale structure and interfacial properties. Chem Mater 34:1245–1258

Andriß C, Kenf A, Donhauser T, Schmeer S (2022) Characterization and modeling of continuous carbon fiber-reinforced polycarbonate under multiaxial loads. Compos Part B Eng 235:109740. https://doi.org/10.1016/j.compositesb.2022.109740

Ann Gan J, Berndt CC (2015) 23—Thermal spray forming of titanium and its alloys. In: Qian M, Froes FH (eds) Titanium powder metallurgy. Butterworth-Heinemann, Boston, pp 425–446

Application of polycarbonate as ultraviolet photoresist material (2020) | Wu Guangpeng. Accessed: Apr. 24, 2024. [Online]. Available: https://typeset.io/papers/application-of-polycarbonate-as-ultraviolet-photoresist-47qxkphzac

Artero-Guerrero JA, Pernas-Sánchez J, Varas D, López-Puente J (2013) Numerical analysis of CFRP fluid-filled tubes subjected to high-velocity impact. Compos Struct 96. https://doi.org/10.1016/j.compstruct.2012.09.020

Aruna Prabha K, Naveen Reddy Ch, Srinivasa Rao L, Sreekesh Reddy S, Yeshwanth Kumar G (2023) Structural and thermal analysis on high-pressure steam turbine blade to determine the optimum material for its manufacturing. Mater Today Proc 92, Part 2:1193–1199. ISSN 2214-7853. https://doi.org/10.1016/j.matpr.2023.05.309

Asif AH, Mahajan MS, Sreeharsha N et al (2022) Enhancement of anticorrosive performance of cardanol based polyurethane coatings by incorporating magnetic hydroxyapatite nanoparticles. Materials (Basel) 15:2308. https://doi.org/10.3390/ma15062308

Bahaskaran R, Palaniswamy N, Rengaswamy NS (2013) A review of differing approaches used to estimate the cost of corrosion, anti-corrosion method and materials. J Appl Sci Res 52:29–41

Baig N, Kammakakam I, Falath W (2021) Nanomaterials: a review of synthesis methods, properties, recent progress, and challenges. Mater Adv 2:1821–1871

Barbagallo R, Mirone G, Landi L, Bua G (2024) Tensile behavior of polycarbonate: key aspects for accurate constitutive modelling and simulation. Appl Eng Sci 18:100183. https://doi.org/10.1016/j.apples.2024.100183

Barshilia HC (2021) Surface modification technologies for aerospace and engineering applications: current trends, challenges and future prospects. Trans Indian Natl Acad Eng 6:173–188. https://doi.org/10.1007/s41403-021-00208-z

Batool SA, Ahmad A, Wadood A, Mateen A, Hussain SW (2018) Development of lightweight aluminum-titanium alloys for aerospace applications. Key Eng Mater 778. https://doi.org/10.4028/www.scientific.net/kem.778.22

Bedi TS, Kumar S, Kumar R (2019) Corrosion performance of hydroxyapaite and hydroxyapaite/titania bond coating for biomedical applications. Mater Res Express 7:015402. https://doi.org/10.1088/2053-1591/ab5cc5

Bhat A et al (2021) Review on nanocomposites based on aerospace applications. Nanotechnol Rev 10(1):237–253. https://doi.org/10.1515/ntrev-2021-0018

Bhat A, Budholiya S, Aravind Raj S, Sultan MT, Hui D, Md Shah AU, Safri SN (2021) Review on nanocomposites based on aerospace applications. Nanotechnol Rev 10(1):237–253. https://doi.org/10.1515/ntrev-2021-0018

Britain G (1985) Wear resistant surfaces in engineering—a guide to their production, properties and selection. London Her Majesty's Stationery Office. Department of Trade and Industry, Great Britain, UK

Buzea C et al (2007) Biointerphases 2, MR17. https://doi.org/10.1116/1.2815690

Cao L, Bradford-Vialva RL, Eckerle RM, Herman F, Jones C, Klosterman DA (2023) Development of carbon nanotube-reinforced nickel-based nanocomposites using laser powder bed fusion. Adv Eng Mater 25(4)

Chaisaenrith P, Taksakulvith P, Pavasupree S (2021) Effect of nano titanium dioxide in intumescent fireproof coating on thermal performance and char morphology. Mater Today Proc 47, Part 12:3462–3467. ISSN 2214-7853. https://doi.org/10.1016/j.matpr.2021.03.461

Chen J, Han J (2020) Effect of hydroxylated carbon nanotubes on the thermal and electrical properties of derived epoxy composite materials. Results Phys 18:103246

Chen Q, Zhang Y-Y, Huang P, Li Y-Q, Fu S-Y (2022) Improved bond strength, reduced porosity and enhanced mechanical properties of 3D-printed polyetherimide composites by carbon nanotubes. Compos Commun 30:101083

Chen J, Wei S, Xie H (2021) A brief introduction of carbon nanotubes: history, synthesis, and properties. J Phys Conf Ser 1948

Chen Y, Zhang J, Li Z, Zhang H, Chen J, Yang W et al (2023) Manufacturing technology of lightweight fiber-reinforced composite structures in aerospace: current situation and toward intellectualization. Aerospace 10. https://doi.org/10.3390/aerospace10030206
Cho G, Park Y, Hong YK, Ha DH (2019) Nano Converg. https://doi.org/10.1186/s40580-019-0187-0
Clausi M, Santonicola MG, Schirone L, Laurenzi S (n.d.) Analysis of ultraviolet exposure effects on the surface properties of epoxy/graphene nanocomposite films on Mylar substrate
Costa P et al (2021) Polycarbonate based multifunctional self-sensing 2D and 3D printed structures for aeronautic applications. Smart Mater Struct 30:085032. https://doi.org/10.1088/1361-665X/ac0cbe
Crawmer DE (2008) Introduction to coatings, equipment, and theory, handbook thermal spray technology, pp 43–46
Crawmer DE (2013) Thermal spray processes, handbook of thermal spray technology. ASM International, pp 54–76
Dhinakaran V, Swapna Sai M, Varsha Shree M (2016) Impact of nanomaterials in aerospace engineering, book nanomaterials and nanocomposites, 1st ed. First Published 2021 Imprint CRC Press Pages 13 eBook ISBN9781003160946, Conference 2016, LNCS, vol 9999. Springer, Heidelberg (2016), pp 1–13
Diniă A, Ripeanu RG, Ilincă CN, Cursaru D, Matei D, Naim RI, Tănase M, Portoacă AI (2024) Advancements in fiber-reinforced polymer composites: a comprehensive analysis. Polymers (Basel) 16:1–49
Dolez P (ed) (2015) Nanoengineering: global approaches to health and safety issues. Elsevier, Amsterdam, Netherland
Ertekin M (2017) Aramid fibers. Elsevier Ltd. https://doi.org/10.1016/B978-0-08-101871-2.00007-2
Evers CE, Vondrasek B, Jolowsky CN, Park JG, Czabaj MW, Ku BE, Thagard KR, Odegard GM, Liang Z (2023) Scalable high tensile modulus composite laminates using continuous carbon nanotube yarns for aerospace applications. ACS Appl Nano Mater 6(13):11260–11268
Fauchais P, Vardelle A (2012) Thermal sprayed coatings used against corrosion and corrosive wear. In: Salimi Jazi H (ed) Advanced plasma spray applications. InTech, pp 3–39. https://doi.org/10.5772/34448
Findik F (2021) Period Eng Nat Sci 62:75. https://doi.org/10.21533/pen.v9i3.1837
Gaffet E (2011) (preprint). http://arxiv.org/abs/1106.2206, https://doi.org/10.1016/j.crhy.2011.06.002
Garg R, Garg R, Sharma R et al (2023) Introduction to nanotechnology and nanomaterials. Nanotechnol Sustain Agric Food Environ 1–15. https://doi.org/10.1201/9781003397861-1
Gawande MB, Goswami A, Felpin FX, Asefa T, Huang X, Silva R, Zou X, Zboril R, Varma RS (2016) Cu and Cu-based nanoparticles: synthesis and applications in catalysis. Chem Rev 116(6):3722–3811. https://doi.org/10.1021/acs.chemrev.5b00482
Girotto C, Rand BP, Genoe J, Heremans P (2009) Exploring spray coating as a deposition technique for the fabrication of solution-processed solar cells. Sol Energy Mater Sol Cell 93(4):454–458
Gu S-Y, Jin S-P, Gao X-F, Mu J (2016) Polylactide-based polyurethane shape memory nanocomposites (Fe_3O_4/PLAUs) with fast magnetic responsiveness. Smart Mater Struct 25:055036. https://doi.org/10.1088/0964-1726/25/5/055036
Guadagno L, Vertuccio L, Naddeo C et al (2019) Self-healing epoxy nanocomposites via reversible hydrogen bonding. Compos Part B Eng 157:1–13
Gürgen S, Kuşhan MC, Diltemiz SF (2016) Fatigue failure in aircraft structural components. Handb Mater Fail Anal Case Stud Aerosp Automot Ind 261–277. https://doi.org/10.1016/B978-0-12-800950-5.00013-2
Hamouz OCSA, Sweileh BA, Al-Salah HA (2006) Synthesis and characterization of polycarbonates by melt-phase interchange reactions with alkylene and arylene diphenyl dicarbonates. J Appl Polym Sci 102(4). https://doi.org/10.1002/app.23996

Han SS, Meng QS, Arabya S, Liu TQ, Demiral M (2019) Mechanical and electrical properties of graphene and carbon nanotube reinforced epoxy adhesives: experimental and numerical analysis. Compos Part A-Appl S 120:116–126

Haque M, Islam SMM, Chowdhury P, Labib KMRU, Bin Rashid A (2024) A comparative study on the influence of MWCNT, GO, and $Al(OH)_3$ gel matrix modification on hierarchical structured composite reinforced with needle-punched jute fiber and glass fiber. Results Eng 21:101671. https://doi.org/10.1016/j.rineng.2023.101671

Hardwicke CU, Lau YC (2013) Advances in thermal spray coatings for gas turbines and energy generation: a review. J Therm Spray Technol 22(5):564–576. https://doi.org/10.1007/s11666-013-9904-0

Heidari M, Thangavel S, Ghafri KA, Kumar A (2024) Future trends and emerging research in nanofluids for aerospace applications. In: Nanofluids technology for thermal sciences and engineering. CRC Press

Hu Z, Zhang D, Lu F et al (2018) Multistimuli-responsive intrinsic self-healing epoxy resin constructed by host-guest interactions. Macromolecules 51:5294–5303. https://doi.org/10.1021/acs.macromol.8b01124

Huang X, Zeng L, Li R, Xi Z, Li Y (2020) Manipulating conductive network formation via 3D T–ZnO: a facile approach for a CNT-reinforced nanocomposite. Nanotechnol Rev 9(1):534–542

Huynh T-P, Haick H (2019) Chapter 10-Self-healing materials for analyte sensing. In: Zenkina OV (ed) Nanomaterials design for sensing applications. Elsevier, pp 325–339

Ibrahim SS, Al Jaafari AA, Ayesh AS (2011) Physical characterizations of three phase polycarbonate nanocomposites. Accessed 15 July 2024 [Online]. Available: https://doi.org/10.1177/8756087911410592

Ichiki M, Zhang L, Yang Z, Ikehara T, Maeda R (2004) Thin film formation on nonplanar surface with use of spray coating fabrication. Microsyst Technol 10(5):360–363

Iijima S, Ichihashi T (1993) Single-shell carbon nanotubes of 1-nm diameter. Nature 363(6430):603–605

Jang J, Kang YH, Lim H-W, Han M (2022) Highly conductive and mechanically robust multiwalled carbon nanotube-polyamide nanocomposites via noncovalent functionalization. J Mater Sci 57(6):4197–4209

Jayakrishna M, Vijay M, Khan B (2023) An overview of extensive analysis of 3D printing applications in the manufacturing sector. J Eng 2023:1–23. https://doi.org/10.1155/2023/7465737

Joudeh N, Linke D (2022) J Nanobiotechnol 20:262. https://doi.org/10.1186/s12951-022-01477-8

Kannaiyan K, Sadr R (2017) The effects of alumina nanoparticles as fuel additives on the spray characteristics of gas-to-liquid jet fuels. Exp Thermal Fluid Sci 87:93–103. ISSN 0894-1777. https://doi.org/10.1016/j.expthermflusci.2017.04.027

Karchin A, Simonovsky FI, Ratner BD, Sanders JE (2011) Melt electrospinning of biodegradable polyurethane scaffolds. Acta Biomater 7(9):3277–3284. https://doi.org/10.1016/j.actbio.2011.05.017

Kausar A (2017) A review of filled and pristine polycarbonate blends and their applications. J Plast Film Amp Sheeting 34(1). https://doi.org/10.1177/8756087917691088

Kember MR, Buchard A, Williams CK (2011) Catalysts for CO_2/epoxide copolymerization. Chem Commun 47(1). https://doi.org/10.1039/c0cc02207a

Khan I, Saeed K, Khan I (2019) Arab J Chem 12:908, 31. https://doi.org/10.1016/j.arabjc.2017.05.011

Khan ZI, Mohamad Z, Rahmat AR, Habib U (2021) Synthesis and characterization of composite materials with enhanced thermo-mechanical properties for unmanned aerial vehicles (Uavs) and aerospace technologies. Pertanika J Sci Technol 29(3). https://doi.org/10.47836/pjst.29.3.15

Khan S, Hossain MK (2022) In nanoparticle-based polymer composites. Woodhead, London, p 15

Kharlamova MV, Kramberger C (2021) Applications of filled single-walled carbon nanotubes: progress, challenges, and perspectives. Nanomaterials-Basel 11(11)

Kiesel L (2021) Advanced air plasma spray TBCs for aerospace and industrial components [Online] www.aerospace.honeywell.com. Available at: https://aerospace.honeywell.com/us/en/learn/about-us/blogs/advanced-air-plasma-spray-tbcs

Kingston CT, Simard B (2003) Fabrication of carbon nanotubes. Anal Lett 36(15):3119–3145

Kohler AR, Som C, Helland A, Gottschalk F (2008) Studying the potential release of carbon nanotubes throughout the application life cycle. J Clean Prod 16:927–937

Koli DK, Agnihotri G, Purohit R (2015) Advanced aluminium matrix composites: the critical need of automotive and aerospace engineering fields. Mater Today Proc 2(4–5):3032–3041. https://doi.org/10.1016/j.matpr.2015.07.290

Krishna TP, Srinivasababu NSN, Babu NR, Adhikari SAS (2011) Assessing the impact performance of PC/ABS blends on addition of PC with Glass. Indian J Appl Res 3(2):121–122. https://doi.org/10.15373/2249555X/FEB2013/41

Krishnakumar B, Prasanna Sanka RVS, Binder WH et al (2020) Catalyst free self-healable vitrimer/graphene oxide nanocomposites. Compos Part B Eng 184:107647. https://doi.org/10.1016/j.compositesb.2019.107647

Krop S, Meijer HEH, van Breemen LCA (2016) Global and local large-deformation response of sub-micron, soft- and hard-particle filled polycarbonate. J Mech Phys Solids 87:51–64. https://doi.org/10.1016/j.jmps.2015.11.005

Kukovecz A et al (2013) Multi-walled carbon nanotubes. Springer Handbook of Nanomaterials, pp 1–30. https://doi.org/10.1007/978-3-642-20595-8_5

Kumar S (2023) Investigating effect of CNT agglomeration in CNT/polymer nanocomposites using multiscale finite element method. Mech Mater Mech Mater 183:104706

Kumar S, Kumar R (2021) Influence of processing conditions on the properties of thermal sprayed coating: a review. Surf Eng 37(11):1339–1372. https://doi.org/10.1080/02670844.2021.1967024

Kumar S, Kumar M, Handa A (2019) High temperature oxidation and erosion-corrosion behaviour of wire arc sprayed Ni–Cr coating on boiler steel. Mater Res Express 6:125533. https://doi.org/10.1088/2053-1591/ab5fae

Kumar S, Kumar M, Jindal N (2020a) Overview of cold spray coatings applications and comparisons: a critical review. World J Eng 17(1):27–51. https://doi.org/10.1108/WJE-01-2019-0021

Kumar M, Bansal M, Garg R (2020b) An overview of beneficiary aspects of zinc oxide nanoparticles on performance of cement composites. Mater Today Proc 43:892–898. https://doi.org/10.1016/j.matpr.2020.07.215

Kumar S, Handa A, Chawla V, Grover NK, Kumar R (2021) Performance of thermal-sprayed coatings to combat hot corrosion of coal-fired boiler tube and effects of process parameters and post coating heat treatment on coating performance: a review. Surf Eng 37(7):833–860. https://doi.org/10.1080/02670844.2021.1924506

Kumar K, Kumar S, Gill HS (2024) Role of carbide-based thermal-sprayed coatings to prevent failure for boiler steels: a review. J Failure Anal Prevent. https://doi.org/10.1007/s11668-024-01974-y

Laurenzi S, Zaccardi F, Semeraro M, Santonicola MG (2018) Nanocomposite layers for spacecraft protection: from multiscale numerical model to experimental data. In: 18th European conference on composite materials. Athens, Greece (2018)

Lee H, Kim MN, Jang HG, Jang J-U, Kim J, Kim SY (2022) Phenyl glycidyl ether-based non-covalent functionalization of nano-carbon fillers for improving conductive properties of polymer composites. Compos Commun 33:101237

Li Y, Xie S, Zhou W, Tang D, Liu Z, Zou X, Wang G (2001) Small diameter carbon nanotubes synthesized in an arc-discharge. Carbon 39(9):1429–1431

Li J, Hu L, Wang L, Zhou Y, Grüner G, Marks TJ (2006) Organic light-emitting diodes having carbon nanotube anodes. Nano Lett 6(11):2472–2477

Li Y, Shen B, Yi D, Zhang L, Zhai W, Wei X, Zheng W (2017) The influence of gradient and sandwich configurations on the electromagnetic interference shielding performance of multilayered thermoplastic polyurethane/graphene composite foams. Compos Sci Technol 138:209–216. https://doi.org/10.1016/j.compscitech.2016.12.002

Li Z, Liu F, Yang G et al (2018) Enhanced energy storage performance of ferroelectric polymer nanocomposites at relatively low electric fields induced by surface modified $BaTiO_3$ nanofibers. Compos Sci Technol 164:214–221. https://doi.org/10.1016/j.compscitech.2018.05.052

Li H, Wang L, Zhu Y et al (2021) Tailoring the polarity of polymer shell on $BaTiO_3$ nanoparticle surface for improved energy storage performance of dielectric polymer nanocomposites. Chinese Chem Lett 32:2229–2232. https://doi.org/10.1016/j.cclet.2020.12.032

Liang C, Song P, Ma A et al (2019) Highly oriented three-dimensional structures of Fe_3O_4 decorated CNTs/reduced graphene oxide foam/epoxy nanocomposites against electromagnetic pollution. Compos Sci Technol 181:107683. https://doi.org/10.1016/j.compscitech.2019.107683

Liao YC, Li TH, Tsai PH, Jang JSC, Hsieh KC, Chen CY, Tsao IY (2020) Designing novel lightweight, high-strength and high-plasticity Tix (AlCrNb) 100-x medium-entropy alloys. Intermetallics 117, Article 106673

Liu Y, Du H, Liu L, Leng J (2014) Shape memory polymers and their composites in aerospace applications: a review. Smart Mater Struct 23(2):023001. https://doi.org/10.1088/0964-1726/23/2/023001

Liu T, Guo S (2010) Properties of polyurethane/carbon nanotube anocomposites Polymer nanotube nanocomposites: synthesis, properties and applications. Scrivener, Salem, MA, pp 141–76

Lopez NR (2020) Resistant Project: How can nanotechnology reduce environmental impacts of aircrafts? [Online] www.openaccessgovernment.org Available at: https://www.openaccessgovernment.org/nanotechnology-reduce-environmental-impacts-of-aircrafts/85311

Maamoun A, Xue Y, Elbestawi MA, Veldhuis SC (2018) Effect of selective laser melting process parameters on the quality of Al alloy parts: powder characterization, density, surface roughness, and dimensional accuracy. Materials 11(12). https://doi.org/10.3390/ma11122343

Masrafee MR, Rashid AB, Hoque E (n.d.) Innovations in 3D printing-assisted biopolymer composites for aerospace applications

Mauer G, Jarligo MO, Mack DE, Vaßen R (2013) Plasma-sprayed thermal barrier coatings: new materials, processing issues, and solutions. J Therm Spray Technol 22(5):646–658

Mendoza C, Nirwan VP, Fahmi A (2023) Nanofabrication of hybrid nanomaterials: Macroscopically aligned nanoparticles pattern via directed self-assembly of block copolymers. J Appl Polym Sci 140. https://doi.org/10.1002/app.53409

Meure S, Wu DY, Furman S (2009) Polyethylene-co-methacrylic acid healing agents for mendable epoxy resins. Acta Mater 57:4312–4320. https://doi.org/10.1016/j.actamat.2009.05.032

Mirone G, Barbagallo R, Bua G (2024) Static and dynamic tensile behaviour of polycarbonate connected to the deforming specimen shape. IOP Conf Ser Mater Sci Eng 1306(1):012006. https://doi.org/10.1088/1757-899X/1306/1/012006

Mrinmoy D, Ghosh PS, Rotello VM (2008) Adv Mater 20:4225. https://doi.org/10.1002/adma.200703183

Muhammed NS, Haq MdB, Al-Shehri D, Rahaman MM, Keshavarz A, Hossain SMZ (2020) Comparative study of green and synthetic polymers for enhanced oil recovery. Polymers 12(10):2429. https://doi.org/10.3390/polym12102429

Nagarajan R (2008) Ann Cardiothorac Surg 10:341

Naghdi S (n.d.) Thin films-definition, deposition techniques, and applications. https://doi.org/10.13140/RG.2.2.35029.96480/1

Nan Z, Wei W, Lin Z et al (2023) Flexible nanocomposite conductors for electromagnetic interference shielding. Nano-Micro Lett 15:172. https://doi.org/10.1007/s40820-023-01122-5

National Research Council (2009) Advancing risk assessment

Nguyen TNM, Yoo D-Y, Kim JJ (2020) Cementitious material reinforced by carbon nanotube-Nylon 66 hybrid nanofibers: mechanical strength and microstructure analysis. Mater Today Commun 23:100845

Nhiem LT, Oanh DTY, Hieu NH (2023) Nanocoating toward anti-corrosion: a review. VJCH 61:284–293. https://doi.org/10.1002/vjch.202300025

Olaniyan OT, Adeyomoye O, Adetunji CO, Oloruntoba A (2024) Biocomposites for aerospace engineering applications. In: Advances in biocomposites and their applications. Elsevier, pp 285–298. https://doi.org/10.1016/B978-0-443-19074-2.00010-1

Pal SL, Jana U, Manna PK, Mohanta GP, Manavalan R (2011) JAPS 228:234

Paolillo S, Bose RK, Santana MH, Grande AM (2021) Intrinsic self-healing epoxies in polymer matrix composites (PMCs) for aerospace applications. Polymers (Basel) 13. https://doi.org/10.3390/polym13020201

Parveez B, Kittur MI, Badruddin IA, Kamangar S, Hussien M, Umarfarooq MA (2022) Scientific advancements in composite materials for aircraft applications: a review. Polymers (Basel) 14(22):5007. https://doi.org/10.3390/polym14225007

Polyakov MS, Ivanova VN, Basova TV, Saraev AA, Köksoy B, Şenocak A, Demirbaş E, Durmuş M (2020) 3D, covalent and noncovalent hybrid materials based on 3-phenylcoumarin derivatives and single walled carbon nanotubes as gas sensing layers. Appl Surf Sci 504:144276

Polycarbonate composition with excellent wear resistance and low-temperature impact resistance and preparation method of polycarbonate composition (2019) | Liu Weiwei. Accessed 24 Apr 2024 [Online]

Polycarbonate-based composite material and preparation method and application thereof (2019) Accessed 24 Apr 2024 [Online]

Prater T (2014) Friction stir welding of metal matrix composites for use in aerospace structures. Acta Astronaut 93:366–373. https://doi.org/10.1016/j.actaastro.2013.07.023

Pultrusion Market: Trends, Opportunities and Competitive Analysis [2023–2028] 2023

Qadir D, Sharif R, Nasir R et al (2024) A review on coatings through thermal spraying. Chem Pap 78:71–91. https://doi.org/10.1007/s11696-023-03089-4

Raha S, Ahmaruzzaman M (2022) ZnO nanostructured materials and their potential applications: progress, challenges and perspectives. Nanoscale Adv 4:1868–1925. https://doi.org/10.1039/D1NA00880C

Rai P, Mehrotra S, Priya S, Gnansounou E, Sharma SK (2021) Recent advances in the sustainable design and applications of biodegradable polymers. Biores Technol 325:124739. https://doi.org/10.1016/j.biortech.2021.124739

Rajak DK, Pagar DD, Menezes PL, Linul E (2019) Fiber-reinforced polymer composite: manufacturing, properties, and applications. Polymers (Basel) 11

Ramachandran K, Boopalan V, Bear JC, Subramani R (2021) Multi-walled carbon nanotubes (MWCNTs)-reinforced ceramic nanocomposites for aerospace applications: a review. J Mater Sci 57(6):3923–3953

Ramesh M, Arun Ramnath R, Khan A et al (2020) 1-Electrically conductive self-healing materials: preparation, properties, and applications. In: Khan A, Jawaid M, Raveendran SN, Ahmed Asiri AMBT-S-HCM (eds) Series in composites science and engineering. Woodhead Publishing, pp 1–13

Rana S, Fangueiro R (2016) Advanced composites in aerospace engineering. In: Advanced composite materials for aerospace engineering. Elsevier, pp 1–15. https://doi.org/10.1016/B978-0-08-100037-3.00001-8

Rao CNR, Seshadri R, Govindaraj A, Sen R (1995) Fullerenes, nanotubes, onions and related carbon structures. Mat Sci Eng R 15(6):209–262

Rashid AB, Hoque ME (2022) Polymer nanocomposites for defense applications. In: Advanced polymer nanocomposites. Elsevier, pp 373–414. https://doi.org/10.1016/B978-0-12-824492-0.00015-5

Rashid AB, Haque M, Islam SMM, Uddin Labib KMR (2024) Nanotechnology-enhanced fiber-reinforced polymer composites: Recent advancements on processing techniques and applications. Heliyon 10(2):e24692. https://doi.org/10.1016/j.heliyon.2024.e24692

Raunika A, Raj SA, Jayakrishna K, Sultan MTH (2017) Carbon nanotube: a review on its mechanical properties and application in aerospace industry. IOP Conf Ser Mater Sci Eng 270:012027

Rokaya D, Srimaneepong V, Sapkota J, Qin J, Siraleartmukul K, Siriwongrungson V (2018) Polymeric materials and films in dentistry: an overview. J Adv Res 14. https://doi.org/10.1016/j.jare.2018.05.001

Safety glass with improved impact resistance and safety strength and method for manufacturing the same 2020. Accessed 24 Apr 2024 [Online]. Available: https://typeset.io/papers/safety-glass-with-improved-impact-resistance-and-safety-4cp2p3gsum

Samuel MS, Ravikumar M, Ashwini JJ, Selvarajan E, Patel H, Chander PS, Soundarya J, Vuppala S, Balaji R, Chandrasekar N (2022) A review on green synthesis of nanoparticles and their diverse biomedical and environmental applications. Catalysts 12:459

Sanchez-Romate XF, Jimenez-Suarez A, Campo M, Urena A, Prolongo SG (2021) Electrical properties and strain sensing mechanisms in hybrid graphene nanoplatelet/carbon nanotube nanocomposites. Sensors 21(16)

Santos CSC, Gabriel B, Blanchy M, Menes O, García D, Blanco M, Arconada N, Neto V (2014) 5th international conference on advanced nano materials. https://doi.org/10.1016/j.matpr.2015.04.056

Shaikh SF (2022) Nanomaterials applied in coatings: synthesis, structures, properties, and applications. Coatings 12(11):1773. https://doi.org/10.3390/coatings12111773

Sharma V, Kumar S, Kumar M, Deepak D (2020) High temperature oxidation performance of Ni–Cr–Ti and Ni–5Al coatings. Mater Today Proc 26(3):3397–3406. https://doi.org/10.1016/j.matpr.2019.11.048

Sharma R, Garg R, Bali M, Eddy NO (2023) Potential applications of green-synthesized iron oxide NPs for environmental remediation. Environ Monit Assess 195. https://doi.org/10.1007/s10661-023-12035-6

Shi H, Magaye R, Castranova V et al (2013) Titanium dioxide nanoparticles: a review of current toxicological data. Part Fibre Toxicol 10:15. https://doi.org/10.1186/1743-8977-10-15

Siengchin S (2023) A review on lightweight materials for defence applications: present and future developments. Def Technol 24:1–17. https://doi.org/10.1016/j.dt.2023.02.025

Singh P, Campidelli S, Giordani S, Bonifazi D, Bianco A, Prato M (2009) Organic functionalisation and characterisation of single-walled carbon nanotubes. Chem Soc Rev 38(8):2214–2230

Singh NP, Gupta VK, Singh AP (2019) Graphene and carbon nanotube reinforced epoxy nanocomposites: a review. Polymer 180

Singh AK, Shishkin A, Koppel T, Gupta N (2018) A review of porous lightweight composite materials for electromagnetic interference shielding. Compos B Eng 149:188–197

Song HJ, Zhang ZZ, Men XH (2007) Surface-modified carbon nanotubes and the effect of their addition on the tribological behavior of a polyurethane coating. Eur Polym J 43:4092–4102

Steirer KX, Reese MO, Rupert BL, Kopidakis N, Olson DC, Collins RT, Ginley DS (2009) Ultrasonic spray deposition for production of organic solar cells. Sol Energy Mater Sol Cell 93(4):447–453

Stephen AJ, Rees NV, Mikheenko I, Macaskie LE (2019) Platinum and palladium bio-synthesized nanoparticles as sustainable fuel cell catalysts. Front Energy Res 7:66. https://doi.org/10.3389/fenrg.2019.00066

Su J-F, Zhang X-L, Guo Y-D et al (2019) Experimental observation of the vascular self-healing hollow fibers containing rejuvenator states in bitumen. Constr Build Mater 201:715–727. https://doi.org/10.1016/j.conbuildmat.2019.01.001

Sumey JL, Sarver JA, Kiran E (2019) Foaming of polystyrene and poly (methyl methacrylate) multilayered thin films with supercritical carbon dioxide. J Supercrit Fluids 145:243–252. https://doi.org/10.1016/j.supflu.2018.12.001

Tao Y, Hadigheh SA, Wei Y (2023) Recycling of glass fibre reinforced polymer (GFRP) composite wastes in concrete: a critical review and cost benefit analysis. Structures 53:1540–1556. https://doi.org/10.1016/j.istruc.2023.05.018

Thakur A, Kaya S, Kumar A (2023) Recent trends in the characterization and application progress of nano-modified coatings in corrosion mitigation of metals and alloys. Appl Sci 13(2):730. https://doi.org/10.3390/app13020730

Vasilev C (2021) Applying nanocoatings to aviation: a review. AZoNano. Accessed 06 Nov 2024 from https://www.azonano.com/article.aspx?ArticleID=5843
Vela-Cano M, Garcia-Fontana C, Osorio F et al (2021) Silver-derived antimicrobial coatings for the prevention of microbial biofilms in metal pipes. Water Air Soil Pollut 232:313. https://doi.org/10.1007/s11270-021-05264-3
Venkataramanan S, Selvaraj K, Ponnusamy P (2021) A brief review of carbon nanotube reinforced metal matrix composites for aerospace and defense applications. Int Nano Lett 11(4):321–345
Vertuccio L, Guadagno L, Spinelli G et al (2018) Smart coatings of epoxy based CNTs designed to meet practical expectations in aeronautics. Compos Part B Eng 147:42–46. https://doi.org/10.1016/j.compositesb.2018.04.027
Vidales-Herrera J (2020) Nanomaterials in coatings: an industrial point of view. In: Handbook of nanomaterials for manufacturing applications, micro and nano technologies, pp 51–77
Wang S-D, Chang M-H, Lan KM-D, Wu C-C, Cheng J-J, Chang H-K (2005) Synthesis of carbon nanotubes by arc discharge in sodium chloride solution. Carbon 43(8):1792–1795
Wang Y, Pham DT, Zhang Z et al (2016) Sustainable self-healing at ultra-low temperatures in structural composites incorporating hollow vessels and heating elements. R Soc Open Sci 3:160488. https://doi.org/10.1098/rsos.160488
Wang H, Zhou H, Huang Z, Zhang Y, Zhao X (2017) Constitutive modeling of polycarbonate over a wide range of strain rates and temperatures. Mech Time-Depend Mater 21(1):97–117. https://doi.org/10.1007/s11043-016-9320-1
Wilson S (2013) Abradable thermal spray applications and technology, ASM handbook, Volume 5A: thermal spray technology. ASM International, pp 287–295
Wu Q, Miao W-S, Zhang Y-D, Gao H-J, Hui D (n.d.) Mechanical properties of nanomaterials: a review. J Nanotechnol Rev. https://doi.org/10.1515/ntrev-2020-0021
Xia H, Song M (2005) Preparation and characterization of polyurethane–carbon nanotube composites. Soft Matter 1:386–394
Xie J, Fan L, Yao D et al (2022) Ultra-robust, self-healable and recyclable polyurethane elastomer via a combination of hydrogen bonds, dynamic chemistry, and microphase separation. Mater Today Chem 23:100708. https://doi.org/10.1016/j.mtchem.2021.100708
Xu S, Wen M, Li J, Guo S, Wang M, Du Q, Shen J, Zhang Y, Jiang S (2008) Structure and properties of electrically conducting composites consisting of alternating layers of pure polypropylene and polypropylene with a carbon black filler. Polymer 49:4861–4870. https://doi.org/10.1016/j.polymer.2008.08.056
Xu W, Kasar AK, Menezes PL, Dieringa H, Varma RS, Gupta M (2019) Metals 9(3):330. https://doi.org/10.3390/met9030330
Yadav R, Tirumali M, Wang X et al (2020) Polymer composite for antistatic application in aerospace. Def Technol 16:107–118. https://doi.org/10.1016/j.dt.2019.04.008
Yadav L, Sihmar A, Kumar S et al (2024) Review of nano-based smart coatings for corrosion mitigation: mechanisms, performance, and future prospects. Environ Sci Pollut Res. https://doi.org/10.1007/s11356-024-33234-9rtyuyujjqaezdy5bqz5ybzybzaybzyyzyzy
Yu Z, Xue P, Yao P, Zahran MS (2019) Analytical determination of the critical impact location for wing leading edge under birdstrike. Lat Am J Solids Struct 16(1). https://doi.org/10.1590/1679-78255352
Zai L et al (2020) Laser powder bed fusion of precipitation-hardened martensitic stainless steels: a review. Metals. https://doi.org/10.3390/met10020255
Zhang X, He Q, Gu H et al (2013) Polyaniline stabilized barium titanate nanoparticles reinforced epoxy nanocomposites with high dielectric permittivity and reduced flammability. J Mater Chem C 1:2886. https://doi.org/10.1039/c3tc30129j
Zhang W, Gao Y, Ye N, Huang W, Li D (2017) Experimental study on dynamic mechanical behaviors of polycarbonate, presented at the shock compression of condensed matter. In: 2015: proceedings of the conference of the american physical society topical group on shock compression of condensed matter. Tampa Bay, Florida, USA, p 110014. https://doi.org/10.1063/1.4971677

Zhang C, Ling YQ, Zhang XQ, Liang M, Zou HW (2022) Ultra-thin carbon fiber reinforced carbon nanotubes modified epoxy composites with superior mechanical and electrical properties for the aerospace field. Compos Part A-Appl S 163

Zhang J, Lin G, Vaidya U, Wang H (2023) Past, present and future prospective of global carbon fibre composite developments and applications. Compos Part B Eng 250

Zhu Y, Cao K, Chen M, Wu L (2019) Synthesis of UV-responsive self-healing microcapsules and their potential application in aerospace coatings. ACS Appl Mater Interfaces 11:33314–33322. https://doi.org/10.1021/acsami.9b10737

Chapter 2
Nanomaterials Tribological Coatings for Aerospace Applications

Santosh Kumar and Rakesh Kumar

Abstract Tribological coating is a special coating that applied to a surface of parts to improve the performance under friction, wear and lubrication. Tribological coatings act as a protective barrier, reducing friction between contracting surfaces, which in turn increase part life span, smoother operation, lower energy consumption in machinery and require less frequent replacement. There are different types of tribological coating that can applied in wide range of coating materials and each provide specific advantages. The examples of different coatings are metal coatings (Nickel, Chromium, etc.), Ceramic coatings, polymer coating, diamond like carbon coating, and nanomaterial coatings. The tribological coating is not only used in aerospace applications (landing gear, engine parts) but also used in automotive (engine parts, transmission, gear), manufacturing (tool and machine parts), medical devices (surgical instruments), and consumer goods (doorknobs, etc.). The diamond like carbon coatings, ceramic oxide coating, etc., are the examples of tribological coating in aerospace application. Overall, tribological coatings are a valuable technology in aerospace industry because it reduces friction, wear, maintenance need. Further it helps to improve aircraft performance, fuel efficiency, and operational cost. This chapter presents an overview of the various tribological coatings, advantages and disadvantages, nanomaterial tribological coatings for aerospace application. Finally, challenges and future direction were discussed.

Keywords Tribological coatings · History · Nanomaterial · Aerospace applications · Future scope · Challenges

---

S. Kumar
Department of Applied Science, CGC College of Engineering, Chandigarh Group of Colleges, Landran, Mohali 140307, Punjab, India
e-mail: santoshdgc@gmail.com

R. Kumar (✉)
Department of Regulatory Affairs & Quality Assurance, Kaushik Orthopaedic Pvt. Ltd., Sonipat, Haryana, India
e-mail: rk05691@gmail.com

Visakh P. M. (ed.), *Nanomaterials for Defense, Aeronautical and Aerospace Applications*, Advanced Structured Materials 236,
https://doi.org/10.1007/978-3-031-92723-2_2

2.1 Introduction

In the 1960s, the term "tribology" was used to refer to the engineering and science of rubbing surfaces. It involves learning and using the concepts of wear, lubrication, and friction. Tribology is hence interdisciplinary in nature and has tight ties to a number of other fields, including physics, chemistry, and material mechanics. To put it simply, when two bodies move relative to one another, rubbing occurs at the interface, which leads to friction, which in turn wears down one or both of the materials. Consequently, the function of lubricant is introduced. Lubricants can be used to reduce wear and friction between two bodies. Water and air can function as lubricants in the environment to increase wear resistance and lower friction. Various external factors significantly influence the friction and wear performance of structures at the nano, micro, and macroscale. Key parameters such as relative speed, temperature, and contact pressure play a crucial role in determining tribological behavior. Several organizations have made significant contributions to the field of tribology by advancing research, standards, and best practices. Notable institutions include the American Society of Mechanical Engineers (ASME), founded in 1880, and the Institution of Mechanical Engineers (IMechE), established in London in 1847. Other influential organizations include the Japanese Society of Tribologists (JAST), founded in 1956, the American Society for Testing and Materials (ASTM), established in 1898, the American Gear Manufacturers Association (AGMA), founded in 1916, and the Society of Tribologists and Lubrication Engineers (STLE), which was originally established as the American Society of Lubrication Engineers (ASLE) in 1944. These institutions continue to play a vital role in shaping the future of tribology through research, innovation, and standardization.

The International Tribology Council is home to a number of these groups (Pisano 2015; Rathaur et al. 2020; Stachowiak and Batchelor 1993; Williams 2005).

2.1.1 History of Tribological Coatings

To enhance lubricity and extend wear life in demanding tribological applications, innovative solid lubricants and advanced lubrication techniques have been developed in recent years. Traditionally, solid lubricants were primarily formulated as metal, ceramic, or polymer-matrix composites, demonstrating effectiveness across various technical applications. Recent advancements in thin-film deposition techniques have led to the development of next-generation adaptive, self-lubricating coatings with composite or multilayered structures. These coatings are created using duplex or multiplex surface treatments, improving their performance and adaptability to extreme conditions. As a result, these advanced self-lubricating coatings are gradually entering the commercial market, meeting the growing demands of high-performance applications. In extreme operating conditions, where conventional lubrication methods fail, solid lubricants often become the only viable solution for minimizing wear and friction in tribological systems (Farfan-Cabrera et al. 2022; Cheah et al. 2020; Xu et al. 2014; Markandan et al. 2024). A general design evaluation of the tribological needs on a coated-surface contact may be created as shown in Fig. 2.1.

The use of coatings to enhance the tribological characteristics of parts, including machine parts and metal cutting and shaping tools. Combining materials with varied qualities to create composites is a common way to improve the functioning of contemporary materials. There are several kinds of composites, the majority of which focus on better mechanical qualities including stiffness, strength, toughness, and fatigue resistance. Surface-engineered materials, also known as coating composites, are made to precisely enhance tribological, electrical, optical, electronic, chemical, and magnetic properties. Therefore, applying coatings to tools and machine components is a very effective method of enhancing their resistance to wear and friction. Obviously, the goal of using tribological coatings is to extend their lifespan. However, as Fig. 2.2 illustrates, there are a number of additional benefits.

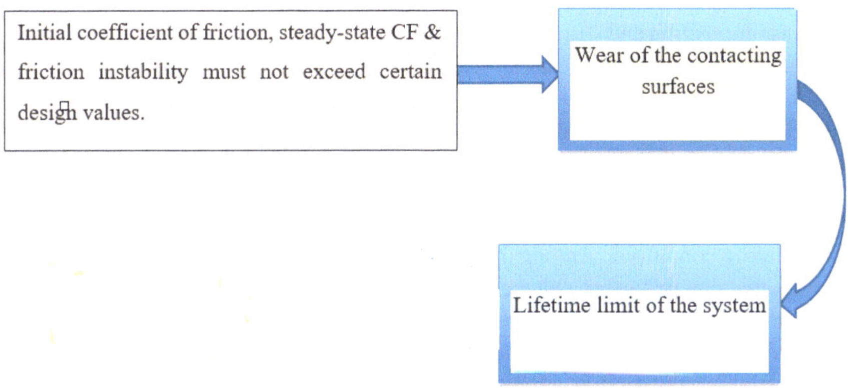

Fig. 2.1 Evaluation of tribological needs for coated surfaces to enhance durability and performance

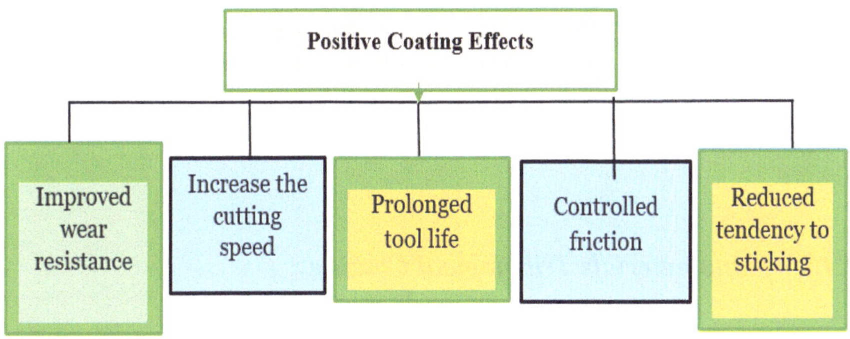

Fig. 2.2 Represent the distinct positive coating effects

Detachment, cracking, etc., of the coating material can cause a coating (tribological) to fail too soon. The coating should be chosen based on the desired application's life-limiting surface damage processes and tribological conditions. It is easy to classify these processes into three groups, such as damage without material exchange, damage with material loss and damage with material pick-up. As long as the coating covers the contact region, it mostly determines the resistance against wear of a coated part.

While tribological requirements vary across applications, coating adhesion is often the most critical factor. Differentiating between practical adhesion and true adhesive forces—such as atomic bonding or physical adhesion between the substrate and coating—provides valuable insights into coating performance. When certain work materials are machined at high cutting speeds, for example, tribochemical layers may occur. The thickness of the coatings varies from a few microns to several millimeters. The thinnest coatings are usually produced using atomistic techniques. Certain techniques use high deposition temperatures, which might cause the coated component to undergo undesirable phase transitions, soften, or alter shape.

PVD and CVD coating techniques offer significant advantages due to their flexibility in composition and structure, making them suitable for a wide range of mechanical components. Common coating materials include nitrides (TiN, CrN), carbides (TiC, CrC, W_2C, WC/C), oxides (such as alumina), and combinations of these materials. Additionally, diamond-like carbon (DLC), molybdenum disulfide (MoS_2), and diamond coatings have gained widespread success in various applications.

Coatings such as WC/C, MoS_2, and DLC are particularly notable for their low friction properties, with friction coefficients typically ranging between 0.05 and 0.25 in dry sliding conditions. Among these, amorphous DLC represents a broad class of coatings that can be further enhanced by doping with metals, nitrides, or carbides to improve their mechanical and tribological performance, as well as adhesion to the substrate. When applied under optimal conditions, DLC coatings provide excellent wear resistance, low friction, and strong corrosion protection. However, their primary limitation is poor thermal stability, as they begin to degrade at temperatures exceeding 300 °C. In contrast, nitrides, carbides, and oxides generally offer superior wear resistance compared to low-friction coatings, though they typically exhibit higher friction values ranging from 0.4 to 0.9 in dry sliding conditions. A notable exception is CVD diamond coatings, which combine exceptionally low friction with outstanding wear resistance, making them highly effective in demanding applications (Blomberg 1994; Hirvonen 1996; Hogmark et al. 1992; Holmberg and Matthews 1994; Nordgren and Melander 1988).

2.1.2 Nanomaterials Tribological Coatings

Carbon nanoparticles have attracted significant attention due to their important tribological applications and proven friction and wear properties. Their diverse morphologies, high thermal conductivity, excellent mechanical properties, and superior corrosion resistance make them highly valuable in various fields. Carbon materials have been extensively researched for their applications in optics, electrochemistry, mechanics, and tribology. Among them, carbon nanomaterials exhibit exceptional electrochemical properties, including one-dimensional (1D) carbon nanotubes (CNTs), two-dimensional (2D) graphene, three-dimensional (3D) nanodiamonds, and zero-dimensional (0D) fullerenes. Allotropes of carbon with sp^2 hybridization, such as graphene and CNTs, have attracted significant attention due to their exceptional mechanical flexibility, electrochemical stability, and high charge carrier mobility. These exceptional qualities have led to the widespread usage of CNTs and graphene in a variety of applications, including electronics, chemical separation, catalysis support, and super lubrication (Avouris et al. 2007; Dürkop et al. 2004; Park et al. 2013; Chae et al. 2013; Mochalin et al. 2011; Won et al. 2016; Rogers et al. 2010; Wu et al. 2011; Raccichini et al. 2015; Sekitani et al. 2008; Lipomi et al. 2011; Wang et al. 2010; Umeda et al. 2015; Mittendorff 2014). In 1991, arc-discharge evaporation was used to create CNTs for the first time. CNT-based coatings exposed to wear and friction have been the focus of several studies (Hentour et al. 2016; Keshri et al. 2010). Regarding the tribological performances of CNT-reinforced composite coatings material and their wear rate (Lee et al. 2013; Nie et al. 2015; Koumoulos et al. 2015), as seen in Fig. 2.3, the different studies employed materials of varying thicknesses.

Because of graphene, it is a viable option for tribological applications at both the Nano and macro scales. Nano diamonds have extremely low surface roughness in comparison to microcrystalline diamonds, which is crucial for the wear and friction behaviors of self-mated tribosystems (Marko et al. 2015; Shenderova et al. 2014; Kim et al. 2013, 2016; Shirvani et al. 2016; Dolmatov et al. 2003; Novak et al. 2014; Sharma et al. 2011). The creation and tribological characteristics of the Nano diamond coatings applied to the silicon substrate by microwave plasma assisted chemical vapor deposition were documented by Abdeen et al. (2019). Corrosion and tribological problems with wear and friction have a negative impact on component performance and operational life in industrial systems. These factors cause significant material and energy losses in engineering equipment, which results in significant financial losses. This has prompted the development of many tactics for the creation

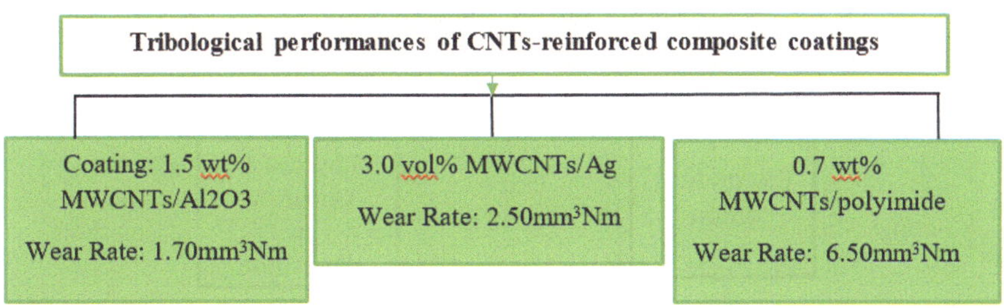

Fig. 2.3 CNTs-reinforced composite coatings materials and their wear rate

of surface protection that is anticorrosive and antifriction/wear. The use of coatings is one such effective and tested approach. Because there are so many contributing factors, including thickness, grain size, coating adherence to the substrate, hardness, etc., it can be difficult to design and create coatings for a certain substrate, application, and operating environment. Therefore, it is difficult and multidisciplinary to comprehend the physical and chemical phenomena for a coating–substrate system since it entails. Coatings undergo mechanical loads and thermal strains when in use, which result in the development of micro cracks and, ultimately, the fracturing or peeling off of the coatings from their substrates. In real-world applications, coatings are subjected to a broad variety of temperatures, including up to 2700 °C for IC engines, in addition to different mechanical stresses. Solar applications require 80–200 °C, gas turbines require 2100 °C, and aircraft components require 100–300 °C. This has an impact on component service life because the underlying substrate materials are either directly exposed to the corrosive environment around them and deteriorate, or in the case of tribological conditions, they come into direct contact with their counter face, increasing wear and friction. The effective creation of nanostructured coatings, which contain at least one component at the nanometer scale, is a result of the special qualities that nanomaterials provide over micro sized materials. Because they offer better surface protection, nanostructured coatings are being used in a variety of industries, including electronics, healthcare, food packaging, and shipping (Tzanakis et al. 2012; Sørensen et al. 2009; Hogmark et al. 2000; Hirvonen et al. 1996; Khan et al. 2018; Voevodin et al. 1999a; Nazir et al. 2017, 2016; Gu et al. 2020; Safakish 2012; Du et al. 2018; Nicholls 2003; Aliofkhazraei 2011a, b). Furthermore, Fig. 2.4 displays the various techniques employed by various researchers to create nanostructured coatings.

The quality of a coating on a substrate is highly dependent on its surface conditions, including cleanliness (absence of contaminants, dirt, oxides, or scales), surface defects (such as pores and scratches), and roughness. Therefore, proper substrate preparation, which involves cleaning and chemical modification, is essential before applying coatings (Gu et al. 2020; Kumar et al. 2019a, 2021, 2020, 2019b, 2021, 2024; Kumar and Kumar 2021; Yilmaz and Yorgancioglu 2019; Kumar 2024; Basile et al. 2010; Bedi et al. 2019; Sharma et al. 2020; Shishkovsky and Lebedev 2011;

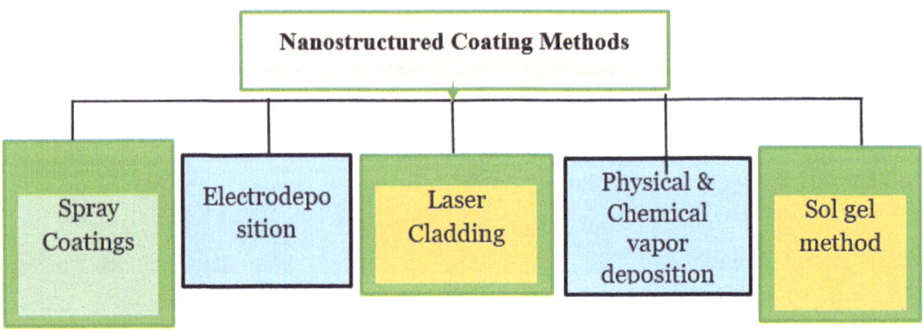

Fig. 2.4 Nanostructured coating methods

Callister and Rethwisch 2020). Several factors influence the properties of nanostructured coatings, including thickness, microstructure, grain size, nanoparticles, and the substrate material. Ceramic nanostructured coatings, in particular, are widely used due to their superior resistance to corrosion, oxidation, and wear compared to metals, especially in high-temperature environments. These coatings are commonly applied in engine valves, boiler components, automotive body parts, orthopedic implants, and various other applications. Additionally, they exhibit excellent electrical and thermal insulation properties. Various ceramic materials are utilized for nanocoatings, including metallic nanostructured coatings, tantalum nitride (Ta_2N), titanium oxide (TiO_2), alumina (Al_2O_3), and tantalum pentoxide (Ta_2O_5). These materials enhance the durability and functionality of coated components, making them suitable for demanding industrial and biomedical applications (Mersagh Dezfuli and Sabzi 2018; Deyab and Keera 2014).

2.1.3 Merits and Demerits

The merits of nanomaterial tribological coatings include improved wear resistance, reduced friction, enhanced corrosion resistance, and a thinner coating that lowers the overall weight of aerospace parts. These benefits contribute to reducing the hazard of part failure, decreasing maintenance and downtime, and increasing the lifespan of aerospace parts. However, demerits includes high cost to develop, complicated deposition process, limited standardization, probability of health related risk, and threat in coating thickness control, etc. (Fairbrother 1967; Ma et al. 2016; Global Nanocoatings Market 2020–2030).

2.2 Nanomaterials Tribological Coatings for Aerospace Applications

Without modifying bulk materials, nanostructured coatings offer an efficient way to prevent corrosion, wear, and friction by protecting the surface. As a result, there are several uses for nanostructured coatings. Marine, aerospace, automotive, sports, construction, defense, energy, food packaging, and medical (dental and orthopedic) are just a few of the engineering systems that use nanostructured coatings. The market for nanostructured coatings is predicted to grow by about 20% between 2020 and 2030, per a research. Sensors; robotics (magnetic and dielectric); medical devices (biocompatibility); surgical equipment; solar cells (optical); aerospace, space, and automobile (mechanical); textile (wetting); space, turbine (cossosion & heat properties), etc., are among the most notable application areas for nanostructured coatings (Abdel-Karim and Waheed 2013; Holmberg et al. 1997). The most prevalent solid lubricants are transition-metal dichalcogenides (MoS_2, WS_2, $NbSe_2$, etc.), which are an example of composite coatings for aerospace applications. Additionally, vacuum deposition offered a range of sophisticated hard coatings for wear protection (Kim et al. 2013, 2016; Shirvani et al. 2016; Dolmatov et al. 2003), as well as a number of advanced hard coatings for wear protection (Holleck and Schier 1995; Petlyuk et al. 1989). The lubrication by WS2-graphite particles with an inorganic binder studied for aerospace applications ten years ago (Aliofkhazraei 2011a, b) has a comparable impact on friction control with environmental References cycling. Nevertheless, these powders are gentle and comfortable to wear, and they are unable to precisely measure thickness when applying sub-mm coatings. The ideal combination of hardness, low wear rate, and low friction in high/low humidity is possessed by the comparatively harder WC/DLC/WS_2 coatings. On the nm scale, they may be applied precisely in terms of thickness. When combined, these offer the possibility of using these coatings in a variety of aerospace applications. The low wear, friction, and unique environmental adaptability of WC/DLC/WS_2 coatings open new possibilities for aerospace applications, particularly for similar carbide/DLC/dichalcogenide nanocomposites.

Furthermore, aluminum alloys, titanium alloys, glass fiber reinforced aluminum, ceramic composites, glass fiber reinforced polymers, shape memory, polymers, and ceramic composites are examples of lightweight materials and their applications in space (Voevodin et al. 1999b; Murthy 2023).

2.3 Nanomaterials Coatings for Defense Applications

Tribological coatings are essential for improving the functionality and lifespan of engineering systems and parts. In defense applications, thermal spray techniques are commonly used for coating applications. This article explores the development of low-friction coatings for high-temperature environments, high-temperature abatable coatings, and wear-resistant carbide coatings. Due to the partial dissolution of carbide in the NiCr binder, chromium carbide-based coatings applied using high-velocity spray techniques exhibit a reduced primary carbide content while forming a dense and hard protective layer. Heat treatment causes precipitation in the supersaturated metallic binder, which raises the volume percentage of carbides. There is also a simultaneous alteration in the binder phase's composition and microstructure. Heat treatment of the coating causes microstructural changes that impact its mechanical characteristics. Various authors have examined various tribological applications in defense systems, namely in missile and aero-engine systems (Rhys-Jones 1990; Abed and Jawad 2022). Nowadays, nanotechnology is being used in research, projects, and applications by the majority of advanced nations' armies. Stronger and lighter weapons, silver-packed meals that are antiviral and antibacterial, gas and biological sensors, and nanomedicines and bandages that promote wound healing and halt bleeding are all possible thanks to nanotechnology.

Even while nanotechnology has advanced and improved, there are still hazards associated with its creation and even use (Bhatnagar 2016). Because of their strength-to-weight ratio, lower carbon emissions, superior mechanical performance, fatigue endurance, damage resistance, and strong thermal and radiation resistance, lightweight composite materials have been widely used in the space industry. The use of lightweight composites in aircraft has enhanced manufacturability and decreased energy consumption, production costs, and industrial waste. These benefits have led the scientific community to develop a number of viewpoints for investigating potential ways to improve the performance of aircraft and aerospace materials. MMC, PMC, CMC, and fiber composites are some of the types of lightweight composites that fall under this category. Lightweight composites were frequently used in the aerospace industry for components of rocket engines, such as fairings, fuselage skin, upper/lower wing coverings, vertical/horizontal stabilizers, interior panels, passenger seats, and the spacecraft body (Polmear et al. 2017; Joshi and Chatterjee 2016; Kumar and Dixit 2019; Kar 2016). The biggest rise in the past ten years was the 4% increase in global defense spending in 2019. The rise in Chinese and American military spending was a prerequisite for this (Fontanel 2019 , 2022). Lockheed Martin Corp., Boeing, Leonardo, Thales, Dassault, and others are among the world's defense corporations (https://www.3ds.com/industries). In order to provide lightweight resistance to severe temperatures, researchers in the military and defense industry have begun employing magnesium in generator housings and helicopter gears. Despite having comparatively greater densities, titanium and nickel may be used in a variety of military and naval-related components due to their superior corrosion and crash resistance. High-performance ceramic materials are becoming essential in contemporary armament and defense systems (van Wegen and Carton 2008; Bao et al. 2021; Misra and Misra 2022)

2.4 Future Scope

Tribology has become an integral part of various fields, including science, engineering, design, manufacturing, and life cycle management. The extensive knowledge available in this field can be leveraged to enhance the performance and longevity of machines and mechanical systems. Understanding the difficult interactions between materials, and operating conditions is essential for minimizing friction and preventing failures caused by tribological effects. Given the scale of industrial applications, even minor improvements in the efficiency of a single machine component can lead to substantial global cost savings. At the nanoscale, known as nanotribology, friction phenomena have been explored, where atomic forces significantly influence system behavior. Key advancements in tribology include the development of innovative lubricants and additives, improvements in surface characteristics (such as microgeometry and coatings), and expanding the applications of tribometry and bio tribology for tissue engineering and prosthetic development. In the defense sector, there is a growing demand for ultra-lightweight, high-strength materials that can be integrated into large structures to enhance the protection of combat vehicles against emerging threats. However, the high cost of these advanced materials remains a significant challenge, limiting their widespread adoption. Research and development efforts are crucial to reducing the manufacturing costs of lightweight materials, improving their recyclability, and enhancing their integration into final products. Strength is a critical factor for lightweight materials in defense applications, but traditional destructive testing methods are expensive and time-consuming. The use of simulation software mainly reduce time and cost of designing the new materials. Additionally, nanotechnology plays a major role in distinct defense applications, intelligent systems, and autonomous technologies for missiles, artillery, and mortars (Spikes 2001; Carnes 2005).

2.5 Conclusion

AS the aerospace industry continues to evolve and demand more efficient, sustainable and high performance solutions, nanomaterial tribological coatings are likely to play an increasingly important role. Thus development and use of nanomaterial tribological coatings for aerospace use have shown tremendous promise in improving the performance, life of aerospace parts, and efficiency. These coatings provide resistance against corrosion, wear resistance, making them ideal for use in harsh aerospace environments. Further research and development are needed to fully realize the potential of these coatings but the future of aerospace tribology looks promising with nanomaterials at the forefront.

Acknowledgements Authors are greatly thankful to CGC Landran for motivation during this research.

References

Abdeen DH, el Hachach M, Koc M, Atieh MA (2019) A review on the corrosion behaviour of nanocoatings on metallic substrates. Materials 12:210

Abdel-Karim R, Waheed AF (2013) Nanocoatings modern surface engineering treatments. InTech, London, UK

Abed MS, Jawad ZA (2022) Nanotechnology for defence applications. In: Mubarak NM, Gopi S, Balakrishnan P (eds) Nanotechnology for electronic applications. Materials horizons: from nature to nanomaterials. Springer, Singapore. https://doi.org/10.1007/978-981-16-6022-1_10

Aliofkhazraei M (2011a) Nanocoatings: size effect in nanostructured films; springer science and business media. Amsterdam, The Netherlands

Aliofkhazraei M (2011b) Synthesis, processing and application of nanostructured coatings nanocoatings. Springer, Amsterdam, The Netherlands, pp 1–28

Avouris P, Chen Z, Perebeinos V (2007) Carbon-based electronics. Nat Nanotechnol 2:60e615

Bao Y, Gao X, Wu Y, Sun M, Li G (2021) Research progress of armor protection materials. J Phys Conf Ser 1855, Article 012035 No. 1. IOP Publishing

Basile F, Benito P, Fornasari G, Monti M, Scavetta E, Tonelli D, Vaccari A (2010) A novel electrochemical route for the catalytic coating of metallic supports. Stud Surf Sci Catal 175:51–58

Bedi TS, Kumar S, Kumar R (2019) Corrosion performance of hydroxyapaite and hydroxyapaite/titania bond coating for biomedical applications. Mater Res Express 7:015402. https://doi.org/10.1088/2053-1591/ab5cc5

Bhatnagar A (ed) (2016) Lightweight ballistic composites: military and law-enforcement applications. Woodhead Publishing

Blomberg A (1994) Wear mechanisms and tribo mapping of Al_2O_3 and SiC in dry sliding. Wear 171(1–2):77–89. https://doi.org/10.1016/0043-1648(94)90350-6

Callister WD Jr, Rethwisch DG (2020) Fundamentals of materials science and engineering: an integrated approach. Wiley, Hoboken, NJ, USA

Carnes K (2005) The ten greatest events in tribology history: as chosen by the readers of TLT magazine. Tribol Lubr Technol 61(6):38–47

Chae SH, Yu WJ, Bae JJ, Duong DL, Perello D, Jeong HY et al (2013) Transferred wrinkled Al_2O_3 for highly stretchable and transparent graphene-carbon nanotube transistors. Nat Mater 12:403e409

Cheah MY, Ong HC, Zulkifli NWM, Masjuki HH, Salleh A (2020) Physicochemical and tribological properties of microalgae oil as biolubricant for hydrogen-powered engine. Int J Hydrogen Energy 45:22364–22381. https://doi.org/10.1016/j.ijhydene.2019.11.020

Deyab MA, Keera ST (2014) Effect of nano-TiO2 particles size on the corrosion resistance of alkyd coating. Mater Chem Phys 146:406–411

Dolmatov VY, Fujimura T, Burkat GK, Orlova EA (2003) Preparation of wearresistant chromium coatings using different types of nanodiamonds. Powder Metall Met Ceram 42:587e591

Du K, Calautit J, Wang Z, Wu Y, Liu H (2018) A review of the applications of phase change materials in cooling, heating and power generation in different temperature ranges. Appl Energy 220:242–273

Dürkop T, Getty S, Cobas E, Fuhrer M (2004) Extraordinary mobility in semiconducting carbon nanotubes. Nano Lett 4:35e39

Fairbrother F (1967) The chemistry of niobium and tantalum. Elsevier, Amsterdam, The Netherlands

Farfan-Cabrera LI, Franco-Morgado M, González-Sánchez A, Pérez-González J, Marín-Santibáñez BM (2022) Microalgae biomass as a new potential source of sustainable green lubricants. Molecules 27(4):1205. https://doi.org/10.3390/molecules27041205.PMID:35208995;PMCID:PMC8875479

Fontanel J (2019) New global and national military expenditure series developed by SIPRI, Economie de la sécurité internationale

Fontanel J (2022) Is the SIPRI estimate of military expenditure a reliable indicator of the power of states? Defence economics

Global Nanocoatings Market by Type (Anti-fingerprint, Antimicrobial, Easy-to-Clean & Anti-Fouling, Self-Cleaning {Bionic & Photocatalytic}, Anti-Icing & Deicing, Anticorrosion, Conductive, UV, Abrasion, & Wear Resistant), by Application (Electronics, Energy, Food & Packaging, Construction, Marine Industry, Military & Defense, Automotive, Aerospace, Healthcare)-Global Opportunity Analysis and Industry Forecast, 2020–2030. Available online: https://www.vantagemarketresearch.com/industry-report/nanocoatings-market-1096. Accessed 28 Jan 2022

Gu Y, Xia K, Wu D, Mou J, Zheng S (2020) Technical characteristics and wear-resistant mechanism of nano coatings: a review. Coatings 10:233

Hentour K, Marsal A, Turq V, Weibel A, Ansart F, Sobrino JM et al (2016) Carbon nanotube/alumina and graphite/alumina composite coatings on stainless steel for tribological applications. Mater Today Commun 8:118e126

Hirvonen J-P (1996) Present progress in the development of low friction coatings. Surf Coat Technol 80(1–2):139–150. https://doi.org/10.1016/0257-8972(95)02701-7

Hirvonen JP, Koskinen J, Jervis JR, Nastasi M (1996) Present progress in the development of low friction coatings. Surf Coat Technol 80:139–150
Hogmark S, Jacobson S, Larsson M (2000) Design and evaluation of tribological coatings. Wear 246:20–33
Hogmark S, Jacobson S, Vingsbo O (1992) Friction lubrication and wear technology. ASM Handbook, vol 18, p 176
Holleck H, Schier V (1995) Surf Coat Technol 76–77:328
Holmberg K, Ronkainen H, Matthews A (1997) Proceedings of 1st tribology congress on new directions in tribology. Institute of Mechanical Engineers, London, Paper 251
Holmberg K, Matthews A (1994) Coatings tribology, tribology series, 28. Elsevier, Amsterdam
https://www.3ds.com/industries
Joshi M, Chatterjee U (2016) Polymer nanocomposite: an advanced material for aerospace applications advanced composite materials for aerospace engineering. Woodhead Publishing, pp 241–264
Kar KK (ed) (2016) Composite materials: processing, applications, characterizations. Springer
Keshri AK, Huang J, Singh V, Choi W, Seal S, Agarwal A (2010) Synthesis of aluminum oxide coating with carbon nanotube reinforcement produced by chemical vapor deposition for improved fracture and wear resistance. Carbon 48:431e442
Khan ZA, Latif J, Hammad Nazir AS, Stokes K (2018) Predictive and prognostic modelling and simulation of coatings subject to corrosion and mechanical failures. Mater Charact 6:487–498
Kim HS, Park JW, Park SM, Lee JS, Lee YZ (2013) Tribological characteristics of paraffin liquid with nanodiamond based on the scuffing life and wear amount. Wear 301:763e767
Kim ST, Woo JY, Lee YZ (2016) Friction, wear and scuffing characteristics of marine engine lubricants with nanodiamond particles. Tribol T 59:1098e1103
Koumoulos EP, Jagadale P, Lorenzi A, Tagliaferro A, Charitidis CA (2015) Evaluation of surface properties of epoxy-nanodiamonds composites. Compos Part B-Eng 80:27e36
Kumar S (2024) Comprehensive review on high entropy alloy-based coating. Surf Coat Technol 477:1–30. https://doi.org/10.1016/j.surfcoat.2023.130327
Kumar S, Kumar R (2021) Influence of processing conditions on the properties of thermal sprayed coating: a review. Surf Eng 37(11):1339–1372. https://doi.org/10.1080/02670844.2021.1967024
Kumar S, Kumar M, Handa A (2019a) High temperature oxidation and erosion-corrosion behaviour of wire arc sprayed Ni–Cr coating on boiler steel. Mater Res Express 6:125533. https://doi.org/10.1088/2053-1591/ab5fae
Kumar S, Handa A, Kumar R (2019b) Overview of wire arc spray process: a review. J Compos Theory 12(7):900–907
Kumar S, Kumar M, Jindal N (2020) Overview of cold spray coatings applications and comparisons: a critical review. World J Eng 17(1):27–51. https://doi.org/10.1108/WJE-01-2019-0021
Kumar S, Handa A, Chawla V, Grover NK, Kumar R (2021) Performance of thermal-sprayed coatings to combat hot corrosion of coal-fired boiler tube and effects of process parameters and post coating heat treatment on coating performance: a review. Surf Eng 37(7):833–860. https://doi.org/10.1080/02670844.2021.1924506
Kumar N, Dixit A (2019) Nanotechnology for defence applications. Springer International Publishing
Kumar K, Kumar S, Gill HS (2024) Role of carbide-based thermal-sprayed coatings to prevent failure for boiler steels: a review. J Failure Anal Prevent. https://doi.org/10.1007/s11668-024-01974-y
Kumar R et al (2021) The role of additive manufacturing for biomedical applications: acritical review. J Manuf Process. https://doi.org/10.1016/j.jmapro.2021.02.022
Lee H-D, Penkov OV, Kim D-E (2013) Tribological behavior of dual-layer electroless-plated Ag-carbon nanotube coatings. Thin Solid Films 534:410e416

Lipomi DJ, Vosgueritchian M, Tee BC, Hellstrom SL, Lee JA, Fox CH et al (2011) Skin-like pressure and strain sensors based on transparent elastic films of carbon nanotubes. Nat Nanotechnol 6:788e792

Ma JJ, Xu J, Jiang S, Munroe P, Xie Z-H (2016) Effects of pH value and temperature on the corrosion behavior of a Ta2N nanoceramic coating in simulated polymer electrolyte membrane fuel cell environment. Ceram Int 42:16833–16851

Markandan K, Nagarajan T, Peter AP et al (2024) Recent progress in tribological and thermophysical behaviour of bio-based nanolubricant: a review. Emergent Mater. https://doi.org/10.1007/s42247-024-00923-7

Marko M, Branson B, Terrell E (2015) Tribological Improvements of dispersed nanodiamond additives in lubricating mineral oil. J Tribol 137:011802

Mersagh Dezfuli S, Sabzi M (2018) A study on the effect of presence of CeO_2 and benzotriazole on activation of self-healing mechanism in ZrO_2 ceramic-based coating. Int J Appl Ceram Technol 15:1248–1260

Misra RDK, Misra KP (2022) Fundamentals of ceramics: introduction, classification, and applications, ceramic science and engineering. Elsevier, pp 5–20

Mittendorff M, Winzer T, Malic E, Knorr A, Berger C, De Heer WA et al (2014) Anisotropy of excitation and relaxation of photogenerated charge carriers in graphene. Nano Lett 14:1504e1507

Mochalin VN, Shenderova O, Ho D, Gogotsi Y (2011) The properties and applications of nanodiamonds. Nat Nanotechnol 7:11e23

Murthy JKN (2023) Development of tribological coatings for aerospace and missile applications. Def Sci J 73(2):212–219. https://doi.org/10.14429/dsj.73.18718

Nazir MH, Khan ZA, Saeed A, Stokes K (2016) Modeling the effect of residual and diffusion-induced stresses on corrosion at the interface of coating and substrate. Corrosion 72:500–517

Nazir MH, Khan ZA, Saeed A (2017) A novel non-destructive sensing technology for on-site corrosion failure evaluation of coatings. IEEE Access 6:1042–1054

Nicholls JR (2003) Advances in coating design for high-performance gas turbines. MRS Bull 28:659–670

Nie P, Min C, Song HJ, Chen X, Zhang Z, Zhao K (2015) Preparation and tribological properties of polyimide/carboxyl-functionalized multi-walled carbon nanotube nanocomposite films under seawater lubrication. Tribol Lett 58:1e12

Nordgren A, Melander A (1988) Tool wear and inclusion behaviour during turning of a Ca-treated quenched and tempered steel using coated cemented carbide tools. In: Proceedings of 1st international conference on the behaviour of materials in machining, Stratford-upon-Avon, UK, Paper 20

Novak C, Kingman D, Stern K, Zou Q, Luan G (2014) Tribological properties of paraffinic oil with nanodiamond particles. Tribol.. T 57:831e837

Park S, Vosguerichian M, Bao Z (2013) A review of fabrication and applications of carbon nanotube film-based flexible electronics. Nanoscale 5:1727e1752

Petlyuk AM, Lazovskaya OV, Shelentsova NN, Seliva-Nova MK, Mashinostroenija V (1989) 3:62

Pisano R (2015) History of mechanism and machine science. In: A bridge between conceptual frameworks Volume 27, tribology: a historical overview of the relation between theory and application, pp 135–154. https://doi.org/10.1007/978-94-017-9645-3

Polmear I, StJohn D, Nie JF, Qian M (2017) Light alloys: metallurgy of the light metals. Butterworth-Heinemann

Raccichini R, Varzi A, Passerini S, Scrosati B (2015) The role of graphene for electrochemical energy storage. Nat Mater 14:271e279

Rathaur AS, Katiyar JK, Patel VK (2020) Introduction and applications of tribology. In: Katiyar J, Ramkumar P, Rao T, Davim J (eds) Tribology in materials and applications. materials forming, machining and tribology. Springer, Cham. https://doi.org/10.1007/978-3-030-47451-5_1

Rhys-Jones TN (1990) Thermally sprayed coating systems forsurface protection and clearance control applications inaero engines. Surf Coat Technol 43(44):402–415. https://doi.org/10.1016/0257-8972(90)90092-q

Rogers JA, Someya T, Huang Y (2010) Materials and mechanics for stretchable electronics. Science 327:1603e1607

Safakish GR (2012) Temperature estimation in the combustion chamber of an internal combustion engine. Adv Mech Eng 4:931584

Sekitani T, Noguchi Y, Hata K, Fukushima T, Aida T, Someya T (2008) A rubberlike stretchable active matrix using elastic conductors. Science 321:1468e1472

Sharma V, Kumar S, Kumar M, Deepak D (2020) High Temperature Oxidation Performance of Ni–Cr–Ti and Ni–5Al Coatings. Mater Today Proc 26(3):3397–3406. https://doi.org/10.1016/j.matpr.2019.11.048

Sharma N, Kumar N, Sundaravel B, Panda K, David W, Kamarrudin M et al (2011) Effect of CH_4/H_2 plasma ratio on ultra-low friction of nano-crystalline diamond coating deposited by MPECVD technique. Tribol Int 44:980e986

Shenderova O, Vargas A, Turner S, Ivanov DM, Ivanov MG (2014) Nanodiamondbased nanolubricants: investigation of friction surfaces. Tribol T 57:1051e1057

Shirvani KA, Mosleh M, Smith ST (2016) Nanopolishing by colloidal nanodiamond in elastohydrodynamic lubrication. J Nanopart Res 18:248

Shishkovsky IV, Lebedev PN (2011) Chemical and physical vapor deposition methods for nanocoatings. In; Nanocoatings and ultra-thin films. Woodhead Publishing, Sawston, UK, pp 55–77

Sørensen PA, Kiil S, Dam-Johansen K, Weinell CE (2009) Anticorrosive coatings: a review. J Coat Technol Res 6:135–176

Spikes HA (2001) Tribology research in the twenty-first century. Tribol Int 34:789–799

Stachowiak GW, Batchelor AW (1993) Engineering tribology: abrasive, erosive and cavitation wear. Tribol Ser 24:557–612

Tzanakis I, Hadfield M, Thomas B, Noya SM, Henshaw I, Austen S (2012) Future perspectives on sustainable tribology. Renew Sustain Energy Rev 16:4126–4140

Umeda J, Fugetsu B, Nishida E, Miyaji H, Kondoh K (2015) Friction behavior of network-structured CNT coating on pure titanium plate. Appl Surf Sci 357:721e727

Voevodin AA, O'Neill JP, Zabinski JS (1999a) Nanocomposite tribological coatings for aerospace applications. Surf Coat Technol 116–119:36–45

Voevodin AA, O'Neill JP, Zabinski JS (1999) Nanocomposite tribological coatings for aerospace applications. 116–119(none):36–45. https://doi.org/10.1016/s0257-8972(99)00228-5

Wang H, Luo J, Robertson A, Ito Y, Yan W, Lang V et al (2010) High-performance field effect transistors from solution processed carbon nanotubes. ACS Nano 4:6659e6664

van Wegen FTM, Carton EP (2008) New lightweight metals for armors. In: Proceedings of the 24th international symposium on Ballistics, pp 830–837

Williams JA (2005) Wear and wear particles—some fundamentals. Tribol Int 38(10):863–870

Won S, Jang J-W, Choi H-J, Kim C-H, Lee SB, Hwangbo Y et al (2016) A graphene meta-interface for enhancing the stretchability of brittle oxide layers. Nanoscale 8:4961

Wu B, Kuang Y, Zhang X, Chen J (2011) Noble metal nanoparticles/carbon nanotubes nanohybrids: synthesis and applications. Nano Today 6:75e90

Xu Y, Hu X, Yuan K, Zhu G, Wang W (2014) Friction and wear behaviors of catalytic methylesterified bio-oil. Tribol Int 71:168–174. https://doi.org/10.1016/j.triboint.2013.11.018

Yilmaz O, Yorgancioglu A (2019) Nanocoatings: preparation, properties, and biomedical applications. In: Polymeric nanomaterials in nanotherapeutics. Elsevier, Amsterdam, The Netherlands, pp 299–331

Chapter 3
Nanomaterials Coatings for Aerospace and Defense Applications: Current Prospects and Future Trends

Santosh Kumar and Rakesh Kumar

Abstract The nanomaterial coating trend is rapidly developing owing to their incredibly small size, high strength to weight ratio, improving functionality and exhibiting unique properties that would not be seen in bulk materials. Nanomaterial coatings have wide range of applications in aerospace and defense. Aircraft and military vehicles faces harsh environments. Nano coating significantly improve resistance to wear and tear, enhances durability and corrosion resistance. In addition, in defense and aerospace weight reduction is crucial. Hence, Nano coatings can provide the same level of protection as traditional, heavier coatings, allowing for light construction of vehicles and aircraft. This translates to better fuel efficiency for airplanes and increased maneuverability for military vehicles. This chapter provides an overview of coatings and their types, merits and demerits, focusing on current prospects and future trends along with aerospace and defense applications.

Keywords Nano coatings · Types · Merits and demerits · Current and Future · Challenges · Applications

3.1 Introduction

The term "nano" originates from the Greek word "nanos," meaning "dwarf" or something extremely small. Over the past decade, the use of the prefix "nano" has significantly expanded in scientific literature, becoming a standard term in various fields of research. The defining feature of nanoparticles is their size, which falls within

S. Kumar (✉)
Department of Applied Science, CGC College of Engineering, Chandigarh Group of Colleges, Landran, Mohali, Punjab, India
e-mail: santoshdgc@gmail.com

R. Kumar
Department of Quality Assurance & Regulator Affairs, Kaushik Orthopaedic Pvt. Ltd., Sonipat, Haryana, India

Visakh P. M. (ed.), *Nanomaterials for Defense, Aeronautical and Aerospace Applications*, Advanced Structured Materials 236,
https://doi.org/10.1007/978-3-031-92723-2_3

the range of 1–100 nm in at least one dimension (Buzea et al. 2007; Findik 2021; Gaffet 2011; Pal et al. 2011; Buzea and Pacheco 2007; Afolalu et al. 2019; Nagarajan 2008; Abdalkreem 2018; Mrinmoy et al. 2008). Many scientific publications classify nanomaterials as substances with at least one dimension within this nanoscale range. Nanotechnology has gained widespread attention due to its innovative applications across multiple industries. Nanomaterials are now essential in fields such as biomedicine, energy, pollution control, food engineering, mechanical engineering, coating (Santos et al. 2014; Joudeh and Linke 2022; National Research Council 2009). Their unique properties enable advancements that were previously unattainable with bulk materials. Today, nanotechnology plays a vital role in global progress. For instance, the adoption of green technologies that rely solely on nanomaterials presents a promising solution to combat climate change and reduce global warming. Research has shown that nanomaterials often exhibit superior efficiency compared to conventional bulk materials, making them a key component of sustainable technological advancements. Second, technologies for diagnosing and managing the global pandemic of illnesses are being developed using nanomaterials. In 2019, nanoparticles were utilized in the detection and treatment of COVID-19, a disease that had a significant global impact and caused widespread fatalities. Additionally, the application of nanomaterials allowed for the diagnosis that is now afflicting people worldwide. It is anticipated that nanotechnology, nanoscience, and nanomaterials will be crucial to global growth in the future. Everyone must thus possess sufficient knowledge and comprehension of nanomaterials. Nanoparticles and materials have been the subject of several investigations throughout the past 20 years. Nonetheless, a number of recent research suggest that they concentrate on the qualities, applications, preparation technique, and categorization of every nanomaterial. Very little research has been done on preparation, categorization, etc. (Lu and Ozcan 2015; Baig et al. 1821; Editorial 2021; Talebian et al. 2020a; Kumar et al. 2018; Nonymous 2020; Barhoum and Makhlouf 2018).

3.1.1 Historical Development

The use of nanotechnology dates back to the ninth century, making it a very old technology. If a material has at least one dimension in the 1–100 nm nanoscale range, it is referred to as a nanomaterial. Nanomaterials are widely recognized for their diverse and valuable applications across various industries. To put their scale into perspective, a nanometer—equivalent to one-billionth of a meter—can be illustrated by aligning five silicon atoms or ten hydrogen atoms in a row. Any material with at least one dimension ranging from 1 to 100 nm is classified as a nanomaterial. Tracing the exact history of nanoscale material usage is challenging; however, their application is not a recent development. Throughout history, humans have unknowingly utilized nanomaterials for different purposes, such as reinforcing ceramic mixtures with asbestos nanofibers approximately 4500 years ago. Nanomaterials have a propensity to change their basic characteristics as their size increases within the nanometer range. The intriguing nanomaterials have a wide range of uses in research and technology and show great basic characteristics. However, the idea of nanotechnology was there in primordial times without anybody realizing it. As a result, the historical overview, usage, and evolution are covered in this chapter. Compared to micro particles, nanoparticles have a larger surface area and more particles per unit mass. At first, tiny particle measurements were connected to gold. There are several synthetic approaches for creating nanoparticles, including top-down, hydro/solvothermal, and bottom-up. One of the most popular and widely used techniques for creating nanostructured materials, etc., is the hydrothermal process. The bottom-up techniques include seeded growth, 2D template-confined growth, mediated synthesis of small molecules and ions, and organic ligand-assisted growth.

Higher wear, erosion, high temperature oxidation, and corrosion resistance are among the specific enhancements that are desired. These materials fall into two primary types in the field of hard coatings: (1) hard nanocomposite coatings and (2) hard multilayer coatings. The carbides, and oxides of transition metals and elements (III and IV group) are mostly used nanostructured coatings in hardness applications. Because of their flexibility, which allows for the customization of nano coating characteristics, the use of a range of substrates, and the assurance of high repeatability of deposition conditions, these coatings are commonly manufactured using plasma-assisted procedures. Examples of nano coatings include WC–CrNi coatings, WC–CoCr coatings, WC–Co coatings, WC–Co cermet's, WC-based MMC coatings, and others. Nanomaterial coatings, particularly those applied through thermal spray techniques, play a crucial role in aerospace and defense applications by enhancing surface properties such as wear resistance, thermal stability, and corrosion protection. The future of thermal spray coatings in these industries is evolving toward nanostructured materials, offering superior performance, extended durability, and advanced functionalities tailored for extreme operational conditions (Amen et al. 2021; Keten and Buehler 2010; Valladas et al. 2001; Dai et al. 2002; Yetisen et al. 2016; Polette-Niewold et al. 2007; Kagdada et al. 2023; Sudha et al. 2018; Baig 2021; Trotta and Mele 2019; Heiligtag 2013; Singh et al. 2024; Kumar et al. 2021, 2020, 2019; Bedi et al. 2019).

3.1.2 Merits and Demerits of Nano Material Coatings

The following are the benefits and drawbacks of nanomaterial coatings (Saji et al. 2012; Aliofkhazraei 2015) (Fig. 3.1).

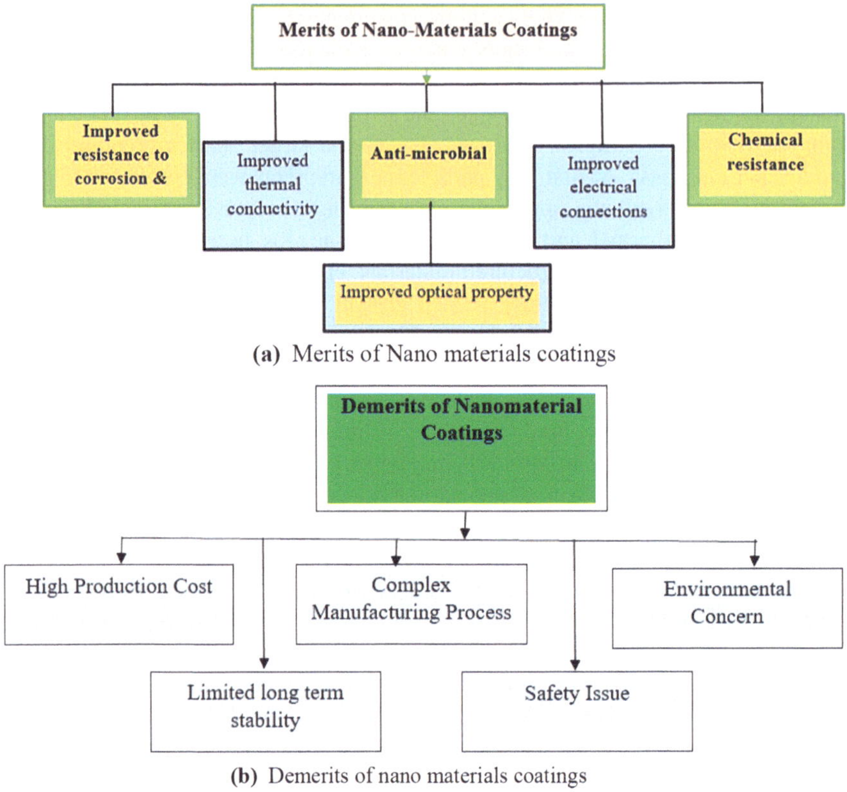

Fig. 3.1 **a** Merits of Nano materials coatings. **b** Demerits of nano materials coatings

3.2 Different Types of Nanomaterials and Nanomaterial Coatings for Aerospace Applications

Nanomaterials have a wide range of applications across various industries, including the chemical and cosmetic sectors, where they are used in nanoscale chemicals, compounds, paints, and coatings. In the field of nanomedicine, they play a crucial role in tissue engineering, medical devices, and the development of nanodrugs.

Nanotechnology also contributes significantly to multiple disciplines, such as nanoparticles, carbon nanotubes, and biopolymers. In food science, it is utilized in food processing, nutraceuticals, and nano-encapsulation techniques. Environmental and energy-related applications include fuel cells, photovoltaics, and advanced filtration systems for air and water purification. Additionally, nanomaterials are extensively used in military and defense technologies, enhancing biosensors, weapon systems, and sensory devices. In electronics, they are crucial for semiconductors, memory storage, photonics, and optoelectronics. Moreover, they have significantly advanced scientific research tools, such as atomic force microscopy and scanning tunneling microscopy, enabling precise analysis at the nanoscale (Shaikh 2022; Baig et al. 2021; Samuel et al. 2022; Prakash Sharma et al. 2018). Nanomaterials are playing an increasingly vital role in technological advancements and sustainable development. One significant application is in green technology, where nanomaterials contribute to reducing environmental impact by enhancing energy efficiency and minimizing pollution. Compared to bulk materials, nanomaterials have been shown to offer superior performance in various applications, making them essential in efforts to combat climate change. Additionally, nanomaterials have proven valuable in medical diagnostics and treatment, particularly in managing infectious diseases. In recent years, they have been utilized to develop advanced detection and treatment methods for viral outbreaks, such as COVID-19. Researchers are also exploring their potential in identifying and treating other diseases, including monkeypox. Looking ahead, the continued advancement of nanotechnology, nanoscience, and nanomaterials is expected to drive significant progress in multiple fields, shaping the future of global innovation. Everyone must so possess sufficient knowledge and comprehension of nanomaterials (Baig et al. 1821; Talebian et al. 2020b; Anonymous 2020). The group of nanomaterials as per number of criteria, including dimensions, origin. Composition, as seen in Fig. 3.2 below, nanomaterials may be categorized into five groups based on their size, origin, structural design, pore size, and possible toxicity.

Nanomaterials can be classified into two categories based on their origin: natural and artificial nanoparticles. Natural nanomaterials are widely present in the environment and include viruses, protein molecules, and minerals such as clay. They can also be found in natural colloids like milk and blood (liquid colloids), fog (aerosol type), and gelatin (gel type). Additionally, mineralized biological structures such as shells, corals, bones, and certain insect wings exhibit nanoscale features. Other naturally occurring nanomaterials include opals, spider silk, lotus leaves, gecko feet, volcanic ash, and ocean spray. In contrast, artificial nanomaterials are intentionally engineered using precise mechanical and manufacturing techniques. Examples include carbon nanotubes and semiconductor nanoparticles such as quantum dots (QDs),

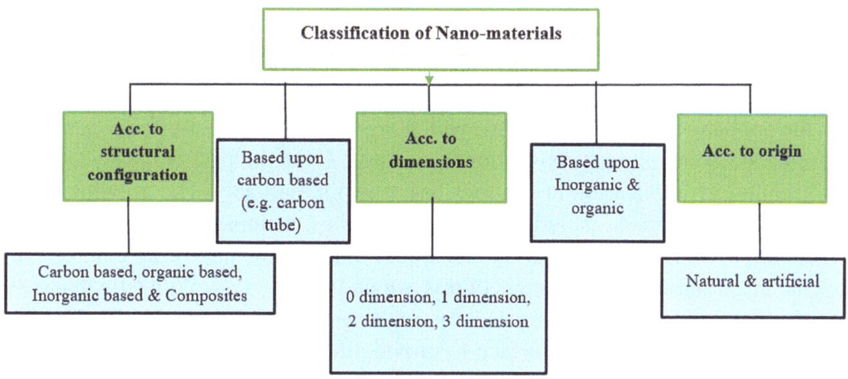

Fig. 3.2 General types of nanomaterials

which are designed for various technological applications. These synthetic nanoparticles are tailored to exhibit specific properties that make them highly valuable in fields like electronics, medicine, and material science. The classified of nanomaterial as dendrimers, which are materials based on metal. When it comes to structural composition and arrangement, the options include C based, inorganic, organic and composite (Barhoum and Makhlouf 2018; Zheng 2012; Khan and Hossain 2022). Examples of organic nanomaterials include ferritin, liposomes, dendrimers, and micelles. However, metal-based or metal oxide-based nanomaterials make up inorganic nanoparticles. Furthermore, according to their size dimensions, nanomaterials are divided into 4 groups: 0D, 1D, 2D, and 3D respectively (Joudeh and Linke 2022; Cho et al. 2019; Khan and Saeed 2019). In addition, the primary classification of nanomaterials is based on pore diameters (microporous, microporous), origin (natural and manufactured), possible toxicity (biopwesistent, CMAR), etc. (Findik 2021; Ijaz et al. 2020). A substrate is coated in order to provide various purposes, such as coloration, protection, or other uses. Resins, solvents, pigments, fillers, and additives make up its composition. While fillers affect the mechanical characteristics and provide the substrate certain extra properties, pigments give the coating color or corrosion protection. Since they enhance and add new qualities to paint, nanoparticles have recently attracted a lot of attention from the coatings sector. Scientists from all over the world and from a variety of fields, including biology, physics, and chemistry, have been interested in nanotechnology. Considering its many uses, advantages, and benefits, it is a promising technology. The coatings business is one of the areas where nanoparticles are anticipated to have a major impact.

The ability of nanomaterials to serve several purposes is well-known. Researchers in the automobile, automotive, and aerospace sectors have lately been interested in the latest developments in nanotechnology. High strength, low weight, and enhanced sensing and actuation capabilities are some of the benefits of nanomaterials. Because of their delamination toughness, fatigue resistance, impact resistance, and self-healing properties, nanoparticles have also recently been used in health monitoring apparel. In many industrial applications, coatings serve as an interface between environment and substrate material, improving the substance's longevity and performance. The aerospace and aviation industries make extensive use of thin films and functional surface coatings to enhance the qualities of cutting-edge building materials. An increasing number of aviation and aerospace businesses are looking into the application of nano coatings, which provide notable cost-effectiveness and performance benefits over conventional coatings. Over the past twenty years, the nanotechnology application in coatings has grown remarkably. The growing availability of nanomaterials, including carbon nanotubes, nanoparticles, and others, as well as improvements in deposition techniques that provide nanoscale control over the coating structure, are the causes of this astounding progress. With a thickness of 100 nm or less, Nano coatings are usually single- or multi-phase solid structures that are applied to an undelaying surface to provide the substrate material a particular characteristic or purpose. At the nanoscale, a variety of materials display distinctive characteristics. The particle size of a material plays a crucial role in determining its fundamental properties, including melting point, electrical conductivity, magnetic permeability, hydrophobicity, and chemical reactivity. For example, boron nitride is widely utilized as a coating material in the aerospace industry due to its exceptional thermal conductivity, chemical and thermal stability, and hydrophobic nature. In fiber form, it enhances composite materials by providing continuous and uniform properties. Additionally, boron nitride exhibits excellent insulation characteristics, as its nanotubes maintain a consistent wide band gap of over 5.2 eV, regardless of their diameter, chirality, or the number of tubular walls (Rai and Nguyen 2021; Dolez 2015; Vidales-Herrera 2020; Lopez 2020; Umaira et al. 2021; Barshilia 2021; Kiesel 2021).

3.3 Nanomaterials Coatings for Defense Applications

Because lightweight materials enable components to be lighter without compromising strength, they are essential in defense applications. Every nation's military forces made an effort to improve the agility of their vehicles and soldiers without sacrificing the capacity of their protective gear to withstand severe impacts. Aluminum, magnesium, titanium, and nickel are a few common lightweight metals and alloys. Because of their monolithic character, these lesser weight materials were considered unsuitable for use in military and defense uses in the past, when steel was the sole common metal. The hurdle to employing the aforementioned metals as steel substitutes has been broken by the advent of research on alloys and metal coating in the modern age. These lightweight metal alloys all have superior corrosion resistance and high specific strengths. The defense industry comprises significant industrial and economic components. The economic, social, and technical well-being of the populace and the environment have a direct impact on the nation's defense budget. On the other hand, the country's chances both at home and abroad are secured by the steady growth of all these factors. The biggest rise in the past ten years was the 4% increase in global defense spending in 2019. The rise in Chinese and American military spending was a prerequisite for this (Vasilev 2021; Abbasi 2020). Spending on defense has a variety of effects on economic growth. In addition to directly creating jobs, defense spending can indirectly increase economic production by transferring human capital and technology to the civilian sector. Because it takes funds away from government initiatives that may do more to spur growth, defense spending also has opportunity costs. As a result, we also take into account the relative impacts of infrastructure expenditures against defense spending. Numerous studies on the connection between defense expenditure and economic growth have been published by the defense and peace economics literature, but there is still disagreement about whether or not this link is favorable (Lizcano et al. 2021; Fontanel. 2019, 2022; Heo and Ye 2016; Kumar and Kumar 2021).

Lockheed Martin Corp., Boeing, BAE Systems, Leonardo, and others are defense-based businesses worldwide (Pieroni 2009; Dunne and Tian 2013; Castillo et al. 2001; https://www.boeing.com/defense/unmanned-little-bird-h-6u/index.page). Figure 3.3

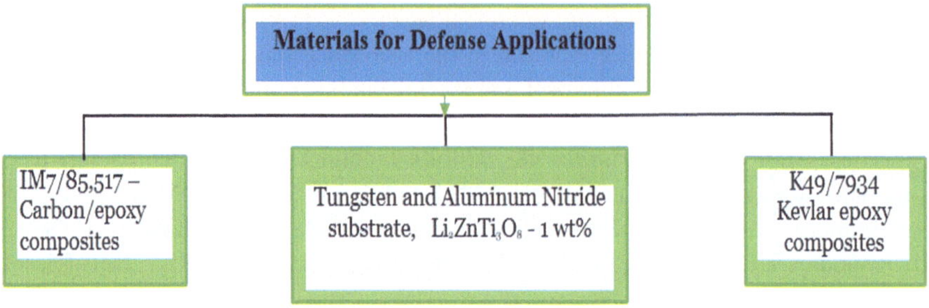

Fig. 3.3 Lightweight materials applications

illustrates that many lightweight materials are utilized for military purposes (https://electronics.leonardo.com/en/home; https://www.boeing.com/defense/support/a-10-wing-replacement-program/index.page; Sharon 2019; Peng and Li 2024; Shah and Maiaru 2021).

Liao and his colleagues developed a novel lightweight alloy for application in transportation. These alloys exhibit desirable mechanical properties, including enhanced compression plasticity exceeding 30% at room temperature, a high yield strength of approximately 1500 MPa, and a fracture strength compression type of around 1800 MPa (Mourad et al. 2020). Recent research has also highlighted the use of magnesium-based alloys, in the production of thin-walled aircraft profiles. These alloys enable the manufacturing of complex cross-sectional shapes, which are essential for lightweight extruded products in the aviation industry, ultimately contributing to reduced costs and energy consumption. Lowering the weight of aircraft has demonstrated significant improvements in material selection and production efficiency (Thomas et al. 2013). Syntactic foams, which are lightweight composite materials consisting of hollow particles dispersed within a matrix, are widely used in weight-sensitive applications such as pipeline thermal insulation, aerospace structures, and marine vessels. Metal-reinforced polymer foams have also shown excellent electromagnetic interference (EMI) shielding capabilities at relatively low densities (See et al. 2017). Metal matrix composites (MMCs) have demonstrated superior performance compared to traditional alloys in the defense sector. Carbon-based nanomaterials, such as graphene and carbon nanotubes (CNTs), are frequently used as reinforcements in MMCs, enabling their use at higher operating temperatures. These materials exhibit exceptional wear resistance, making them ideal for vehicle armor and lightweight protective applications. For example, carbon nanotubes are incorporated into an aluminum alloy matrix to produce Kryron Terminator, a ballistic body & vehicle armor. Similarly, Israel's Light Improved Ballistic Armor (LIBA) can withstand up to 20 strikes from 14.5 mm caliber projectiles, showcasing the effectiveness of metal matrix nanocomposites (Liao et al. 2020). The rapid advancement of military and aerospace communication systems is largely driven by the precision required for telecommunication applications. Researchers have explored the use of ceramic paints and inks, ultralow-temperature ceramics, etc., in military telecommunications systems. These materials are particularly beneficial for active electronically steered antennas (AESA), which consist of thousands of transmitter and receiver modules operating across various platforms and systems (Śliwa et al. 2017). Polymer materials are most efficient choice for manufacturing durable, lightweight components in military and defense applications. Due to their numerous advantages, such as non-conductive properties, and water resistance, polymers have increasingly replaced metal components in military applications. Polymers are widely regarded as secure and reliable materials for diverse defense applications, ranging from compass boxes to precision weapon components. Graphene nanofiller-reinforced polymer composites have been used in the development of advanced protective gear, such as the high-tech armor inspired by "Star Wars." This armor exhibits superior ballistic resistance, exceptional strength-to-weight ratio, hydrophobicity, and remarkable electrical and thermal conductivity. The suit also includes additional components such as a padded khaki uniform, gloves, a rifle, a radio wire, an exoskeleton layer, and a helmet with tinted night vision capabilities (Śliwa et al. 2017; Singh et al. 2018; Venkataramanan et al. 2021).

Polymers offer significant advantages in large-scale manufacturing, reducing both production time and costs due to their ease of assembly. Injection molding techniques have been widely adopted in the production of military components, i.e., hydraulic fittings in tanks, etc. These polymer-based components exhibit outstanding resistance to heat, chemicals. Unlike metals and ceramics, polymers do not require additional processing steps, further lowering production costs. Polymers used in military and defense include polypropylene, PEEK, polycarbonate, etc. (Schuldies and Nageswaran 2010; Oladele et al. 2020; Bird and Ravindra 2021).

3.4 Current Prospects and Future Trends

With their numerous advantages, nanomaterial coatings are reentering the aerospace and military sectors. At the moment, these coatings provide remarkable surface hardness, adhesion strength, and corrosion resistance, making them perfect for tough environments. Additional developments in nanomaterial coatings for defense and aerospace include advanced insulating materials, self-healing coatings, chemical agent detection, and safety enhancement. The worldwide aerospace and military coatings market is expected to develop at a compound yearly growth rate of 6.62% between 2023 and 2029. Certain materials, including magnesium alloy with an outstanding strength to weight ratio and Ti–Al-based alloy (for jet engine components), are being investigated for potential future uses. Defense technology is now headed toward greatness and will mostly concentrate on these areas for the industry's expansion. Artificial intelligence in defense greatly enhances computational military thinking, which is suitable for surveillance, reconnaissance missions, and intelligence. The application of computer vision to enhance the capabilities of autonomous weapon systems & also manage the security of force (military) equipment directly reduces personnel injuries. The defense is also improving its defensive technologies to make it more creative and efficient in the face of contemporary threats. Current technical developments include space militarization, directed energy weaponry, and supersonic flight (Elehinafe and Ayeni 2019; Reghunadhan et al. 2020; https://www.afmc.af.mil/News/Article-Display/Article/1879395/afrl-produces-lighter-thinner-transparent-armor; Thakur et al. 2023).

3.5 Conclusion

All things considered, nanomaterial coatings have the potential to significantly influence how the aerospace and military sectors develop in the future. They are a fascinating field for study and development because of their special qualities and possible uses. Further research and collaborative efforts will be crucial to fully harness their transformative potential, leading to more durable, efficient, and safer systems in these critical industries.

Acknowledgements The Chandigarh Group of College Landran Mohali Punjab & Kaushik Orthopaedic Pvt. Ltd. is acknowledged by the authors for providing the opportunity to conduct this study and motivation.

Conflict of Interest There is no conflict of interest, according to the authors.

References

Abbasi S (2020) Functionalized nanomaterials for the aerospace, vehicle, and sports industries. In: Micro and nano technologies, pp 795–825

Abdalkreem TM (2018) Master thesis, Sudan University of Science and Technology

Afolalu SA, Soetan SB, Ongbali SO, Abioye AA, Oni AS, Conf IOP (2019) Ser Mater Sci Eng 640:012065. https://doi.org/10.1088/1757-899X/640/1/012065

Aliofkhazraei M (2015) Anti-abrasive nano coatings, current and future applications, pp 210–280. https://doi.org/10.1016/C2013-0-16186-7

Amen R et al (2021) Chapter 1—History and development of nanomaterials. In: Nanomaterials: synthesis, characterization, hazards and safety micro and nano technologies, pp 1–14

Anonymous (2020) Nanomedicine and the COVID-19 vaccines. Nat Nanotechnol 15:963. https://doi.org/10.1038/s41565-020-00820-0

Bahaskaran R, Palaniswamy N, Rengaswamy NS (2013) A review of differing approaches used to estimate the cost of corrosion, anti-corrosion method and materials. J Appl Sci Res 52:29–41

Baig N (2021) Nanomaterials: a review of synthesis methods, properties, recent progress, and challenges. Mater Adv 2(6):1821–1871

Baig N, Kammakakam I, Falath W (1821) Mater. Adv 2021(2):71. https://doi.org/10.1038/s41565-020-0751-0

Baig N, Kammakakam I, Falath W (2021) Nanomaterials: a review of synthesis methods, properties, recent progress, and challenges. Mater Adv 2:1821–1871

Barhoum A, Makhlouf ASH (2018) Emerging applications of nanoparticles and architecture nanostructures. Elsevier, Amsterdam, The Netherlands. https://doi.org/10.1016/B978-0-323-51254-1.00012-9

Barshilia HC (2021) Surface modification technologies for aerospace and engineering applications: current trends, challenges and future prospects. Trans Indian Natl Acad Eng 6:173–188. https://doi.org/10.1007/s41403-021-00208-z

Bedi TS, Kumar S, Kumar R (2019) Corrosion performance of hydroxyapaite and hydroxyapaite/titania bond coating for biomedical applications. Mater Res Express 7:015402. https://doi.org/10.1088/2053-1591/ab5cc5

Bhat A, Budholiya S, Aravind Raj S, Sultan MT, Hui D, Md Shah Umaira A, Safri SNA (2021) Review on nanocomposites based on aerospace applications. Nanotechnol Rev 10(1):237–253. https://doi.org/10.1515/ntrev-2021-0018

Bird DT, Ravindra NM (2021) Additive manufacturing of sensors for military monitoring applications. Polymers 13(9):1455

Buzea C, Pacheco II (2007) Robbie, Biointerphases 2, MR17. https://doi.org/10.1116/1.2815690

Buzea C et al (2007) Biointerphases 2, MR17. https://doi.org/10.1116/1.2815690

Castillo J, Lowell J, Tellis AJ, Munoz J, Zycher B (2001) Military expenditures and conomic growth. Rand Arroyo Center Santa Monica CA

Cho G, Park Y, Hong YK, Ha DH (2019). Nano Converg. https://doi.org/10.1186/s40580-019-0187-0

Dai ZR, Bradley JP, Joswiak DJ, Brownlee DE, Hill HGM, Genge MJ (2002) Nature 418:157–159

Dolez P (ed) (2015) Nanoengineering: global approaches to health and safety issues. Elsevier, Amsterdam, Netherland

Dunne JP, Tian N (2013) Military expenditure and economic growth: a survey. Econ Peace Secur J 8(1)

Editorial (2021) Let's talk about lipid nanoparticles. Nat Rev Mater 6:99. https://doi.org/10.1038/s41578-021-00281-4

Elehinafe FB, Ayeni AO (2019) Processing of polymer-based nanocomposites in advanced engineering and military application polymer nanocomposites for advanced engineering and military applications. IGI Global, pp 1–9

Findik F (2021) Period Eng Nat Sci 62:75. https://doi.org/10.21533/pen.v9i3.1837

Fontanel J (2022) Is the SIPRI estimate of military expenditure a reliable indicator of the power of states? Defence Econ

Fontanel J (2019) New global and national military expenditure series developed by SIPRI Economie de la sécurité internationale

Gaffet E (2011) http://arxiv.org/abs/1106.2206, https://doi.org/10.1016/j.crhy.2011.06.002

Heiligtag FJ, Niederberger M (2013) Mater Today 16:262–271 https://doi.org/10.1016/j.mattod.2013.07.004

Heo U, Ye M (2016) Defense spending and economic growth around the globe: the direct and indirect link. Int Interact 42(5):774–796

https://electronics.leonardo.com/en/home

https://www.afmc.af.mil/News/Article-Display/Article/1879395/afrl-produces-lighter-thinner-transparent-armor

https://www.boeing.com/defense/support/a-10-wing-replacement-program/index.page

https://www.boeing.com/defense/unmanned-little-bird-h-6u/index.page

Ijaz I, Gilani E, Nazir A, Bukhari A (2020) Green Chem Lett Rev 13(223):45. https://doi.org/10.1080/17518253.2020.1802517

Joudeh N, Linke D (2022) J Nanobiotechnol 20:262. https://doi.org/10.1186/s12951-022-01477-8

Kagdada HL, Bhojani AK, Singh DK (2023) An overview of nanomaterials: history, fundamentals, and applications. In: Singh DK, Singh S, Singh P (eds) Nanomaterials. Springer, Singapore. https://doi.org/10.1007/978-981-19-7963-7_1

Keten S, Buehler MJ (2010) Appl Phys Lett 96:153701

Khan I, Saeed K, Khan I (2019) Arab J Chem 12:908, 31. https://doi.org/10.1016/j.arabjc.2017.05.011

Khan S, Hossain MK (2022) Nanoparticle-based polymer composites. Woodhead, London, p 15

Kiesel L (2021) Advanced air plasma spray TBCs for aerospace and industrial components [Online] www.aerospace.honeywell.com. Available at: https://aerospace.honeywell.com/us/en/learn/about-us/blogs/advanced-air-plasma-spray-tbcs

Kumar S, Kumar M, Handa A (2018) Combating hot corrosion of boiler tubes-a study. Eng Fail Anal 94:379–395. https://doi.org/10.1016/j.engfailanal.2018.08.004

Kumar S, Kumar M, Handa A (2019) Comparative study of high temperature oxidation behavior of wire arc sprayed Ni–Cr and Ni–Al coatings. Eng Fail Anal 106:104173–104189. https://doi.org/10.1016/j.engfailanal.2019.104173

Kumar S, Kumar M, Handa A (2020) Erosion corrosion behavior and mechanical property of wire arc sprayed Ni–Cr and Ni–Al coating on boiler steels in actual boiler environment. Mater High Temp 37(6):370–384. https://doi.org/10.1080/09603409.2020.1810922

Kumar S, Handa A, Chawla V, Grover NK, Kumar R (2021) Performance of thermal-sprayed coatings to combat hot corrosion of coal-fired boiler tube and effects of process parameters and post coating heat treatment on coating performance: a review. Surf Eng 37(7):833–860. https://doi.org/10.1080/02670844.2021.1924506

Kumar S, Kumar R (2021) Influence of processing conditions on the properties of thermal sprayed coating: a review. J Surf Engi 37(11):1339–1372. https://doi.org/10.1080/02670844.2021.1967024

Liao YC, Li TH, Tsai PH, Jang JSC, Hsieh KC, Chen CY, Tsao IY (2020) Designing novel lightweight, high-strength and high-plasticity Tix (AlCrNb) 100-x medium-entropy alloys. Intermetallics 117, Article 106673

Lizcano M et al (2021) A lecture on nanotechnology applications for aerospace. Available on https://ntrs.nasa.gov/api/citations/20210014264/downloads/Final%20Draft-A%20Lecture%20in%20Nano%20Technology%20for%20Aerospace%20Applications.pdf

Lopez NR (2020) Resistant project: how can nanotechnology reduce environmental impacts of aircrafts? [Online] www.openaccessgovernment.org. Available at: https://www.openaccessgovernment.org/nanotechnology-reduce-environmental-impacts-of-aircrafts/85311

Lu Y, Ozcan S (2015) Nano Today 10(417):20. https://doi.org/10.1016/j.nantod.2015.04.010

Mourad AH, Cherupurakal N, Hafeez F, Barsoum I, A. Genena F, S. Al Mansoori M, A. Al Marzooqi L (2020) Impact strengthening of laminated kevlar/epoxy composites by nanoparticle reinforcement. Polymers 12(12):2814

Mrinmoy D, Ghosh PS, Rotello VM (2008) Adv Mater 20:4225. https://doi.org/10.1002/adma.200703183

Nagarajan R (2008) Ann Cardiothorac Surg 10:341

National Research Council (2009) Advancing risk assessment

Nonymous (2020) Nanomedicine and the COVID-19 vaccines. Nat Nanotechnol 15:963. https://doi.org/10.1038/s41565-020-00820-0

Oladele IO, Omotosho TF, Adediran AA (2020) Polymer-based composites: an indispensable material for present and future applications. Int J Polym Sci

Pal SL, Jana U, Manna PK, Mohanta GP, Manavalan R (2011) JAPS 228:234

Peng S, Li P (2024) Applications of nanomaterials and nanotechnology in military and aerospace fields. In: Nanomaterials and nanotechnology. Springer, Singapore. https://doi.org/10.1007/978-981-97-8433-2_8

Pieroni L (2009) Military expenditure and economic growth. Defence Peace Econ 20(4):327–339

Polette-Niewold LA, Manciu FS, Torres B, Alvarado M Jr, Chianelli RR (2007) J Inorg Biochem 101:1958–1973

Prakash Sharma V, Sharma U, Chattopadhyay M, Shukla VN (2018) Advance applications of nanomaterials: a review. Mater Today Proc 5:6376–6380

Rai M, Nguyen TA (eds) (2021) Nanomaterials recycling. Elsevier, Amsterdam, Netherland

Randis R et al (2023) The potential of nanocomposite-based coatings for corrosion protection of metals: a review. J Mole Liquids 390, Part A:123067. https://www.sciencedirect.com/journal/journal-of-molecular-liquids

Reghunadhan A, Krishna A, Jose AJ (2020) Polymers in robotics polymer science and innovative applications. Elsevier, pp 393–421

Saji VS et al (2012) The impact of nanotechnology on reducing corrosion cost. In: Corrosion protection and control using nanomaterials, pp 42–72

Samuel MS, Ravikumar M, Ashwini JJ, Selvarajan E, Patel H, Chander PS, Soundarya J, Vuppala S, Balaji R, Chandrasekar N (2022) A review on green synthesis of nanoparticles and their diverse biomedical and environmental applications. Catalysts 12:459

Santos CSC, Gabriel B, Blanchy M, Menes O, García D, Blanco M, Arconada N, Neto V (2014) 5th international conference on advanced nano materials. https://doi.org/10.1016/j.matpr.2015.04.056

Schuldies JJ, Nageswaran R (2010) Ceramic matrix composites for ballistic protection of vehicles and personnel blast protection of civil infrastructures and vehicles using composites. Woodhead Publishing, pp 235–243

See TL, Chantzis D, Royer R, Metsios I, Antar M, Marimuthu S (2017) Ultraviolet-diode pump solid state laser removal of titanium aluminium nitride coating from tungsten carbide substrate. Lasers Manuf Mater Process 4(3):93–107

Shah SP, Maiarù M (2021) Effect of manufacturing on the transverse response of polymer matrix composites. Polymers 13(15):2491

Shaikh SF (2022) Nanomaterials applied in coatings: synthesis, structures, properties, and applications. Coatings 12(11):1773. https://doi.org/10.3390/coatings12111773

Sharon M (2019) Nanotechnology in the defense industry advances, innovation, and practical applications. https://doi.org/10.1002/9781119460503.fmatter

Singh AK, Shishkin A, Koppel T, Gupta N (2018) A review of porous lightweight composite materials for electromagnetic interference shielding. Compos B Eng 149:188–197

Singh H, Kumar M, Singh R, Kumar S (2024) Deposition behaviour of WC-17Co cold-sprayed composite coating on hydro turbine steels. Surf Rev Lett. https://doi.org/10.1142/S0218625X25500209

Śliwa RE, Balawender T, Hadasik E, Kuc D, Gontarz A, Korbel A, Bochniak W (2017) Metal forming of lightweight magnesium alloys for aviation applications. Arch Metall Mater 62

Sudha PN et al (2018) Chapter 12—Nanomaterials history, classification, unique properties, production and market. In: Emerging applications of nanoparticles and architecture nanostructures current prospects and future trends micro and nano technologies, pp 341–384. https://doi.org/10.1016/B978-0-323-51254-1.00012-9

Talebian S, Wallace GG, Schroeder FSA, Conde J (2020a) Nat Nanotechnol 15(618):21. https://doi.org/10.1038/s41565-020-0751-0

Talebian S, Wallace GG, Schroeder FSA, Conde J (2020) Nat Nanotechnol 15:618, 21. https://doi.org/10.1038/s41565-020-0751-0

Thakur A, Kaya S, Kumar A (2023) Recent trends in the characterization and application progress of nano-modified coatings in corrosion mitigation of metals and Alloys. Appl Sci 13(2):730. https://doi.org/10.3390/app13020730

Thomas D, Abhilash P, Sebastian MT (2013) Casting and characterization of $LiMgPO_4$ glass free LTCC tape for microwave applications. J Eur Ceram Soc 33(1):87–93

Trotta F, Mele A (2019) Nanosponges: from fundamentals to applications

Valladas H, Clottes J, Geneste J-M, Garcia MA, Arnold M, Cachier H, Tisnérat-Laborde N (2001) Nature 413:479

Vasilev C (2021) Applying nanocoatings to aviation: a review. AZoNano. Accessed 06 Nov 2024 from https://www.azonano.com/article.aspx?ArticleID=5843

Venkataramanan S, Selvaraj K, Ponnusamy P (2021) A brief review of carbon nanotube reinforced metal matrix composites for aerospace and defense applications. Int Nano Lett 11(4):321–345

Vidales-Herrera J (2020) Nanomaterials in coatings: an industrial point of view. In: Handbook of nanomaterials for manufacturing applications, micro and nano technologies, pp 51–77

Yadav L, Sihmar A, Kumar S et al (2024) Review of nano-based smart coatings for corrosion mitigation: mechanisms, performance, and future prospects. Environ Sci Pollut Res. https://doi.org/10.1007/s11356-024-33234-9rtyuyujjqaezdy5bqz5ybzybzaybzyyzyzy

Yetisen AK, Coskun AF, England G, Cho S, Butt H, Hurwitz J, Yun SH (2016) Adv Mater 28:1724–1742

Zheng S (2012) Reviews, vol 75. University of London, p 625. https://doi.org/10.1016/j.nantod.2008.10.014

Chapter 4
Spray Coating Process of Multi-Walled Carbon Nanotubes (MWCNT) for Aerospace Applications

Santosh Kumar and Rakesh Kumar

Abstract Spray coating is a versatile, simple, efficient, and most widely used technique for applying a thin layer of material to a surface. Spray coating has many advantages such as handle wide range of material and thickness, uniform coating on complex shape, scalability, cost efficient, waste reduction as compared to dipping & brushing, automated for high volume production, reducing labor cost and improving consistency. There are several distinct type of spray processes that can be used in aerospace application such as cold spray. But the selection of spray coating process for an aerospace application depends upon many factors such as target properties (corrosion resistance, wear resistance, etc.), desired coating materials, substrate limitations, and cost considerations. This chapter presents an overview of the various spray coating process, Types, preparation of MWCNT/epoxy nanocomposite films, spray coating process of MWCNT/epoxy nanocomposite films, MWCNT/epoxy nanocomposite films for aerospace applications, future trends and distinct challenges are discussed in detail.

Keywords Spray coatings · Types · Aerospace applications · Preparation of MWCNT/epoxy nanocomposite films · Challenges and opportunities

S. Kumar (✉)
Department of Applied Science, CGC College of Engineering, Chandigarh Group of Colleges, Landran, Mohali, Punjab, India
e-mail: santoshdgc@gmail.com; santosh.3267@cgc.edu.in

R. Kumar
Department of Regulatory Affairs & Quality Assurance, Kaushik Orthopaedic Pvt. Ltd., Sonipat, Haryana, India

Visakh P. M. (ed.), *Nanomaterials for Defense, Aeronautical and Aerospace Applications*, Advanced Structured Materials 236,
https://doi.org/10.1007/978-3-031-92723-2_4

4.1 Introduction

Carbon-based nanocomposite films have been widely researched for their exceptional multifunctional properties. These materials offer remarkable mechanical strength, superior electrical and thermal conductivity, extensive surface area, distinctive optical features, chemical stability, and biocompatibility. Such properties make them highly versatile for various scientific and technological applications. Their potential applications extend across multiple domains, including energy storage systems and emerging biomedical technologies. The synthesis methods used play a crucial role in controlling the composition and structural organization of these nanocomposites. This precise control over material distribution enhances their effectiveness in advanced applications, making them a promising choice for next-generation (Chen et al. 2003; Li et al. 2006; Clausi et al. 2017; Patel et al. 2020; Philip et al. 2003; Qiu et al. 2016; Wang et al. 2009). The main goal of producing nanocomposite films for the aerospace industry is to achieve outstanding tailored material qualities while significantly reducing weight and outperforming conventional materials' electrical, thermal, and mechanical characteristics. UV sensors, radar-absorbing shields, electromagnetic interference shields, radiation-protective and thermo-optical coatings, and more are potential uses for nanocomposite films. Furthermore, Fig. 4.1 illustrates the many applications (Laurenzi et al. 2018; Clausi et al. 2016a; Feng et al. 2021; Munakata et al. 2000; Li et al. 2008; Yang et al. 2007) of carbon-based nano composites (CNCs).

The kind of incorporated carbon nanoparticles determines the functional qualities of the nanocomposite films, and the manufacturing technique used has a significant impact on these properties. In addition to increasing imaging sensitivity and

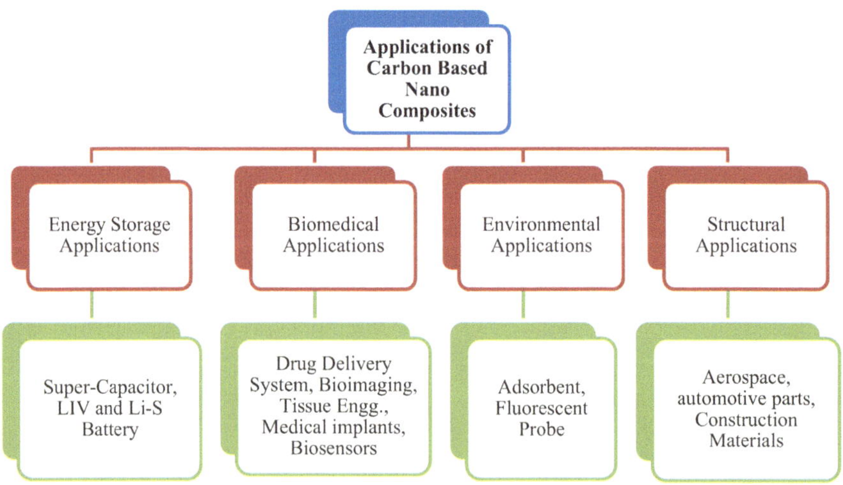

Fig. 4.1 Distinct applications of CNCs

specificity, nanocomposites allow multimodal imaging, offering a thorough understanding of anatomical and functional data (Fernandez et al. 2016; Jeong et al. 2021; Clausi et al. 2016b). Different types of carbon nanoparticles, varying in thickness, surface area, and aspect ratio, have gained significant attention as fillers in modern composite manufacturing. Their incorporation enhances the mechanical strength, thermal & electrical performance of composite materials, making them highly valuable for advanced applications. A crucial aspect of composite fabrication is the effective dispersion of carbon nanoparticles within the polymer matrix, as well as the film deposition process. Proper dispersion ensures uniformity and maximizes the material's functional properties. Compared to other matrices, polymers offer better process ability, making them easier to manipulate and integrate with carbon nanoparticles for improved composite performance.

They are not very structurally reliable for engineering purposes, though. When fillers are successfully added to polymer matrices, their mechanical, electrical, magnetic, optical, chemical, and surface characteristics are often improved in the condition of polymer–polymer blends. But the process of film deposition A thin film is a layer or layers of material that range in thickness from a nanometer (monolayer) to several micrometers (a stack of thin films is termed a multilayer). The thin film is applied on the substrate for a variety of reasons, such as protecting it, altering its appearance (decorative purposes), or altering its optical or electrical characteristics.

Applying a thin coating of any material to a surface—whether it be a substrate or layers that have already been deposited—is known as thin film deposition, and it is a crucial step in many applications. Deposition techniques may be divided into two main groups based on whether the process is largely physical or chemical: this talk will cover the most widely used approaches. A single layer of atoms or molecules can be deposited at a time using a few distinct processes, such as atomic layer deposition (ALD), the Langmuir–Blodgett method, and molecular beam epitaxy (MBE). Unquestionably, thin films are used in many aspects of our daily lives. One straightforward example is the mirror we use every day to view our features.

Therefore, they must be carefully engineered to optimize their functional qualities and produce homogenous and uniform nanocomposite films (Ansari and Giannelis 2009; Eda and Chhowalla 2009; Peng 2008; Song and Youn 2005; Xie et al. 2005; Norrman et al. 2005a; Moshiul 2022; Naghdi n.d.). With the ability to apply coatings on flat surfaces (like glass or plastic) or larger, non-planar surfaces, spray coating has become a popular deposition technique in the industrial sector as an alternative to spin coating. Applying a liquid coating substance, such as polymer solution or photoresist, on a substrate that rotates quickly produces a thin, consistent layer in the case of spin coating. Uniformity, thickness control, fast processing speed, minimal material consumption, and the use of a variety of materials are some benefits of spin coating. Nonetheless, spin coating finds use in energy storage, biological applications (such as biosensors and implantable devices), semiconductor fabrication, and other fields (Naghdi n.d.Ichiki et al. 2004; Girotto et al. 2009; Steirer et al. 2009). Therefore, by regulating the feeding rate, the applied pressure, the spraying distance, etc., the spray coating offers a way to achieve novel shapes by enabling the deposition of precisely calibrated layers. Vak and associates differentiate between wet and dry spray coatings (Vak et al. 2015). The presence of a fluid layer of resin on the substrate during and soon after the spraying process, which may be accomplished by either not heating the substrate or by exposing it to the jet at a slower pace or for a longer period of time, is indicative of a wet spray coating. Conversely, the dry spray coating method uses heated substrates and brief resin jets. After a heat cure cycle, the films were created on a Mylar substrate using epoxy resin supplemented with carbon nanoparticles at varying loadings, ranging from 0.5 to 2 weight percent. Since flexible Mylar membranes are widely utilized in many spacecraft subsystems, particularly as thermal barriers, they were selected as the substrate for the nanocomposites (Cadogan and Lin 1999; Babuscia et al. 2013; Kerzhanovich et al. 2004; Engelhart et al. 2017; Mazzinghi et al. 2018; Fesmire and Johnson 2018). Several negative effects caused by the contact of the subsystem, such as a multilayer insulation (MLI), with the aerospace conditions can be lessened by the use of an electrically conductive coating over a Mylar substrate. The dielectric properties of MLI materials, including Mylar, present significant challenges due to charging phenomena and electromagnetic interference, which can compromise the integrity and functionality of spacecraft. Given the excellent electrical conductivity and electromagnetic absorption capabilities of CNT-based materials, CNT-based nanocomposite films on Mylar provide a highly effective solution, seamlessly integrating with MLIs to mitigate these issues (Rawal et al. 2013).

4.1.1 Historical Development

When Dr. Max Ulrick Schoop of Zurich discovered that "a stream of molten particles impinging upon themselves" could produce a coating, thermal spraying was born. Using high-pressure gas, engineers were experimenting at the time to break up liquids into powders or small particles. This discovery by Dr. Schoop resulted in the creation of the first thermal spray method. In order to create coatings from molten and powdered metals, Dr. Schoop and his German engineering partners, Felix Meyer and Franz Herkenrath, were able to create machinery and procedures. As a result of this progress, the first fixed metal spraying factory was established in 1910. The first device for spraying wire form solid metal was developed in 1912 by Dr. Schoop and his colleagues. The method used to create this instrument was metalizing. The principle of metallizing states that when a wire rod is introduced into an intense, concentrated flame—produced by burning fuel gas with oxygen—it melts. If a stream of compressed gas surrounds the flame, the molten metal is atomized and propelled onto a surface, forming a coating. This process, once known as metallizing, is now commonly referred to as flame spraying or oxy-fuel spraying. Various oxy-fuel processes have since been developed, including ceramic powder spraying, ceramic rod spraying, HVOF spraying, molten metal spraying, wire spraying. The outbreak of World War II in 1939 significantly accelerated advancements in the American thermal spray industry, leading to widespread innovation and adoption of these technologies, and the International Thermal Spray Association's members were instrumental in supplying the "metallizing" that was sorely required for industrial equipment replacement components. Metallizing was acknowledged globally as a significant and extremely useful industrial process by 1945, when the war came to a conclusion. Large raised water tanks, tuna fishing boats, tank vehicles and tanks used in the chemical sector, capacitor castings, and pipes were among the applications during the first phase. ITSA created industry standards for coatings, their uses, and inspection techniques in response to a market that was becoming more complex. These criteria led to more commercial prospects for the whole metallizing sector and were disseminated to engineering firms, designers, and educational institutions worldwide. Fusible alloys, ceramic coatings, plasma spraying, and further specialty coatings were among the many innovations developed by emerging companies. The two primary categories of thermal spraying methods are electrical and flame spray. Powder, wire, detonation, molten metal, ceramic rod, and high velocity oxygen fuel are all included in flame spray. Additionally, there are non-trans electric plasma, wire electric arc, and radiofrequency plasma in the electrical situation. There is a lot of interest in this field since these methods are becoming more popular and thermal spray coating applications are growing quickly. Surface coating is a useful method for improving the mechanical and functional characteristics of certain machine parts or components. However, because of their exceptional qualities in many situations, coatings based on high entropy alloys (HEAs) are becoming more significant than conventional coatings.

These materials possess several attractive characteristics, including distinctive magnetic and electrical properties that make them suitable for both functional and structural applications. They also exhibit high hardness, exceptional fracture toughness, superior strength, resistance to high-temperature oxidation, and protection against hot corrosion. However, despite these advantages, high-entropy alloys come with certain drawbacks, primarily their higher cost compared to conventional alloys. The incorporation of more expensive elements like niobium (Nb), chromium, etc., might be the cause of this increased cost. To withstand erosion, corrosion, tension, and other factors, these HEAs coatings are extensively utilized in a variety of technical and industrial applications (Tucker 2013; Espallargas n.d.; Kumar 2024; Hsu et al. 2005).

4.1.2 Distinct Spray Coating Processes

Thermal spray encompasses a variety of deposition techniques, all of which utilize concentrated heat sources to melt feedstock materials and propel the molten particles onto a surface at different kinetic energy levels. These techniques rely on thermal energy generated from electric discharges, liquid fuels, or the combustion of flammable gases such as hydrogen and hydrocarbons. The term "thermal spray processes" broadly refers to both conventional thermal spraying methods—such as flame spraying as well as deposition techniques driven solely by kinetic energy. Thermal spray methods are typically classified based on the type of energy source used to melt the feedstock material. The most commonly employed energy sources include thermal and kinetic energy from electric discharges like electric arcs or ionized plasma gases, thermal energy from the combustion of gases such as hydrogen or hydrocarbons, and pure kinetic energy utilized in cold spray processes. Thermal spraying is a highly versatile technology with a broad selection of deposition techniques and coating materials, making it valuable across multiple industries. It is widely employed to enhance corrosion and oxidation resistance, provide high-temperature protection, prevent wear and erosion, create abradable coatings, restore dimensional accuracy, and support biomedical and electronic applications. (Qadir et al. 2024; Ann Gan et al. 2015; Agarwal et al. 2022; Britain 1985; Kumar et al. 2019a, 2020, 2021, 2024; Fauchais and Vardelle 2012; Kumar and Kumar 2021; Mauer et al. 2013; Hardwicke and Lau 2013; Wilson 2013; Bediet al. 2019; Sharma et al. 2020). Technologies that use the same concept to deposit metallic or nonmetallic coatings are included in thermal spraying methods. The feedstock material is melted by a heat source, and the molten particles are then given kinetic energy by a jet. They constantly accumulate the required thickness after impacting the substrate surface and quickly cooling to create a solid splat (Crawmer 2013, 2008; Centers for Disease Control and Prevention 2011). The development of functional coatings focused on their functionality, such as their ability to prevent infections or kill pathogens on orthopedic implants, to prevent algae and hard-shell organisms from adhering and growing on water-submerged equipment, and to deposit coatings with electrochemical properties. Bacteria and germs' antimicrobial qualities and surface adherence result in serious consequences such surgical site infections or chronic wounds in the medical area or health problems brought on by tainted and expired food goods in the food business. These factors have contributed to the recent rise in popularity of antimicrobial coatings, which stop or inhibit the growth of harmful microbes (Prudenziati et al. 2006). In contrast to the passive behavior displayed by conventional and functional coatings, smart coatings are thought to represent the next stage in the creation of more competent coatings, explaining their active responsiveness to intrinsic or external stimuli. Sensors, embedded sensors, photoluminescence, self-heating, self-lubrication, and other features are examples of smart coatings (Taniguchi et al. 1982; White et al. 2001; García et al. 2011; Kohler et al. 2008). Antimicrobial compounds can be added to thermal-sprayed coatings, which have a track record of efficacy on surgical implants, to lessen post-operative rejection. Other sectors, such protecting ship hulls and maritime equipment against fouler attachment and growth, would benefit from the new features, which would have a significant effect on productivity and financial turnover. Functional coatings that successfully lessen the influence of detrimental phenomena brought on by environmental circumstances are specifically designed to offer a higher level of protection.

4.1.3 Merits and Demerits

The benefits and drawbacks of spray coating multiwalled carbon nanotubes for aerospace applications are displayed in Fig. 4.2 (a, b). According to Song et al. (2007), PU–CNT coatings have improved electrical and thermal mechanical qualities. When compared to conventional films, Liu and Guo (2010) discovered that PU-MWCNT films had a higher coefficient of friction and a higher level of wear resistance. Research has examined the impacts of PU-MWCNT and plain PU's glass transition temperatures and found significant variations between the two (Xia and Song 2005; Kukovecz et al. 2013). A flux blockage type hypothesis is the most widely accepted idea, even if the processes underlying changes in thermal stability are still mostly understood (Xia and Song 2005; Li et al. 2017).

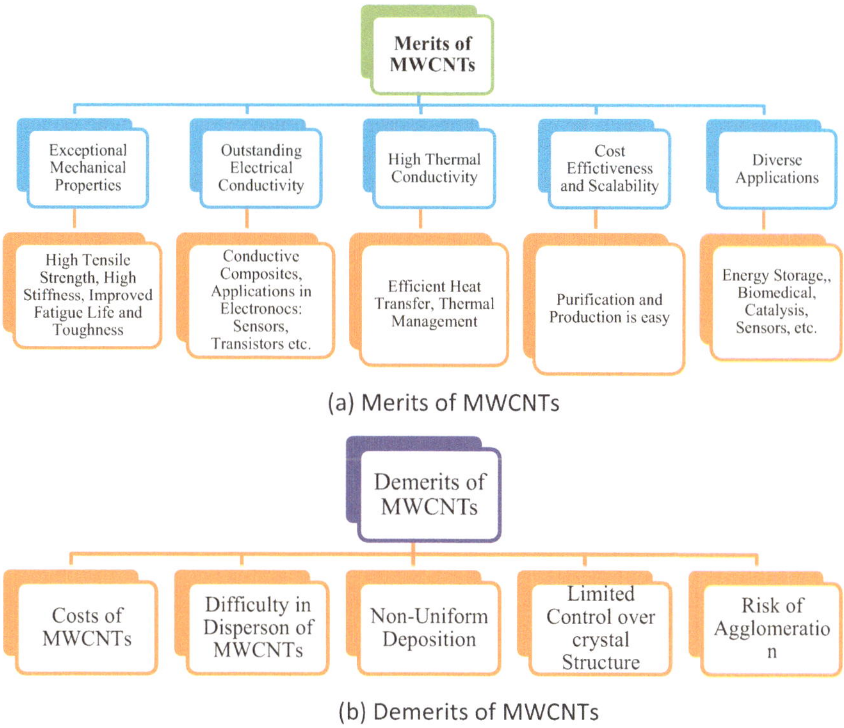

Fig. 4.2 **a** Merits of MWCNTs. **b** Demerits of MWCNTs

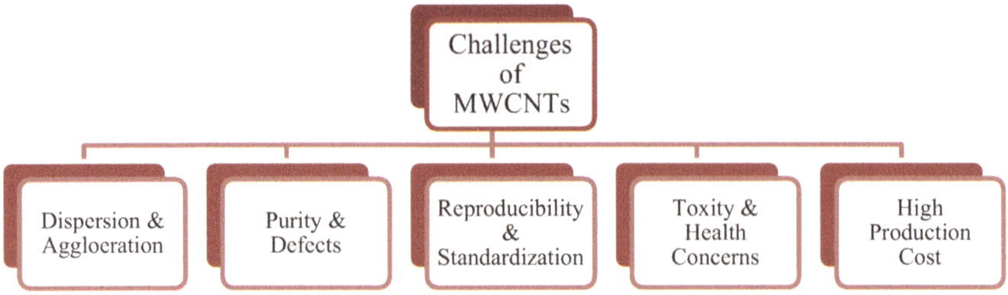

Fig. 4.3 Challenges of MWCNTs

4.2 Preparation of MWCNT/Epoxy Nanocomposite Films

Stacking and foaming play a crucial role in producing multilayered composite foams. To improve the microwave absorption properties of PUG composites, Xu et al. (2008) designed multilayered thermoplastic polyurethane/graphene (PUG) composites by assembling multiple single-layered PUG foams. This approach allows for precise control over each layer's thickness, cell morphology, and graphene concentration. However, its practical application is constrained by the irregular interfaces between adjacent layers. A promising approach for fabricating multilayered foams with continuous interfaces involves stacking multiple sample sheets through hot melt pressing or multilayer coextrusion, followed by the foaming of these laminated structures (Sumey et al. 2019; Zhao et al. 2015). For example, Zhou et al. (2019) utilized a multilayer coextrusion process to develop a poly(ethylene-co-octene) composite featuring alternating foam and film layers. Their study demonstrated that increasing the number of layers led to a reduction in average cell size and enhanced sound absorption performance. Likewise, Yuan et al. (2018) combined physical supercritical CO_2 foaming with melt hot pressing to create poly(methyl methacrylate) (PMMA) foams with a well-defined multilayered cellular morphology. Their results revealed that when the interface spacing was smaller than the critical nucleation size of the cells, the foams exhibited a uniform, continuous, and directionally structured multilayer architecture. Despite these advancements, research on the impact of cell morphology variations across adjacent layers on the overall properties of multilayered composite foams remains limited. This challenge arises primarily due to difficulties in precisely controlling the cell structure in each layer (Li et al. 2016). Epoxy adhesives have been widely employed in the fabrication of conductive composite foams owing to their exceptional characteristics, such as strong adhesion, excellent thermal and chemical stability, and low toxicity (Wang et al. 2019; Fan et al. 2019). For instance, Yang et al. (2017) utilized a batch foaming technique with supercritical CO_2 to produce epoxy/silver nanosheet (EP/AgNS) composite foams. Their findings indicated that incorporating 20 wt% AgNS resulted in impressive material properties, including an electrical conductivity of 89.12 S/m, a specific electromagnetic interference (EMI) shielding efficiency (SE) of 334.59 dB·cm^3/g, and a thermal conductivity of 58.71 mW/m·K. Similarly, Li et al. (2006) employed chemical foaming to develop epoxy/

nickel-coated carbon fiber (EP/NCCF) conductive foams, which achieved a specific EMI SE of up to 77.4 dB·cm^3/g in the X-band at a density of 0.45 g/cm^3. However, research on epoxy-based conductive foams with alternating multilayer cell structures remains largely unexplored. To address this gap, recent studies have examined the foaming behavior of two- and four-layer composite foams with varying concentrations of multiwalled carbon nanotubes (MWCNTs) or chemical foaming agents (CFAs) in adjacent layers. These studies aim to establish correlations between multilayered architectures and key performance metrics, such as mechanical strength, EMI shielding effectiveness, electrical conductivity, and thermal conductivity.

4.3 MWCNT/Epoxy Nanocomposite Films for Aerospace Applications

Flexible membranes coated with nanocomposite materials have a number of intriguing qualities that make them valuable for aerospace applications, including charging mitigation layers and thermal blankets. Because of their remarkable multifunctional qualities, carbon-based nanocomposite films are the subject of much research (Laurenzi et al. 2018; Fan et al. 2019). The main goal of producing nanocomposite films for the aerospace industry is to achieve outstanding tailored material qualities while significantly reducing weight and outperforming conventional materials' electrical, thermal, and mechanical characteristics. Nanocomposite films may be used for electromagnetic interference shields, UV sensors, radar absorption, radiation protection, and thermo-optical coatings (Clausi et al. 2016a). The functional properties of nanocomposite films are largely influenced by the type of carbon nanoparticles incorporated, as well as the manufacturing technique employed. Various carbon nanoparticles, differing in thickness, specific surface area, and aspect ratio, have gained significant attention as fillers in modern composite production. Their inclusion enhances the mechanical strength, electrical conductivity, and thermal performance of the final composite materials (Eda and Chhowalla 2009; Peng 2008; Norrman et al. 2005b). Important production procedures that require careful planning to produce homogenous and uniform nanocomposite films and enhance their functional qualities are the dispersion of carbon nanoparticles in the polymer matrix and the film deposition procedure. With the benefit of being able to apply coatings on larger, non-planar surfaces, spray coating has replaced spin coating as a deposition technology in the industrial sector (Steirer et al. 2009; Babuscia et al. 2013). Spray coating is a technique that involves expelling resin (printing ink) through the nozzle of an aerograph by applying external pressure, typically up to 3.5 bar. Unlike traditional printing or coating methods, which apply a solution in a single step to form a film, spray coating deposits multiple layers of fine droplets. By adjusting factors such as feeding rate, applied pressure, spraying distance, and the composition of the sprayed mixture, this method enables the precise deposition of layers, allowing for the creation of advanced material architectures. Vak and colleagues categorize spray coating into two types: wet and dry. Wet spray coating is characterized by the presence of a fluid resin layer on the substrate during and immediately after spraying, achieved either by not heating the substrate or by applying the spray at a slower rate or for a longer duration. In contrast, dry spray coating employs heated substrates and short bursts of resin jets. Given the widespread use of flexible Mylar membranes in spacecraft subsystems, particularly as thermal barriers, they were selected as the substrate for the nanocomposite coatings. The application of an electrically conductive coating over Mylar can mitigate adverse effects caused by interactions between spacecraft subsystems, such as multilayer insulation (MLI), and the harsh conditions of space. The dielectric nature of MLI materials, including Mylar, makes them susceptible to electromagnetic interference and charging phenomena, which can compromise the spacecraft's integrity and functionality. Due to their excellent electromagnetic absorption and electrical conductivity properties, CNT-based nanocomposite films applied to Mylar present a promising solution for integration with MLIs. This study first investigates the dispersion state and viscosity of MWCNT/epoxy blends at varying carbon nanotube concentrations. It then evaluates how different spray coating parameters influence the quality and uniformity of the deposited coatings. The research examines two commonly used industrial spray gun nozzle sizes, along with varying spraying distances and carbon nanotube concentrations. Additionally, the electrical properties of nanocomposite films coated on Mylar are analyzed concerning spraying parameters. Specifically, the uniformity of surface electrical conductivity is assessed by comparing samples from different regions of the sprayed nanocomposite films. Additionally, electrical measurements taken from various spray-covered locations provide insights into the critical parameters that must be optimized during the manufacturing process to enhance the properties of these films for aerospace applications (Kumar et al. 2018, 2019b). A subset of composite materials called nanocomposites is distinguished by having better mechanical qualities than ordinary composite materials. There is currently a research gap on the usage of nanocomposites in the aerospace sector, namely with regard to determining the potential future applications. The usage of appropriate nanocomposites is responsible for the majority of aerospace sector triumphs. For aerospace applications, materials with a lower material erosion yield factor and the ability to withstand temperature variations of hundreds of degrees are taken into consideration. Aerospace structures require materials with exceptional strength and stiffness to maintain their mechanical properties under high-temperature conditions. The expected outcomes in application areas include enhanced base material strength, improved temperature resistance, and better tribological performance, it was first accomplished using composites that contained a mixture of graphene and epoxy resin as a curing agent. Advanced composites have been used to corrosion resistance, fatigue damage, and lightweight aircraft constructions. Wire arc spray coating (WASC) and MWCNT/epoxy nanocomposite films are two different techniques for improving the qualities of materials, especially in aerospace applications. Despite their disparate methods, they both seek to enhance materials' performance by adding new features or fortifying their current qualities (Binder et al. 1999; Kumar et al. 2019c; Liang et al. 2009 Jan). The vibration dampening coefficients of the polymer nanocomposites reinforced with CNT sheets were much lower. The MWCNT can be utilized in aerospace applications due of its enhanced mechanical, electrical, and thermal characteristics (Pernigoni and Grande 2020 Dec). Nanocomposites are the perfect technology for aircraft applications because of their interconnected sustainable effects and steady technological advancement. Nanocomposites may lower waste production in manufacturing processes and be incorporated into intricate aircraft designs. This may be applied to the design of fuselages and structures that are lightweight and need less maintenance (Kumar et al. 2021). Boostani et al. investigated the use of powder metallurgy in the production of graphene nanosheets and SiC nanoparticles, which accumulated to form thermally active compounds for use in aerospace applications. The aerospace sector was transformed by the fast expansion of additive manufacturing (AM) techniques, although the usage of fused deposition modeling (FDM) was constrained due to its decreased strength. Using glass pellet copolymers, SMP was created by combining SMP pellets in a (DMF) solvent, adding carbon black (CB), and then sonicating the mixture. Solvent casting along with DMF evaporation was used to create the nanocomposites. Because conductive CB amplifies the electrical stimulation, SMP nanocomposites have better toughness capabilities. With their optimal constructions for cube-satellites, they might find value in aeronautical applications. When it comes to creating microelectronics for use in aircraft, MWCNTs are crucial (Rosales et al. 2018; Ramachandran et al. 2022).

4.4 Aerospace Specific Challenges

For long-term stability, MWCNT-based coatings must have exceptional mechanical qualities such strength and toughness, high electrical conductivity, and chemical resistance in addition to being able to tolerate high temperatures and harsh environmental conditions. Selecting nanocomposites over traditional alloys is made simpler by the need for materials with long-lasting qualities in space technology and its ever-growing problems (Ramachandran et al. 2022; Huang et al. 2024).

4.5 Future Scope

Nanocomposites have made significant advancements in the field of aircraft technology. These materials are particularly well-suited for high-performance applications that demand structurally superior components. Due to their exceptional properties, nanocomposites effectively meet the stringent requirements of aerospace engineering. By using these materials, certain components' structural capabilities have been greatly enhanced, and the aerospace industry's strict material and production requirements have been satisfied. Sustainable technology like solar power helped to expand future space exploration (Gola et al. 2020). When it comes to space-based tasks, such as MMOD damage resistance and radiation insulation needs in orbit, the future of nanocomposites appears bright [99, 100].

4.6 Conclusion

In aerospace, nanocomposites have successfully aided in ground-breaking achievements. Because of coatings, the development of nanocomposite materials has led to increased fatigue strength, lightweight components, and radiation control.

Nanocomposite components were designed and processed using a variety of production techniques, with novel methods continuously being examined and created. Each approach has advantages, and the choice of technique can significantly impact the quality of the nanocomposite, depending on the intricacy of the machinery and the complexity of the processes to be followed. Superior mechanical qualities have been demonstrated by graphene, and the molecular dynamics simulation demonstrated that graphene allowed for increased strength stability and decreased fatigue stress. These exhibit outstanding qualities that have the potential to completely transform the aerospace sector. The increase in MWCNTs led to an improvement in hardness and fracture resistance, which might be improved by 76% for aerospace applications.

Acknowledgements The authors would like to thank Chandigarh Group of Colleges Landran Mohali Punjab & Kaushik Orthopaedic Pvt. Ltd. Haryana for the motivation & support in this research.

Conflict of Interest The authors declares that no any conflict of interest.

References

Agarwal H, Breining WM, Sánchez-Velázquez G, Lynn DM (2022) Reactive multilayers and coatings fabricated by spray assembly: influence of polymer structure and process parameters on multiscale structure and interfacial properties. Chem Mater 34:1245–1258

Ann Gan J, Berndt CC (2015) 23—Thermal spray forming of titanium and its alloys. In: Qian M, Froes FH (eds) Titanium powder metallurgy. Butterworth-Heinemann, Boston, pp 425–446

Ansari S, Giannelis EP (2009) Functionalized graphene sheet—poly (vinylidene fluoride) conductive nanocomposites. J Polym Sci B Polym Phys 47(9):888–897

Araújo FA, Freire FN, Pinho DC, Dutra KH, Rocha PA, da Silva ME (2021) Study of surfaces, produced with the use of granite and titanium, for applications with solar thermal collectors. Rev Adv Mater Sci 60(1):47–56

Babuscia A, Corbin B, Knapp M, Jensen-Clem R, Van de Loo M, Seager S (2013) Inflatable antenna for cubesats: motivation for development and antenna design. Acta Astronaut 91:322–332

Bedi TS, Kumar S, Kumar R (2019) Corrosion performance of hydroxyapaite and hydroxyapaite/titania bond coating for biomedical applications. Mater Res Express 7:015402. https://doi.org/10.1088/2053-1591/ab5cc5

Binder K, Gennes PG, Giannelis EP, Grest GS, Hervet H, Krishnamoorti R, et al (1999) Polymers in confined environments, 1st ed. Springer, Berlin, Heidelberg. https://doi.org/10.1007/3-540-69711-X

Britain g (1985) Wear Resistant surfaces in engineering– guide to their production, properties and selection; London Her Majesty's Stationery Office, Department of Trade and Industry, UK

Cadogan D, Lin J (1999) Inflatable solar array technology. In: 37th aerospace sciences meeting and exhibit, pp 1075

Centers for Disease Control and Prevention (2011) Vital Signs: Incidence and Trends of Infection with Pathogens Transmitted Commonly Through Food - Foodborne Diseases Active Surveillance Network, 10 U.S. Sites, 1996–2010, MMWR 60(22):749–755

Chen WX, Lee JY, Liu Z (2003) The nanocomposites of carbon nanotube with Sb and SnSb0.5 as Li-ion battery anodes. Carbon 41(5):959–966. https://doi.org/10.1016/S0008-6223(02)00425-6

Clausi M, Santonicola MG, Laurenzi S (2016b) Fabrication of carbon-based nanocomposite films by spin-coating process: an experimental and modeling study of the film thickness. Compos Appl Sci Manuf 88:86–97

Clausi M, Santonicola MG, Schirone L, Laurenzi S (2017) Analysis of ultraviolet exposure effects on the surface properties of epoxy/graphene nanocomposite films on Mylar substrate. Acta Astronaut 134:307–313

Clausi M, Laurenzi S, La Saponara V, Gabriella Santonicola M (2016) Multifunctional sensors for UV and mechanical damage detection of aerospace structures. In: Proceedings of the international astronautical congress. IAC

Clausi M, Santonicola MG, Schirone L, Laurenzi S (n.d.) Analysis of ultraviolet exposure effects on the surface properties of epoxy/graphene nanocomposite films on Mylar substrate

Crawmer DE (2008) Introduction to coatings, equipment, and theory, handbook thermal spray technology, pp 43–46

Crawmer DE (2013) Thermal Spray processes, handbook of thermal spray technology. ASM International, pp 54–76

Eda G, Chhowalla M (2009) Graphene-based composite thin films for electronics. Nano Lett 9(2):814–818

Engelhart D, Cooper R, Cowardin H, Maxwell J, Plis E, Ferguson D, Barton D, Schiefer S, Hoffmann R (2017) Space weathering experiments on spacecraft materials

Espallargas N (n.d.) Introduction to thermal spray coatings. https://doi.org/10.1016/B978-0-85709-769-9.00001-4

Fan X, Zhang G, Gao Q, Li J, Shang Z, Zhang H, Zhang Y, Shi X, Qin J (2019) Highly expansive, thermally insulating epoxy/Ag nanosheet composite foam for electromagnetic interference shielding. Chem Eng J 372:191–202. https://doi.org/10.1016/j.cej.2019.04.069

Fauchais P, Vardelle A (2012) Thermal sprayed coatings used against corrosion and corrosive wear. In: Salimi Jazi H (ed) Advanced plasma spray applications. InTech, pp 3–39. https://doi.org/10.5772/34448

Feng Z, Adolfsson KH, Xu Y, Fang H, Hakkarainen M, Wu M (2021) Carbon dot/polymer nanocomposites: from green synthesis to energy, environmental and biomedical applications. Sustain Mater Technol 29:e00304. https://doi.org/10.1016/j.susmat.2021.e00304

Fernandez M, Shi H, Barnard AS (2016) Geometrical features can predict electronic properties of graphene nanoflakes. Carbon 103:142–150

Fesmire JE, Johnson WL (2018) Cylindrical cryogenic calorimeter testing of six types of multilayer insulation systems. Cryogenics 89:58–75

García SJ, Fischer HR, White PA, Mardel J, González-García Y, Mol JMC, Hughes AE (2011) Self-healing anticorrosive organic coating based on an encapsulated water reactive silyl ester: synthesis and proof of concept. Prog Org Coat 70(2–3):142–149

Girotto C, Rand BP, Genoe J, Heremans P (2009) Exploring spray coating as a deposition technique for the fabrication of solution-processed solar cells. Sol Energy Mater Sol Cell 93(4):454–458

Gola Y, Kim D, Namilae S (2020) Piezoresistive nanocomposites for sensing MMOD impact damage in inflatable space structures. Compos Commun 1(21):100375

Hardwicke CU, Lau YC (2013) Advances in thermal spray coatings for gas turbines and energy generation: a review. J Therm Spray Technol 22(5):564–576. https://doi.org/10.1007/s11666-013-9904-0

Hsu Y-J, Chiang W-C, Wu J-K (2005) Corrosion behavior of FeCoNiCrCux high-entropy alloys in 3.5% sodium chloride solution. Mater Chem Phys 92(1):112–117

Huang S, Fedorov RG, Ein-Eli Y (2024) Silicon-coated multi-walled carbon nanotube (MWCNT) tissues as flexible free-standing anodes for advanced Li-ion batteries. J Solid State Electrochem 28:2139–2149. https://doi.org/10.1007/s10008-023-05737-x

Ichiki M, Zhang L, Yang Z, Ikehara T, Maeda R (2004) Thin film formation on nonplanar surface with use of spray coating fabrication. Microsyst Technol 10(5):360–363

Jeong S, Yoo SW, Kim HJ, Park J, Kim JW, Lee C et al (2021) Recent progress on molecular photoacoustic imaging with carbon-based nanocomposites. Materials 14(19):5643. https://doi.org/10.3390/ma14195643

Kerzhanovich VV, Cutts JA, Cooper HW, Hall JL, McDonald BA, Pauken MT, White CV, Yavrouian AH, Castano A, Cathey HM, Fairbrother DA, Smith IS, Shreves CM, Lachenmeier T, Rainwater E, Smith M (2004) Breakthrough in mars balloon technology. Adv Space Res 33(10):1836–1841

Kohler AR, Som C, Helland A, Gottschalk F (2008) Studying the potential release of carbon nanotubes throughout the application life cycle. J Clean Prod 16:927–937

Kukovecz A et al (2013) Multi-walled carbon nanotubes. Springer Handbook of Nanomaterials, pp 1–30. https://doi.org/10.1007/978-3-642-20595-8_5

Kumar S (2024) Comprehensive review on high entropy alloy-based coating. Surf Coat Technol 477:1–30

Kumar S, Kumar R (2021) Influence of processing conditions on the properties of thermal sprayed coating: a review. Surf Eng 37(11):1339–1372. https://doi.org/10.1080/02670844.2021.1967024

Kumar S, Kumar M, Handa A (2018) Combating hot corrosion of boiler tubes-a study. Eng Fail Anal 94:379–395. https://doi.org/10.1016/j.engfailanal.2018.08.004

Kumar S, Kumar M, Handa A (2019a) High temperature oxidation and erosion-corrosion behaviour of wire arc sprayed Ni–Cr coating on boiler steel. Mater Res Express 6:125533. https://doi.org/10.1088/2053-1591/ab5fae

Kumar S, Kumar M, Handa A (2019b) Comparative study of high temperature oxidation behavior of wire arc sprayed Ni–Cr and Ni–Al coatings. Eng Fail Anal 106:104173–104189. https://doi.org/10.1016/j.engfailanal.2019.104173

Kumar S, Handa A, Kumar R (2019c) Overview of wire arc spray process: a review. Jf Compos Theory 12(7):900–907

Kumar S, Kumar M, Jindal N (2020) Overview of cold spray coatings applications and comparisons: a critical review. World J Eng 17(1):27–51. https://doi.org/10.1108/WJE-01-2019-0021

Kumar S, Handa A, Chawla V, Grover NK, Kumar R (2021) Performance of thermal-sprayed coatings to combat hot corrosion of coal-fired boiler tube and effects of process parameters and post coating heat treatment on coating performance: a review. Surf Eng 37(7):833–860. https://doi.org/10.1080/02670844.2021.1924506

Kumar K, Kumar S, Gill HS (2024) Role of carbide-based thermal-sprayed coatings to prevent failure for boiler steels: a review. J Failure Anal Prevent. https://doi.org/10.1007/s11668-024-01974-y

Kumar R et al (2021) The role of additive manufacturing for biomedical applications: a critical review . J Manuf Process. https://doi.org/10.1016/j.jmapro.2021.02.022

Laurenzi S, de Zanet G, Santonicola MG (2020) Numerical investigation of radiation shielding properties of polyethylenebased nanocomposite materials in different space environments. Acta Astronaut 1(170):530–538

Laurenzi S, Zaccardi F, Semeraro M, Santonicola MG (2018) Nanocomposite layers for spacecraft protection: from multiscale numerical model to experimental data. In: 18th European conference on composite materials. Athens, Greece

Li J, Hu L, Wang L, Zhou Y, Grüner G, Marks TJ (2006) Organic light-emitting diodes having carbon nanotube anodes. Nano Lett 6(11):2472–2477

Li Y, Chen C, Zhang S, Ni Y, Huang J (2008) Electrical conductivity and electromagnetic interference shielding characteristics of multiwalled carbon nanotube filled polyacrylate composite films. Appl Surf Sci 254(18):5766–5771

Li J, Zhang G, Ma Z, Fan X, Fan X, Qin J, Shi X (2016) Morphologies and electromagnetic interference shielding performances of microcellular epoxy/multi-wall carbon nanotube nanocomposite foams. Compos Sci Technol 129:70–78. https://doi.org/10.1016/j.compscitech.2016.04.003

Li Y, Shen B, Yi D, Zhang L, Zhai W, Wei X, Zheng W (2017) The influence of gradient and sandwich configurations on the electromagnetic interference shielding performance of multilayered thermoplastic polyurethane/graphene composite foams. Compos Sci Technol 138:209–216. https://doi.org/10.1016/j.compscitech.2016.12.002

Liang F, Tang Y, Gou J, Gu HC, Song G (2009) Multifunctional nanocomposites with high damping performance for aerospace structures. ASME Int Mech Eng Congress Exposition 43840(1):267–273. https://doi.org/10.1115/IMECE2009-12542

Liu T, Guo S (2010) Properties of polyurethane/carbon nanotube anocomposites Polymer nanotube nanocomposites: synthesis, properties and applications. Scrivener, Salem, MA, pp 141–176

Mauer G, Jarligo MO, Mack DE, Vaßen R (2013) Plasma-sprayed thermal barrier coatings: new materials, processing issues, and solutions. J Therm Spray Technol 22(5):646–658

Mazzinghi A, Sabbadini M, Freni A (2018) Enhanced RF behavior multi-layer thermal insulation. Sci Rep 8(1):91

Moshiul Alam AKM (2022) Methods of nanoparticle dispersion in the polymer matrix. In: Nanoparticle-based polymer composites woodhead publishing series in composites science and engineering, pp 469–479. https://doi.org/10.1016/B978-0-12-824272-8.00008-7

Munakata N, Makita K, Bolsée D, Gillotay D, Horneck G (2000) Spore dosimetry of solar UV radiation: applications to monitoring of daily irradiance and personal exposure. Adv Space Res 26(12):1995–2003

Naghdi S (n.d.) Thin films-definition, deposition techniques, and applications. https://doi.org/10.13140/RG.2.2.35029.96480/1

Norrman K, Ghanbari-Siahkali A, Larsen N (2005) 6 Studies of spin-coated polymer films. Ann Rep Sect C(Phys Chem) 101:174–201

Patel BR, Imran S, Ye W, Weng H, Noroozifar M, Kerman K (2020) Simultaneous voltammetric detection of six biomolecules using a nanocomposite of titanium dioxide nanorods with multi-walled carbon nanotubes. Electrochim Acta 362:137094. https://doi.org/10.1016/j.electacta.2020.137094

Peng H (2008) Aligned carbon nanotube/polymer composite films with robust flexibility, high transparency, and excellent conductivity. J Am Chem Soc 130(1):42–43

Pernigoni L, Grande AM (2020) Development of a supramolecular polymer based self-healing multilayer system for inflatable structures. Acta Astronaut 1(177):697–706. https://doi.org/10.1016/j.actaastro.2020.08.025

Philip B, Abraham JK, Chandrasekhar A, Varadan VK (2003) Carbon nanotube/PMMA composite thin films for gas-sensing applications. Smart Mater Struct 12(6):935

Prudenziati M, Cirri G, Bo PD (2006) Novel high-temperature reliable heaters in plasma spray technology. J Therm Spray Technol 15(3):329–331. https://doi.org/10.1361/105996306X124293

Qadir D, Sharif R, Nasir R et al (2024) A review on coatings through thermal spraying. Chem Pap 78:71–91. https://doi.org/10.1007/s11696-023-03089-4

Qiu X, Lu L, Leng J, Yu Y, Wang W, Jiang M et al (2016) An enhanced electrochemical platform based on graphene oxide and multi-walled carbon nanotubes nanocomposite for sensitive determination of sunset yellow and Tartrazine. Food Chem 190:889–895. https://doi.org/10.1016/j.foodchem.2015.06.045

Ramachandran K et al (2022) Multi-walled carbon nanotubes (MWCNTs)-reinforced ceramic nanocomposites for aerospace applications: a review. J Mater Sci 57(2):42–58. https://doi.org/10.1007/s10853-021-06760-x
Rawal S, Brantley J, Karabudak N (2013) Development of carbon nanotube-based composite for spacecraft components. In: 6th international conference on recent advances in space technologies (RAST), pp 13–19
Rosales CA, Duarte MF, Kim H, Chavez L, Hodges D, Mandal P et al (2018) 3D printing of shape memory polymer (SMP)/carbon black (CB) nanocomposites with electro-responsive toughness enhancement. Mater Res Express. 5(6):065704. https://doi.org/10.1088/2053-1591/aacd53
Sharma V, Kumar S, Kumar M, Deepak D (2020) High temperature oxidation performance of Ni–Cr–Ti and Ni–5Al coatings. Mater Today Proc 26(3):3397–3406. https://doi.org/10.1016/j.matpr.2019.11.048
Song YS, Youn JR (2005) Influence of dispersion states of carbon nanotubes on physical properties of epoxy nanocomposites. Carbon 43(7):1378–1385
Song HJ, Zhang ZZ, Men XH (2007) Surface-modified carbon nanotubes and the effect of their addition on the tribological behavior of a polyurethane coating. Eur Polym J 43:4092–4102
Steirer KX, Reese MO, Rupert BL, Kopidakis N, Olson DC, Collins RT, Ginley DS (2009) Ultrasonic spray deposition for production of organic solar cells. Sol Energy Mater Sol Cell 93(4):447–453
Sumey JL, Sarver JA, Kiran E (2019) Foaming of polystyrene and poly (methyl methacrylate) multilayered thin films with supercritical carbon dioxide. J Supercrit Fluids 145:243–252. https://doi.org/10.1016/j.supflu.2018.12.001
Taniguchi H, Kenichi H, Shiraishi H (1982) Oxygen Sens
Tucker RC, Jr (2013) A brief history of the development of thermal spray processes and materials, ASM handbook thermal spray technology, pp 348–350. https://doi.org/10.31399/asm.hb.v05a.a0005713
Vak D, van Embden J, Wong WW, Watkins S (2015) Optically monitored spray coating system for the controlled deposition of the photoactive layer in organic solar cells. Appl Phys Lett 106(3):7_1
Wang J, Chen Y, Blau WJ (2009) Carbon nanotubes and nanotube composites for nonlinear optical devices. J Mater Chem 19:7425–7443. https://doi.org/10.1039/B906294G
Wang L, He Y, Jiang T, Zhang X, Zhang C, Peng X (2019) Morphologies and properties of epoxy/multi-walled carbon nanotube nanocomposite foams prepared through the free-foaming and limited-foaming process. Compos Sci Technol 182:107776. https://doi.org/10.1016/j.compscitech.2019.107776
White SR, Sottos NR, Geubelle PH, Moore JS, Kessler MR, Sriram SR, Brown EN, Viswanathan S (2001) Autonomic healing of polymer composites. Nature 409(6822):794–797
Wilson s (2013) Abradable thermal spray applications and technology, ASM handbook, volume 5A: thermal spray technology. ASM International, pp 287–295
Xia H, Song M (2005) Preparation and characterization of polyurethane–carbon nanotube composites. Soft Matter 1:386–394
Xie X-L, Mai Y-W, Zhou X-P (2005) Dispersion and alignment of carbon nanotubes in polymer matrix: a review. Mater Sci Eng R Rep 49(4):89–112
Xu S, Wen M, Li J, Guo S, Wang M, Du Q, Shen J, Zhang Y, Jiang S (2008) Structure and properties of electrically conducting composites consisting of alternating layers of pure polypropylene and polypropylene with a carbon black filler. Polymer 49:4861–4870. https://doi.org/10.1016/j.polymer.2008.08.056
Yang Y, Gupta MC, Dudley KL (2007) Towards cost-efficient EMI shielding materials using carbon nanostructure-based nanocomposites. Nanotechnology 18(34):345701
Yang J, Yang Y, Duan H, Zhao G, Liu Y (2017) Light-weight epoxy/nickel coated carbon fibers conductive foams for electromagnetic interference shielding. J Mater Sci: Mater Electron 28:5925–5930. https://doi.org/10.1007/s10854-016-6266-7

Yuan H, Xiong Y, Shen Q, Luo G, Zhou D, Liu L (2018) Synthesis and electromagnetic absorbing performances of CNTs/PMMA laminated nanocomposite foams in X-band. Compos Part A 107:334–341. https://doi.org/10.1016/j.compositesa.2018.01.024

Zhao T, Yang M, Wu H, Guo S, Sun X, Liang W (2015) Preparation of a new foam/film structure poly (ethylene-co-octene) foam materials and its sound absorption properties. Mater Lett 139:275–278. https://doi.org/10.1016/j.matlet.2014.10.061

Zhou D, Xiong Y, Yuan H, Luo G, Zhang J, Shen Q, Zhang L (2019) Synthesis and compressive behaviors of PMMA microporous foam with multi-layer cell structure. Compos B 165:272–278. https://doi.org/10.1016/j.compositesb.2018.11.118

Chapter 5
Introduction of Fibre Reinforced Polymers-Polymers and Composites for Aeronautical and Aerospace Applications

Sonia S. Raj, B. Harish Kumar, and Edwin Raja Dhas

Abstract With the ever-evolving demands for lightweight yet robust materials in the aerospace industry, Fibre reinforced Polymers (FRPs) have emerged as a pivotal class of materials offering an optimal combination of strength, durability, and weight efficiency. Understanding the fundamental aspects of FRPs is becoming increasingly important for engineers and researchers, aiming to bring value to the evolution of lightweight and durable materials for aerospace applications as the industry continues to push the boundaries of innovation. This chapter offers a thorough introduction to the use of FRPs in the field of aeronautical and aerospace applications. The introductory overview delves into the fundamental characteristics of polymers and composites, highlighting the distinctive characteristics which render them desirable for aeronautical engineering. Furthermore, the paper explores how fibre reinforcement in enhancing the mechanical characteristics of polymers, key manufacturing processes employed in fabricating fibre-reinforced polymer composites for specific aerospace needs. The chapter concludes with a glimpse into the ongoing advancements and future prospects of fibre-reinforced polymers in aeronautics and aerospace.

Keywords Fibre reinforced polymer · Light weight · Aeronautical · Aviation

---

S. S. Raj (✉)
Department of Mechanical Engineering, Marian Engineering College, Kazhakuttam, Trivandrum, India
e-mail: soniasraj05@gmail.com

B. H. Kumar
Department of Mechanical Engineering College of Engineering, Trivandrum, India
e-mail: jazzyhari@gmail.com

E. R. Dhas
Department of Automobile Engineering, Noorul Islam Centre for Higher Education, Kumaracoil, Tamil Nadu, India
e-mail: edwinrajadhas@rediffmail.com

Visakh P. M. (ed.), *Nanomaterials for Defense, Aeronautical and Aerospace Applications*, Advanced Structured Materials 236,
https://doi.org/10.1007/978-3-031-92723-2_5

5.1 Introduction

Fibre-reinforced polymers (FRPs) represent a class of composite materials that have gained widespread use and recognition across various industries due to their exceptional mechanical properties and versatility. In certain applications, these materials combine the strength and flexibility of fibres like carbon, glass, or aramid with the lightweight and mouldability of polymers to create a high-performance material with better mechanical qualities than conventional materials like metals and plastics (Dinită et al. 2024). The continuous development of FRP technology contributes to advancements in various sectors, providing solutions for challenges related to weight reduction, durability, and design optimization.

Historically, metal alloys like steel and aluminium were used to build aircraft. The need to improve aircraft performance, fuel efficiency, and structural integrity has led to the introduction of FRPs in aeronautical and aerospace applications. Composite materials were initially employed by the aerospace industry in the 1960s for additional structures and non-structural elements. Composites have become increasingly important in the main aircraft structural construction due to advances in material research and manufacturing techniques (Parveez n.d.). The aerospace composites market is anticipated to expand from 2022 to 2027 at a compound yearly growth rate (CAGR) of 11.7%, from an estimated USD 29.7 billion in 2022 to USD 51.6 billion, as per a report published in Marketsandmarkets (Rajak et al. 2019). Due to its remarkable performance qualities and ability to endure harsh environments, aerospace composites are experiencing rapid market growth. The industry is also predicted to grow in the near future as a result of the global need for fuel-efficient aircraft on a worldwide scale. According to an analysis of the aerospace composites sector, carbon fibre is the market leader and is expected to remain so for some time to come. The preferred material for aircraft and spacecraft is carbon fibre since it is almost twice as durable than ceramic and glass fibres. (Wazeer et al. 2023). Additionally, the aircraft's weight is reduced by the high-strength carbon fibre, which improves engine performance and reduces fuel consumption. Due to its superior strength compared to ceramic and glass fibres, carbon fibre is the preferred material for spacecraft and aeroplanes. High-strength carbon fibre also reduces aircraft weight, which improves engine performance and reduces fuel consumption (Oku n.d.).

In summary, the chapter provides a comprehensive introduction to the subject, covering the basics of polymers and composites, emphasizing the role of fibre reinforcement, detailing manufacturing processes, showcasing aerospace applications, and offering insights into ongoing advancements. This structure serves to equip readers with a foundational understanding of the significance and potential of fibre-reinforced polymers in the field of aeronautical and aerospace engineering.

5.2 Fibre Reinforced Polymer Composites

The term "fibre reinforced polymer composites," or "FRP," refers to a class of composite material consisting of a hard polymer matrix (thermoset or thermoplastic) bonded with fibres, usually carbon, glass, or natural fibres (Yazdani et al. 2018). Figure 5.1 shows an overview of polymer composites. When compared to one another, each of these polymers has advantages and disadvantages in terms of cost, method of application, and durability. Consequently, the type of polymer to be used in the production of the composite material is determined by linked criteria. The reinforcing fibres might be arranged as particles, discontinuous, or continuous. The fibres exhibit high stiffness and strength, homogeneous cross-sectional properties with minimal material property variation between individual fibres, and high aspect ratios for their geometry (extremely high length to diameter ratio). The resin matrix shields the fibres from the working environment, abrasion, and impact damage while distributing the load applied to the composites among all of the individual fibres (Galinska and Cezary 2020). The latter characteristic is crucial to guarantee the equal distribution and transfer of loads through each fibre in each cross section of the fibre-reinforced polymer composites and to avoid failures too soon because of stress concentrations brought on by uneven material characteristics or geometry.

The anisotropic structure of FRP composites makes them directionally dependant. The type, amount, and method of reinforcing used in the manufacturing process determine the finished composite's qualities. Factors such as the type, quantity, and orientation of reinforcing fibre, resin type, product cost, production volume, and service conditions are the most important ones that affect composite design (Milosevic et al.

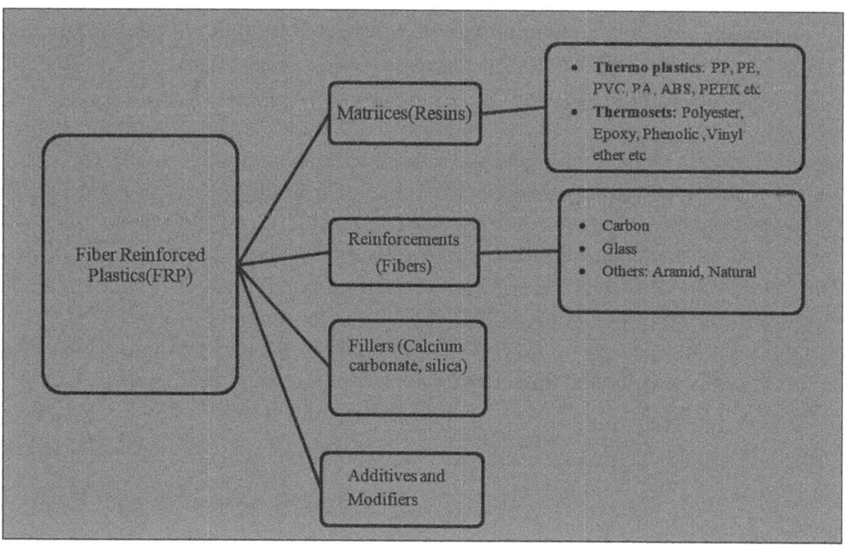

Fig. 5.1 Overview of polymer composites

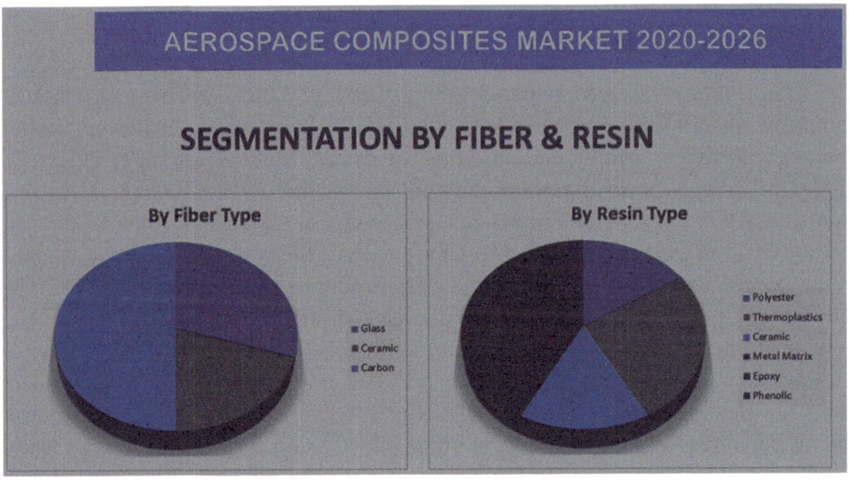

Fig. 5.2 Aerospace composite market segmentation by Fibre type and resin type (Ansari 2019)

2022). Considering these factors in the design phase is essential for developing composite materials that meet specific performance requirements while being cost-effective and manufacturable. The optimization of these parameters contributes to the overall success of composite material applications in various industries. Figure 5.2. shows the aerospace composite market segmentation by Fibre type and resin type (Ansari 2019).

According to the most recent study, the size of the global market for fibre-reinforced polymer (FRP) composites was assessed at USD million in 2022 and is expected to increase at a compound yearly growth rate (CAGR) of percent to reach USD million by 2028 (ABS 2023). Polymer-based composites, more specifically FRP composites, are 25% stronger and 25% lighter than steel. They also offer resistance against deterioration and rust. The car parts are safer and more fuel-efficient since they are stronger, lighter, and more mouldable than aluminium car parts. The industry of transportation makes extensive use of fibre-reinforced polymer (FRP) composites for a range of vehicles, such as cars, boats, ships, aeroplanes, and helicopters. The market position of FRPs by application is displayed in Fig. 5.3 (Fiber Reinforced Polymer (FRP) 2023). Glass Fibre Reinforced Plastics (GFRP), Carbon Fibre Reinforced Plastics (CFRP), Aramid Fibre Reinforced Plastics (AFRP) and natural fibre reinforced plastics (NFRP) are the different categories into which FRP composites can be separated based on the type of reinforcing fibre used.

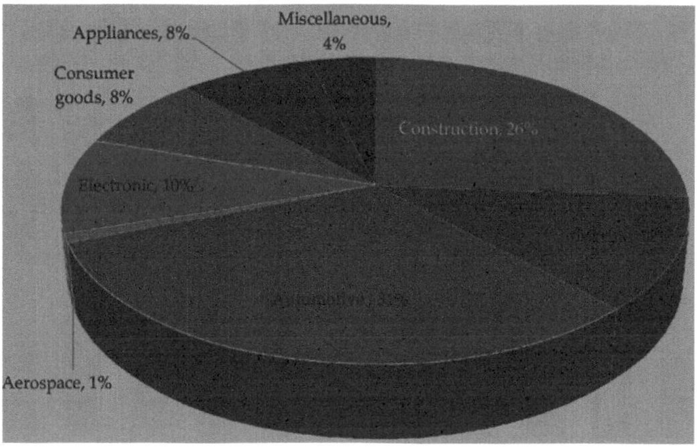

Fig. 5.3 Market share of Fibre Reinforced Polymer (FRP) by application (Fiber Reinforced Polymer (FRP) 2023)

5.2.1 Carbon Fibre Reinforced Plastics (CFRP)

In the first decades and mid-twentieth century, aluminum was the primary material used in the construction of aircraft due to its combination of strength and relatively low weight compared to other metals. While aluminum is strong and has good structural properties, as you mentioned, it also has some drawbacks, particularly in terms of weight. To tackle these fundamental issues, the aerospace sector shifted from using aluminium as the main material for aeroplane building to using carbon fibre. One of the most well-known forms of fibre reinforcement is carbon fibre. The first high-performance fibres for use in aircraft and spacecraft applications hit the market immediately after the creation of carbon fibre in its current form, which dates back to the early 1960s. High tensile strength, strong chemical resistance, high rigidity, minimal thermal expansion, and light weight are all characteristics of carbon fibre. But the most well-known attribute of carbon fibre is its strength-to-weight ratio (Vijayan et al. 2023). The airline industry also reaps significant benefits from the durability of carbon fibre. Lower maintenance costs are another benefit of using carbon fibre as the principal building material for aeroplanes. The two most frequent issues that most metals face are fatigue and corrosion. Aircraft metallic components are prone to strain or stress, particularly when the aircraft is used for many flights. In addition, these parts corrode more quickly than carbon fibres. Carbon fibre is undoubtedly resistant to fatigue and corrosion (Gürgen et al. 2016).

The market for carbon fibre reinforced plastic was estimated to be worth USD 23.6 billion in 2022, and between 2023 and 2030, it is projected to expand at a compound annual growth rate (CAGR) of 7.2% (Carbon Fiber Reinforced Plastic Market Size, Share and Trends Analysis Report n.d.). Government laws to reduce vehicle pollution and the growing need for fuel-efficient vehicles, along with rising

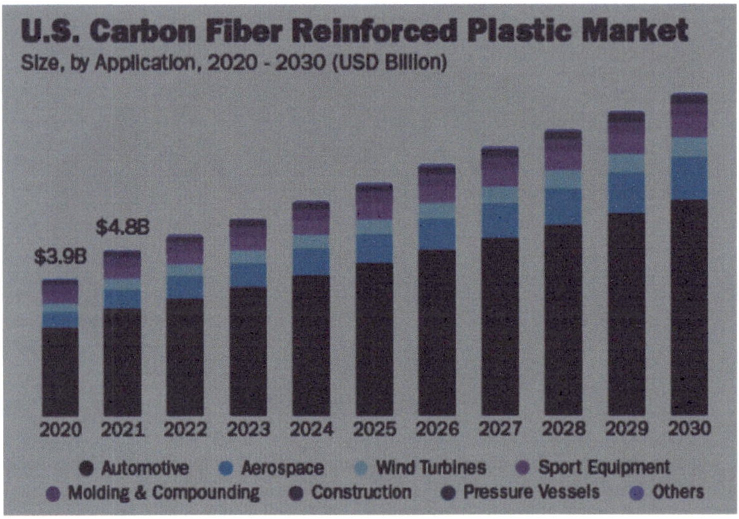

Fig. 5.4 Market for Carbon Fibre Reinforced Plastic (Carbon Fiber Reinforced Plastic Market Size, Share and Trends Analysis Report n.d.)

need for carbon fibre reinforced plastic (CFRP) from the aerospace and automotive sectors, are anticipated to boost the market. It is projected that the increasing demand for lightweight cars will propel market growth in the approaching years. Figure 5.4.shows US market for carbon fibre-reinforced plastic by application.

5.2.2 *Glass Fibre Reinforced Plastics (GFRP)*

Glass fibre is an artificial fibre made from extremely fine glass fibres that is lightweight, strong, and durable. Glass fibres are less brittle than carbon fibres, and they are produced at a comparatively lower cost. Glass fibre reinforced plastic, or GFRP, is a composite material composed of glass fibres and a polymer matrix. Epoxy, vinyl ester, or polyester thermosetting resin are frequently used as the polymer matrix. The product's resistance to chemicals and the environment is attributed to the resin, which also serves as the structural laminate's fibre binder and defines the shape of GFRP components. E-glass, or alumino-borosilicate glass, is the most popular kind of glass fibre used in GRP applications. Electrical/Chemical Resistance, or E-CR-glass, is also frequently utilised in situations that call for extremely strong protection against acidic corrosion (Zhang and Matinlinna 2012). The use of fibreglass reinforcement material has become increasingly common as a result of the increase in corrosion brought on by global warming. One of the components of GFRP is premium vinyl ester resin, which resists corrosion and lengthens the life of structures. When

GFRP rebar is used instead of typical reinforcement materials, it weighs only ¼ the weight of steel and has twice the tensile strength (El-Hassan et al. 2019).

Thermoplastic resins are currently being used more often as raw materials in place of thermoset resins, which are used to make GFRP composites. The amount of time and money needed to produce GFRP composites is decreased when thermoplastic resins are used. In addition, thermoplastic resins provide ultimate composite materials with a surface finish that is superior and comparatively higher strength and impact. Furthermore, it is feasible to mould and recycle thermoplastic resins with simplicity. Utilising these raw materials improves the ultimate quality of GFRP composites, which is one of the main reasons behind the global demand for GFRP composites in a range of applications in industry. Several factors, including the increasing demand for GFRP composites, especially from the building and transportation sectors, are anticipated to propel the global market value for glass fibre reinforced plastic in the near future.

5.2.3 Aramid Fibre Reinforced Plastics (AFRP)

Aramid fibre was the first organic fibre to use as reinforcement in advanced composites because of its high enough tensile modulus and strength. When compared to steel and glass fibres, their mechanical qualities are far superior. The fibres of the aromatic polyamide type that have at least 85% of their amide bonds (–CO–NH–) directly connected to two aromatic rings are referred to as "aramid." Kevlar is another name for aramid fibre. Figure 5.5. shows the structure of Kevlar fibre (Prasad et al. 2014).The polymer is frequently classified based on the arrangement of these bonds, which can be either para or meta. The characteristics of the resulting fibres are significantly altered when aromatic groups replace the aliphatic carbon backbone. The exceptional strength, flexibility, and abrasion resistance of chain molecules are a result of the bonds in their chemical structure being aligned along the fibre axis. The aramid fibre is made of relatively short molecules that form strong bonds with one another.

Aramid fibre, which was initially made available for purchase in the 1960s, is used extensively in ballistic protection. Commercial fibre products are derived from six main aramid compositions. The degree of fibre characteristics that are determined by the molecular weight, content, and manner of fibre creation of the polymers. Body

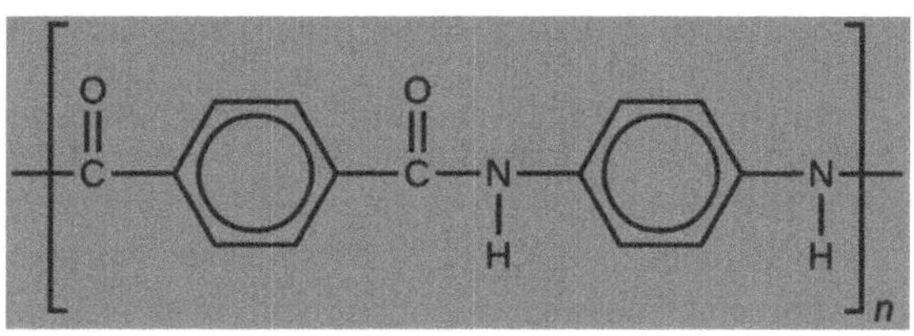

Fig. 5.5 Structure of Kevlar fibre (Prasad et al. 2014)

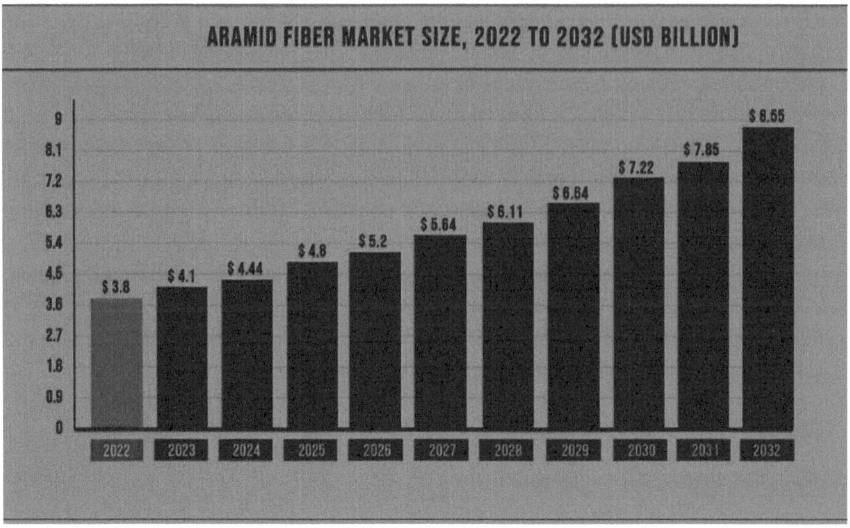

Fig. 5.6 Aramid Fibre market size forecast period 2023 to 2032 (Chemical and Material n.d.)

armour, vehicle armour, military helmets, protective gloves, and firefighter fireproof suits are just a few examples of the many products that use Aramid fibres (Ertekin 2017). One of the high-performance fibres in composite materials, aramid fibre is becoming more and more popular due to its unique properties, which include its low density, excellent strength relative to weight ratio, modulus, and impact resistance. It is anticipated that the AFRP industry will expand significantly in the upcoming years. The market is being driven primarily by the increasing need for lightweight materials in a variety of industries, such as military, automotive, and aerospace. For applications like aviation components, automobile parts, and ballistic protection systems where weight reduction and high performance are essential, AFRP's outstanding strength and stiffness properties make it a great option. According to estimates, the worldwide aramid fibre market was worth USD 3.8 billion in 2022 and is projected to rise at a compound annual growth rate (CAGR) of 8.5% to reach over USD 8.55 billion by 2032 (Chemical and Material n.d.). Figure 5.6 shows the global market for aramid fibre during the forecast period 2022 to 2032.

5.2.4 *Natural Fibre Reinforced Plastics (NFRP)*

Presently, advanced transport structures frequently use reinforced composite materials with high strength fibres including Kevlar, carbon, spectra, glass fibre, aramid, carbon nanotube, and nylon. However, as the world changes due to the depletion of inorganic resources like petroleum and other mineral sources, green materials are taking the lead. Due to the depletion of petroleum-based resources and adverse environmental conditions like pollution, natural fibre-reinforced polymer composites, or ecologically friendly materials, are growing in popularity. The primary constituents of NFRP are the hydrophobic thermoplastic or thermosetting matrix and natural hydrophilic fibres. The carbon dioxide-neutral and renewable materials, such wood or plants, provide the reinforcing fibres that give the composite its strength and rigidity. Glass and mineral fibres are not as environmentally beneficial as natural fibres. Natural fibre reinforced plastic composites, or NFRPs, are a kind of biocomposite that replace glass or carbon fibre with fibres like flax, hemp, jute, sisal, and American bamboo. Unsaturated polyester, polyethylene, and polypropylene are examples of non-biodegradable polymers that are frequently utilised as a matrix to support and transport the charge to the fibres (Vinayagamoorthy 2017). The different biodegradable and non-biodegradable polymers are displayed in Fig. 5.7 (Raj et al. 2020).

The excellent qualities and benefits of these non-synthetic materials have led to a massive rise in field research plant fibre composites. The creation of synthetic materials releases a significant quantity of carbon dioxide into the atmosphere, which is one of its drawbacks. Other drawbacks include their high cost, high energy need, and lack of biodegradability (Thyavihalli Girijappa et al. 2019). Natural fibres have several special properties, like being affordable, lightweight, renewable, and biodegradable. These materials are easily available in nature and are sustainable. Comparing glass fibre with natural fibre in composites is seen in Fig. 5.8 (Ahmad et al. 2015).

The market for natural fibre reinforced composites is anticipated to expand at a compound annual growth rate (CAGR) of 8.39% from 4,081.05 kilo tonnes in

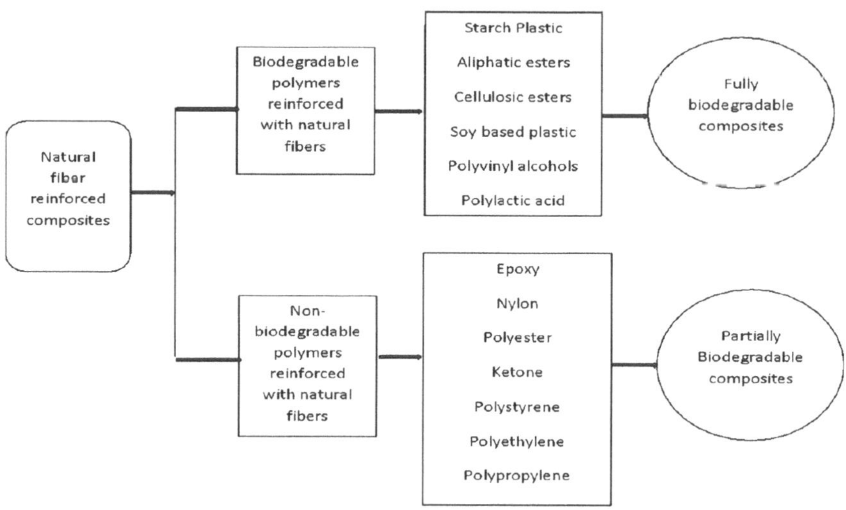

Fig. 5.7 Polymer and composite types (Raj et al. 2020)

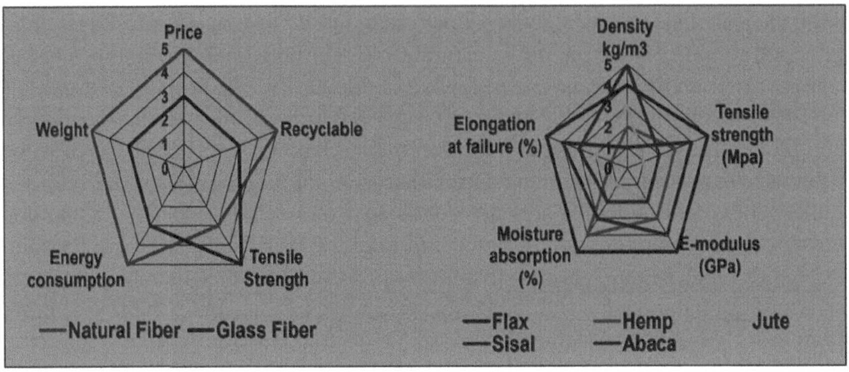

Fig. 5.8 Natural and glass fibre comparison (Ahmad et al. 2015)

2023 to 6,104.85 kilo tonnes by 2028. The automotive industry uses bio-composites bonded with natural fibres to create lightweight, mechanically sound components that reduce CO_2 emissions and improve fuel efficiency (https://www.mordorintelligence.com/team ABS n.d.). It assists in lowering an automobile's manufacturing costs by 20% and its weight by 30%. The use of bio-based materials instead of fossil-based ones has been proposed as a viable mitigation method to reduce the environmental impacts of the aviation sector.

5.3 Preperation of Fibre Reinforced Composites

Since the mid-1990s, the deployment of fibre-polymer composites in helicopters and airliners has grown significantly, and the development of more efficient and affordable manufacturing processes has been a key driver of this rise. The Federal Aviation Agency (FAA) claims that since World War II (WW2), composite materials have been widely used in aircraft manufacturing. Usually, an aircraft's structure contains up to 70% composite material. Nonetheless, the aerospace sector necessitates substantial production costs. As a result, it is crucial to ensure that the composite manufacturing process is economical for use in aerospace applications (Parveez et al. 2022). It's critical to anticipate manufacturing costs from the beginning of the project or throughout. Composites enhance aeroplane performance while consuming less fuel. The structures of aircraft typically bear a great deal of stress, thus it's critical that the composite materials used there have the highest possible strength and creep resistance. Composite materials have apparently been utilised in a range of aircraft types, including general aviation, military, regional, and commercial aircraft. Figure 5.9. shows the proportion of overall structural weight for certain aircraft that is attributable to composites (Rajput and Industry 2018).

Like metal structures, the procedures used to create composite components have a significant effect on material qualities, cost, and quality. This is also true for the

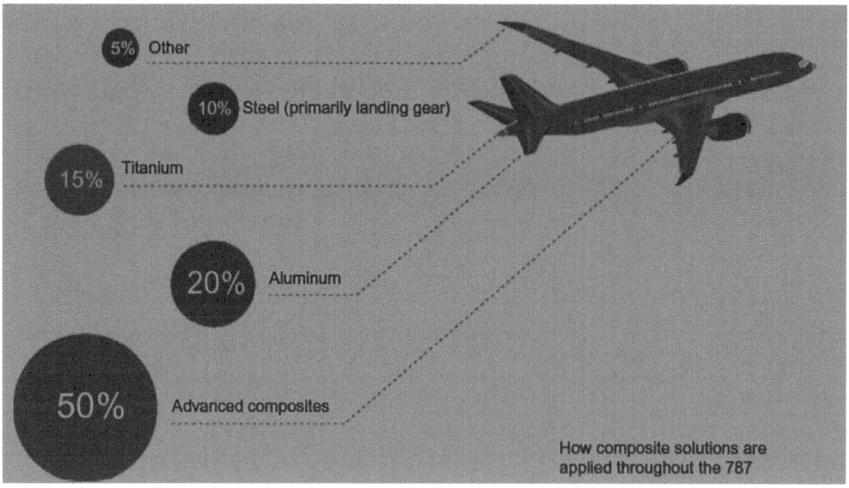

Fig. 5.9 Aerospace composite uses on the Boeing 787 (Rajput and Industry 2018)

casting and forming operations. The design of the composite component is impacted by the manufacturing technique selected; certain procedures work better for intricate, highly curved pieces, while others are better suited for flat or slightly curved structures. The volume proportion and orientation of the reinforcing fibres are two further ways that the process selection impacts the mechanical characteristics of the composite (Baran et al. 2017).

The aerospace sector has made significant investments in the creation of production techniques that can yield a wide variety of composite parts for engines and aircraft structures. Additionally, the industry has created procedures to lower the cost of manufacturing, which normally makes up more than 60–70% of the entire production cost (materials, non-destructive inspection, and other process-related expenses make up the remaining portion) (Blakey-Milner et al. 2021). Processing fibre-reinforced polymer composites naturally involves a variety of multiphysical processes, including solid mechanics, heat transport, resin flow, and state transitions. The process is hard to manage and has many interrelationships among the controlling physics. In composite manufacturing processes, residual stresses will always exist. These stresses have the potential to generate process-induced defects in the part, such as delamination and cracking. Residual shape distortions are more prone to accumulate in the presence of an uneven residual stress distribution, which is undesirable since it affects dimensional accuracy (Chava and Namilae 2021). The objectives of composite production in the aerospace sector are outlined in Fig. 5.10.

Over the past few decades, numerous traditional methods of production have been developed for composite materials. Some of these automated methods employ robots to help with processing, which leads to total automation and a notable boost in output. Creating fibre preforms is the first step in the production of FRP composites. These are then reinforced with the matrix material utilising a number of techniques.

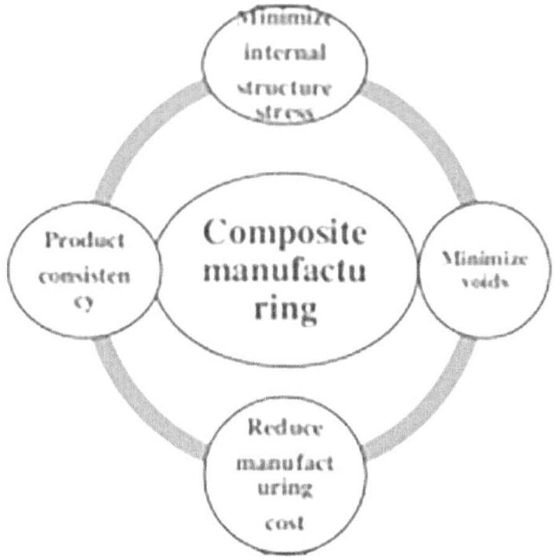

Fig. 5.10 The objectives of composite production in the aerospace sector

Fibre preforms are made by weaving, knitting, braiding, and stitching fibres into long sheets or a mat structure. The primary composite production process is categorised in Fig. 5.11.

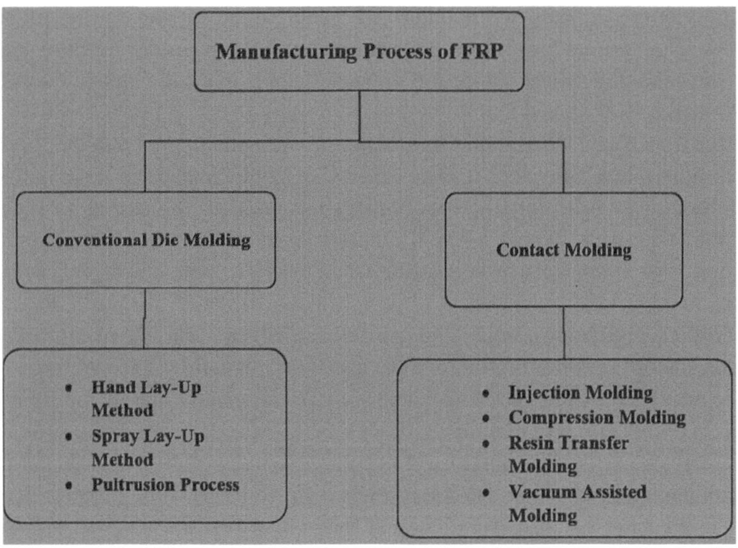

Fig. 5.11 The primary composite production process

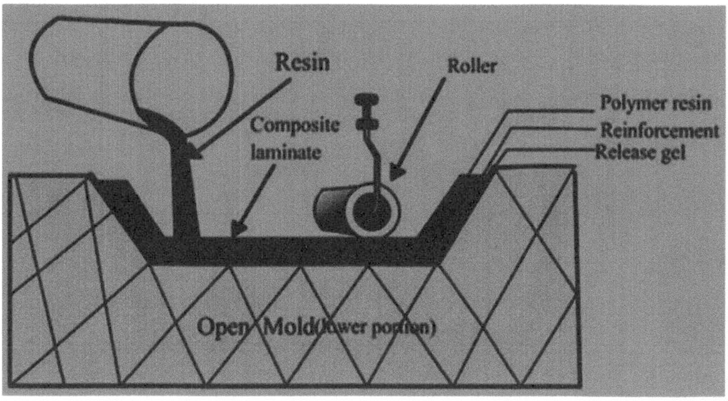

Fig. 5.12 Manual lay-up moulding

5.3.1 Hand Lay-Up Moulding

The conventional manufacturing technique known as "hand layup" entails manually applying layers of fabric and resin to a mould surface in order to create a preform. Although this technique requires little equipment, labour costs are high and it is not very efficient (Al-Furjan et al. 2022). The most popular and extensively utilised method for producing open mould composites is manual lay-up, as depicted in Fig. 5.12. Fibre preforms are initially placed in a mould and lightly coated with an antiadhesive coating to make extraction easier. The resin is applied with a brush or poured onto a reinforcement material. A roller is used to press the resin into the textiles in order to strengthen the contact between the matrix and the increasingly higher layers of reinforcement.

The aerospace industry has gradually seen a decline in the use of the hand layup technique since the products frequently contain several voids, wrinkles, and other faults that cause them to operate below the intended level. Computer and machine-aided Automated Layup Technology (ALT) was introduced in order to speed up the layup procedure and decrease errors in composite constructions while increasing manufacturing efficiency (Chen et al. 2023).

5.3.2 Spray-Up Process

One may consider the spray lay-up technique to be an advancement of the hand lay-up technique. This method involves spraying chopped fibre reinforcement and pressurised resin with a spray cannon. Usually used as reinforcement, glass roving is fed through a spray gun and then cut with a chopper gun. A roller is moved over the sprayed material to liberate trapped air in the lay-ups. Following the spraying of the fibre and resin to the required thickness, the product is cured at room temperature or at a higher temperature. The formed composite part is removed from the mould

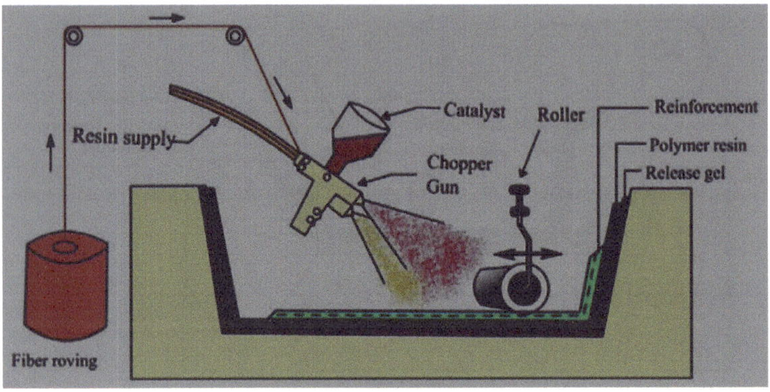

Fig. 5.13 Spray-up open moulding process (Balasubramanian et al. 2018)

and undergoes additional processing when the mould has cured. The duration of the curing process is dependent on the variety of polymer used in composite manufacturing. Figure 5.13 shows how the procedure is carried out (Balasubramanian et al. 2018). Chopped fibres are quieter and more conformable than hand lay-up in an open mould technique.

5.3.3 Vacuum Bag Moulding

The main method for producing laminated structures from composite materials is vacuum bag moulding, which is widely used in the aircraft sector. The portion is enclosed and sealed from the outside air via hoover bag moulding, which uses a flexible film made of nylon polyethylene or polyvinyl alcohol (PVA). The vacuum bag moulding process is frequently used with the hand lay-up technique. The laminate is initially made by hand using a lay-up technique, and it is then placed between the hoover bag and the mould to ensure that the fibres are evenly infused into the matrix material. A vacuum pump then draws out the air between the mould and the vacuum bag, compressing the item as a result of atmospheric pressure. Figure 5.14.provides a clear understanding of the procedure (Yunus et al. 2014). Hierarchical composites were made with multiscale carbon fibre reinforcements using a vacuum bagging process. Flexural and interlinear shear characteristics increased by 15% and 18%, respectively, as a result of the decreased likelihood of observable porosity and inappropriate impregnation of dual reinforcements (Muralidhara et al. 2019).

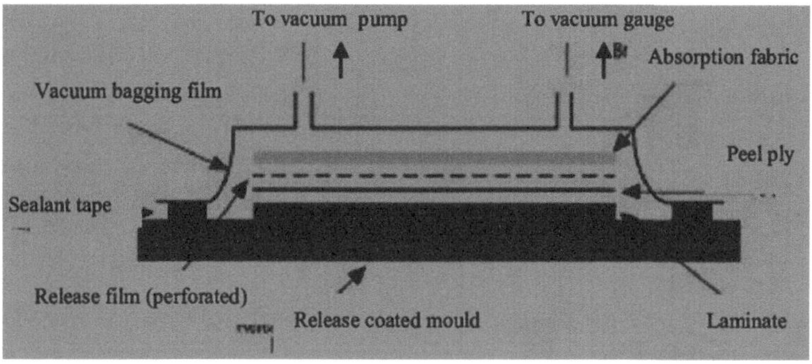

Fig. 5.14 Vacuum bag moulding process (Yunus et al. 2014)

5.3.4 Resin Transfer Moulding (RTM) Process

Using matching male and female moulds, resin transfer moulding is a closed-mould technique. Steel, aluminium, nickel, and composite materials are just a few of the materials that can be used to make moulds for RTM procedures. Before injecting the resin, transfer moulding involves filling the mould with reinforcing fibres, frequently in the form of a matrix. Figure 5.15a shows the RTM process. Glass fibres are the most frequently utilised fibre type in this technique, though carbon fibres, Kevlar, and even certain organic fibres like hemp may also be employed. After inserting the fibres into the enclosing mould, a liquid thermoset resin, like polyurethane, is injected. The fibres are completely saturated with resin, resulting in a robust and long-lasting product that has a superb bond between the two components. Throughout the operation, bolts, toggle clamps, or press are used to keep the mould closed. A liquid heating system is usually utilised to heat the mould, and the typical temperature used during RTM is 104 degrees Fahrenheit.

Vacuum infusion, sometimes referred to as vacuum aided resin transfer moulding (VARTM), is a more recent method that involves placing preform fibres on a mould and a perforated tube between the vacuum bag and the resin container. To solidify the laminate structure, the resin is forced through the perforated tubes over the fibres by vacuum force, as shown in Fig. 5.15b. To improve the strength of textile composites, natural fibres are surface treated. The alkali-treated flax fibre-reinforced epoxy acrylate resin composite, created using the VARTM technique, saw a 19.7% improvement in tensile strength (Summerscales and Searle 2005).

5.3.5 Compression Moulding Process

Achieving product consistency with aerospace composite materials can be difficult from concept to manufacture. When working with intricate, multi-layered materials like carbon fibre reinforced polymers (CFRP), manual procedures might result in

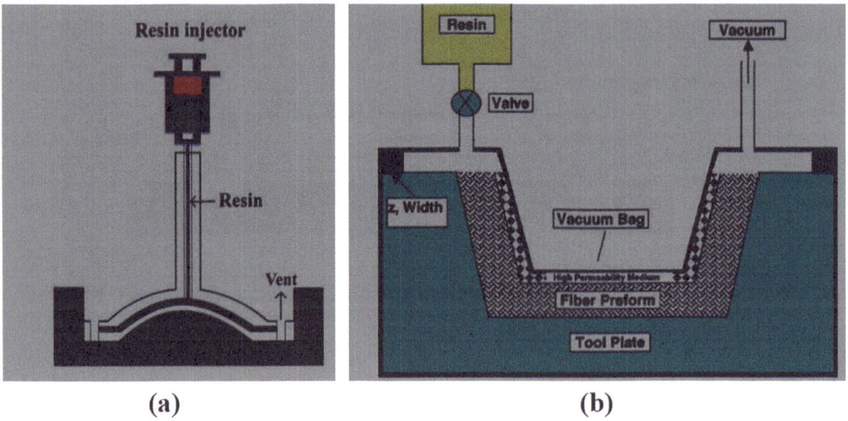

Fig. 5.15 **a** Resin Transfer Moulding, **b** vacuum assisted resin transfer moulding (Summerscales and Searle 2005)

costly and time-consuming rework machining for completed goods. One well-known method for processing polymer composite materials is compression moulding, which can be done with thermoset or thermoplastic matrices. When the fibre-reinforced composite material is made to flow or flex inside the mould cavity, it is frequently referred to as matched-die moulding. Compression moulding is preferred because the technique can create intricate geometries with high fibre volume fractions, resulting in composite products that are reliable and sturdy (Leong et al. 2013).

This process entails pouring a specified quantity of composite material—typically in the form of sheets or pellets—into a mould cavity. The mould is then sealed and heated to a high temperature, which allows the material to expand and take on the shape of the mould. The material compresses and solidifies to create the desired product. Thermoset composite materials are used in close tolerance compression moulding to create lightweight, more rigid, and strong parts compared to metal components. Custom components with stringent dimensional requirements can be made with great ease using this approach. Thermoset compression moulded parts maintain their mechanical properties at elevated temperatures. Thermoset compression-moulded components are a great option for aerospace applications because of all these features (Venugopal et al. 2022) Fig. 5.16. depicts the process of compression moulding (Compression n.d.).

5.3.6 Pultrusion Process

Because of their cost-effectiveness, durability, and lightweight nature, advanced composites are a desirable option for the aerospace sector. Fibre reinforced polymer (FRP) composites are made via an automated closed-moulding technique called

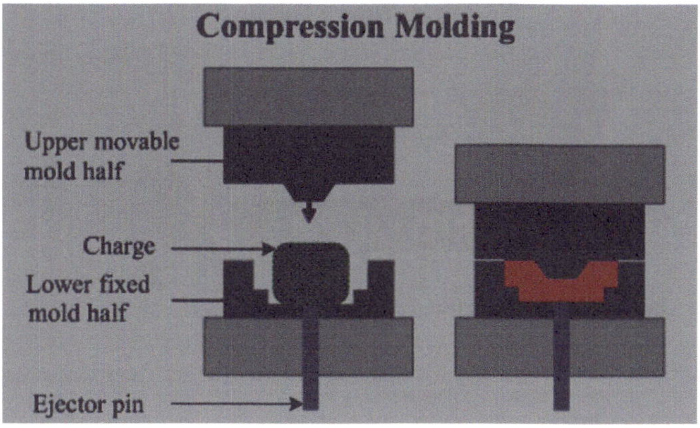

Fig. 5.16 The process of compression moulding (Compression n.d.)

Pultrusion (Volk et al. 2022). Fibreglass rovings are pulled through an impregnator or resin bath until saturated, and then the material hardens from the heated steel. Pultrusion die, producing a robust and lightweight fibre-reinforced plastic (FRP). Pultrusion permits the incorporation of additional resins and reinforcements during the production stages in comparison to other methods of manufacture.

Pultruded FRPs are being used in a variety of industries more and more as businesses become aware of the advantages of pultruded products. Global demand for Pultrusion products is predicted at a cumulative annual growth rate (CAGR) of 4.0%, will reach $3.4 billion by 2024, according to a report by market research firm Lucintel (Pultrusion Market: Trends, Opportunities and Competitive Analysis [2023–2028] 2023). Pultruded FRP composites are inexpensive and have great performance characteristics, making them essential for aerospace innovation. Figure 5.17 provides an explanation of the Pultrusion process (Acquah et al. 2006).

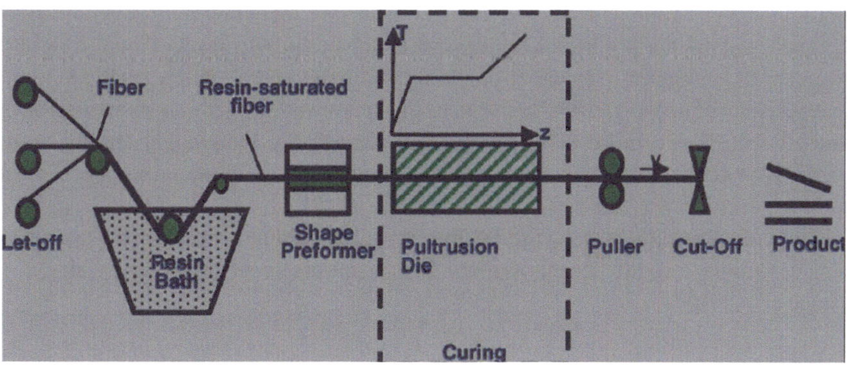

Fig. 5.17 The pultrusion process (Acquah et al. 2006)

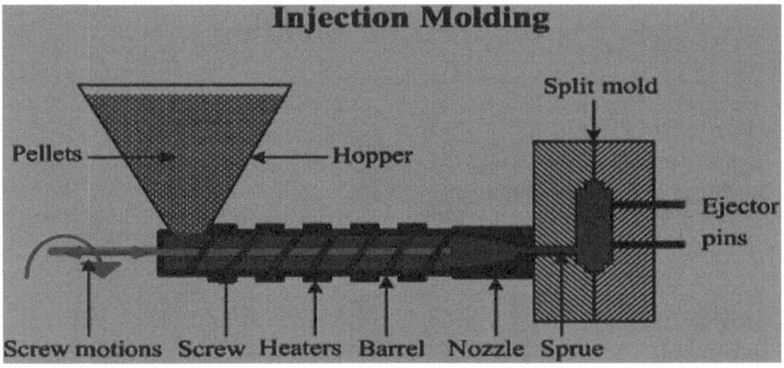

Fig. 5.18 Injection moulding process (Agrawal et al. 1987)

5.3.7 *Injection Moulding Process*

Composite parts can be fabricated by injection moulding at very short cycle times and with excellent precision. As seen in Fig. 5.18, fibre composites are fed as pellets through a hopper in a typical injection moulding process, and they are subsequently conveyed using a screw and heated barrel. (Agrawal et al. 1987). The material is injected into the mould by the screw using a nozzle once the necessary amount has melted in a barrel where it cools and takes on the required form. It has been discovered that injection moulding works incredibly well for thermoplastic encapsulations of electrical devices used in the medical field. Enhancements in fibre-matrix compatibility and uniformity in the fibre dispersion inside the matrix material are achieved through surface treatment of biocomposite.

Due to its economical production, wide selection of materials, and adaptable design, aerospace injection moulding is one such cutting-edge manufacturing technique that is quite useful. In comparison to die casting and machining processes, the setup costs for this mould are comparatively cheap, and it is less costly than traditional die casting moulds (Jayakrishna et al. 2023).Aerospace plastic injection moulding can be done with a broad range of materials, including reinforced composites and standard thermoplastics. Because injection moulding may work with a variety of materials, engineers can select materials that meet certain needs, such strength, heat resistance, and corrosion resistance. Plastics used in injection moulding have strengths similar to those of metals but are remarkably lightweight (Rabbi et al. 2021). An excellent strength-to-weight ratio makes injection moulding a preferred option.

Numerous materials can be used with injection moulding. An extensive overview of the various aircraft injection moulding materials is given in this Table 5.1 (Nagavally 2016).

Table 5.1 Several injection-moulded aircraft materials and their characteristics (Nagavally 2016)

Moulding materials	Material properties
High-Density Polyethylene (HDPE)	Toughness, Flexibility, Weatherproofing, Translucent/Waxy Appearance at Low Temperatures
Thermoplastic Polyurethane (TPU)/ Thermoplastic Vulcanizate (TPV)	Durable, High Ductility resistant to abrasion, and has good compression resistance
Polypropylene (PP)	Durability and Chemical Resistance. Translucency and Thermal Resistance
Acrylonitrile Butadiene Styrene (ABS)	Dimensional stability, tensile strength, hardness, chemical resistance, and abrasion resistance
Polyether Ether Ketone (PEEK) Often reinforced with Glass Fibre or Carbon Fibre	Resistance to Heat, Chemicals
High Impact Polystyrene (HIPS)	Low Cost, Impact Strength, Thermal Resistance, and Dimensional Stability

5.4 Mechanical Characteristics of FRP Composites

Concrete and steel are the main building materials utilised in the contemporary construction sector. But in other cases, a different type of material is required. FRPs are seen to be a viable alternative to steel reinforcement, especially in concrete structures subjected to extreme weather conditions or electromagnetic field effects. A range of materials are used to create FRP reinforcement. Although carbon fibre reinforced polymers (CFRP) provide the best mechanical properties among other FRP composites, the raw materials required to produce them are limited. In terms of mechanical properties and production complexity, aramid (AFRP) and basalt (BFRP) bars are in the middle of the pack, despite their infrequent use in practice. Glass fibre reinforced polymer (GFRP) bars are the most popular type of FRP because they combine affordability for structural fibres with environmental resilience. GFRP bars are extremely durable and can withstand stresses up to 5.6 times greater than those of structural steel (Tao et al. 2023).

The orientation of the fibres inside the matrix, the manufacturing process, the mechanical characteristics of the component materials, and the ratio of fibres to matrix materials all affect the mechanical properties of FRP composites. Fibres, resins, interface, fillers, and additives are the components that make up FRP composite (Abbood et al. 2021). The high elastic modulus of the fibre affects the mechanical characteristics of FRP. However, the resins aid in the distribution of tensions among the fibres to shield them from environmental and mechanical harm. Fibre-matrix interface is known to safely affect the performance of fibre-reinforced polymers. Furthermore, fillers help to lower shrinkage and costs in addition to the three primary components. Beyond making composites more workable, additives help with their mechanical and physical characteristics. Figure 5.2 shows the stress–strain relationship and typical tensile strengths of steel and fibre-reinforced polymers (Balla et al. 2019).

The most widely used structural fibre reinforced polymer (FRP) is e-glass fibre reinforced polymer (GFRP) since it is the least expensive. Nowadays, basalt fibre reinforced polymer (BFRP) can be made for almost the same price as graphene fibre reinforced polymer (GFRP), and it has a resistance to alkalies, somewhat superior strengths, and almost infinite resources. Since the fibres have a lower compressive strength perpendicular to the fibre direction and are more expensive, aramid fibre reinforced polymer, or AFRP, is less frequently used as a structural reinforcement. This characteristic, however, makes aramid fibre the preferred material for ballistic-resistant fabrics since the fibres effectively absorb impact (Altin 2021). Among FRP materials, carbon fibre reinforced polymer (CFRP) offers the highest strength and widest range of strengths. The production processes and carbon source are to blame for the range. When it comes to fatigue failure and creep rupture, CFRP is more resilient than the other FRPs. Its superior strength and resilience to fatigue and cyclic failures more than make up for its higher cost. The stress–strain relationship and average tensile strengths of fibre-reinforced polymers are displayed in Fig. 5.19.

FRP composites, as opposed to steel, are immune to chloride-induced corrosion because of their inherent non-metallic character, which could greatly improve a structure's corrosion resistance (Gopu and Joseph 2022). The features of FRP materials in contrast to conventional steel are shown in Table 5.2. Fibre-reinforced plastics are much stronger and lighter than traditional steel. However, its elongation rates and failure strain are decreased because to its linear elastic mechanical features, which lack an obvious yielding stage. Besides, FRP often has a lower elastic modulus than steel, except for some CFRP that has a high Young's modulus (Ma et al. 2021).

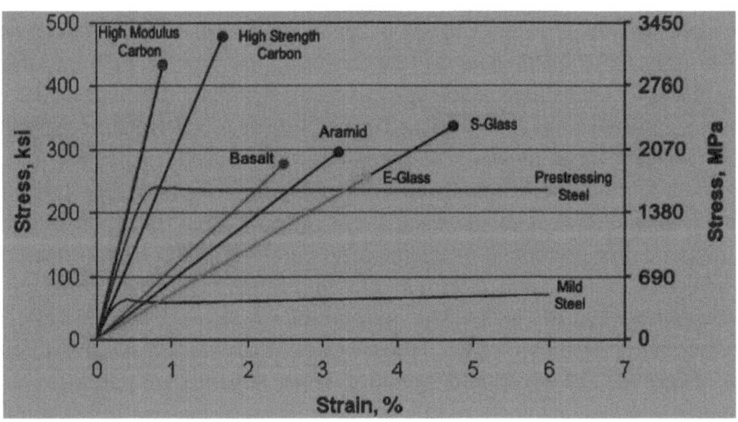

Fig. 5.19 The stress–strain relationship and typical tensile strengths of fibre-reinforced polymers and steel (Balla et al. 2019)

Table 5.2 Features of fibre reinforced polymer materials in relation to conventional steel (Ma et al. 2021)

Material type	Density (gm/cm^3)	Tensile strength (MPa)	Young's modulus (GPa)	Elongation (%)	Coefficient of linear expansion (10^{-6}/C)
CFRP	1.50–2.10	600–3920	37–784	0.5–1.8	–9.0–0.0
GFRP	1.25–2.50	483–4580	1720–3620	600–1500	483–690
AFRP	1.25–1.45	1720–3620	41–175	1.4–4.4	–6.0–2.0
BFRP	1.90–2.10	600–1500	50–65	1.2–2.6	9.0–12.0
Steel	7.85	483–690	200	6.0–12.0	11.7

The reinforcing fibres listed in Tables 5.3, constitute the foundation of the materials systems that have been thought to be beneficial in the aerospace sector. The majority of aviation composites are fabricated via prepregs as their basic ingredients, and autoclave moulding is a common method of manufacture. Shell-like parts, such as the rocket engine enclosures seen in spacecraft and missiles, frequently use filament winding. Glass fibre composites used in small, slow-moving aircraft are mostly cured in an oven or at room temperature (Rajak et al. 2021). Composite tooling is frequently used in situations with low to moderate production rates; nevertheless, metallic conventional tooling is recommended when a high number of components are needed. Additionally, resin injection moulding is used in unique parts like radomes.

5.5 Fibre Reinforced Composites-Aeronautical and Aerospace Applications

Composites have been used more frequently in recent years. Many industries, including aerospace, automotive, industrial, and construction, have employed composites to produce cutting-edge assemblies and products. Many variables, as illustrated in Fig. 5.20, drive the usage of plastics and composite materials in aviation structural development. Weight reduction, durability, design elements, reduced maintenance needed, noise reduction, increased fuel efficiency, and corrosion resistance are some of these factors.

Fibre-reinforced polymers are one type of composite that has grown in popularity (FRPs). When compared to the metals or polymers they are replacing, FRPs have a few special advantages. Personalised material qualities, particular strength, resistance to corrosion, fatigue, and other benefits are a few of these benefits. For the next ten years, it is predicted that the utilisation of composites—especially FRPs—will rise by up to 10–15% annually (Karim et al. 2023).

Table 5.3 Shows the reinforcing fibres used in aviation industry

Fibre	Density (g/cc)	Modulus (GPa)	Strength (GPa)	Application domains
Glass				
E-glass	2.55	65–75	2.2–2.6	Radomes, motor casings for rockets, aircraft interiors, small passenger air conditioning parts, and auxiliary fragments
S-glass	2.47	85–95	4.4–4.8	Components of a high-loading tiny passenger air conditioner
Aramid				
Low modulus	1.44	80–85	2.7–2.8	Fairings; non-load bearing components
Intermediate	1.44	120–128	2.7–2.8	Radar domes, various structural components, casings for rocket motors
High modulus modulus	1.48	160–170	2.3–2.4	Parts that are extremely loaded
Carbon				
Standard modulus	1.77–1.80	270–300	3.0–3.5	Extensively utilised for nearly every kind of component in satellites, antenna dishes, missiles, air conditioners, etc.
Intermediate modulus	1.77–1.81	220–240	5.4–5.7	The main structural components of high-performance fighters Space structures, control surfaces in a/c
High modulus	1.77–1.80	390–450	2.8–3.0	Space structures, control surfaces in a/c
Ultra-high strength	1.80–1.82	290–310	4.0–4.5 7.0–7.5	Key structural components of spacecraft and high-performance fighters

Conventional buildings made of monolithic materials, including metals (such as steel and aluminium), are usually heavy, which raises fuel consumption and construction costs, especially for the road, rail, marine, and aviation transportation industries (Cheng et al. 2017). Furthermore, when exposed to corrosive environments, certain materials, such as aluminium, degrade, decreasing the structure's strength and useful life and raising the expense of repairs and maintenance (Sukiman et al. 2005). In order to increase mechanical properties like strength and lifetime service, as well as to reduce energy consumption and the cost of structural repair and maintenance, companies and research institutions have been working hard to replace monolithic materials with composites. The use of f FRP composites has drawn a lot of attention in the transportation sector, particularly as metal alternatives since stable and fuel-efficient vehicles, ships, and aeroplanes must be built (Parveez et al. 2022).

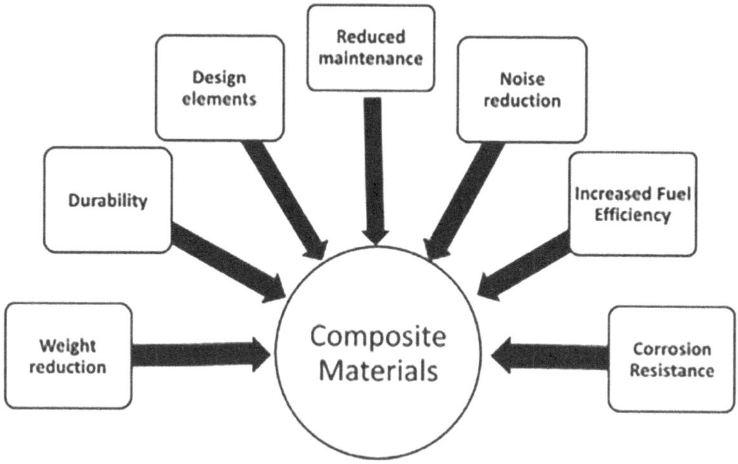

Fig. 5.20 Several elements drive the building of aeronautical structures

Among the desirable qualities that would promote the use of FRP composites in the field of transportation are excellent strength, resistance to corrosion, lightweight design, and appropriate fatigue endurance. The development of improved high performance structural materials is driven by the aerospace industry's constant pursuit of improving the efficiency of both military and commercial aircraft. Composites are widely utilised among these materials in both current and upcoming aviation components. Composite materials are especially desirable for use in aviation and aerospace applications due to their exceptional strength, stiffness-to-density ratios, and better physical attributes. With a projected value of USD 93.69 billion in 2022, the worldwide composites market is projected to expand at a compound annual growth rate (CAGR) of 7.2% between 2023 and 2030 (Composites Market Size, Share & Trends Analysis Report 2022). Table 5.4 lists the essential specifications for an aircraft structure along with how they affect the structure's design (Mangalgiri 1999).

The growing need for lightweight components in the automotive and transportation sectors is the reason for this rise. The composite materials utilised in various aircraft industry segments from 2013 to 2022 are shown in Fig. 5.21 as a pie chart (Al-Furjan et al. 2022). The necessity to integrate and maximise the composite manufacturing sector in our daily lives is highlighted by the fact that about sixty-three percent of all parts and pieces of equipment used in commercial aviation are made of composite materials.

The military was the first to use composites before commercial aircraft adopted the technology. Prior to internal components and secondary constructions, radomes had the earliest military uses. Because of the low modulus of glass compared to metals, the development of primary composite structures was postponed until carbon reinforcements were added. Since confidence in FRPs has grown, integrated composite designs have been developed, leading to the widespread use of composites today. This is the outcome of a progressive direct replacement of metal components. The

Table 5.4 Requirements for aerospace structures (Mangalgiri 1999)

SI no.	Need	Relevance	Impact
1	Light-weight	Every Programme in Aerospace	• Semi-monocoque design using materials with low densities • Stiffened structures or thin-walled boxes • Wood, Al-alloys Composites • High strength/weight; High stiffness/weight
2	A high degree of reliability	Every Programme in Aviation	• Stringent quality assurance comprehensive testing for trustworthy data • Authenticity: Validation of the design
3	Passenger security	Vehicles for passengers	• Utilising fire-resistant materials • Comprehensive testing: Reliability
4	Resilience Corrosion and wear deterioration	Aircraft Spacecraft	• Extensive testing and analysis of fatigue • There is no fatigue limit on aluminium alloys • Plans for preventing corrosion • Concerns about damage and safe life, life extension • Thorough testing in the necessary conditions • High-integrity thin materials
5	Performance in aerodynamics	Aircraft: Reusable satellites	• Extremely intricate loading • Thin, flexible surfaces and wings • Deformed form due to aeroelasticity • Intricately curved shapes • Capability of Manufacturing: NC Machining; Moulding
6	Multiple functions or roles	Every Programme in Aerospace	• Effective design • Application: functionally-grade composites
7	Fly-by-wire	Aeroplanes, primarily fighters but also some passenger aircraft	• Interactions between structure and control • Aero-servo-elasticity • Heavy reliance on electronics and computers and EMI shielding
8	Stealth	Particular Uses in Military Aerospace	• Particular Aircraft Shape and Surface • Stealth Coatings
9	All Weather Conditions	Aircraft	• Defence against lightning and resilience to erosion

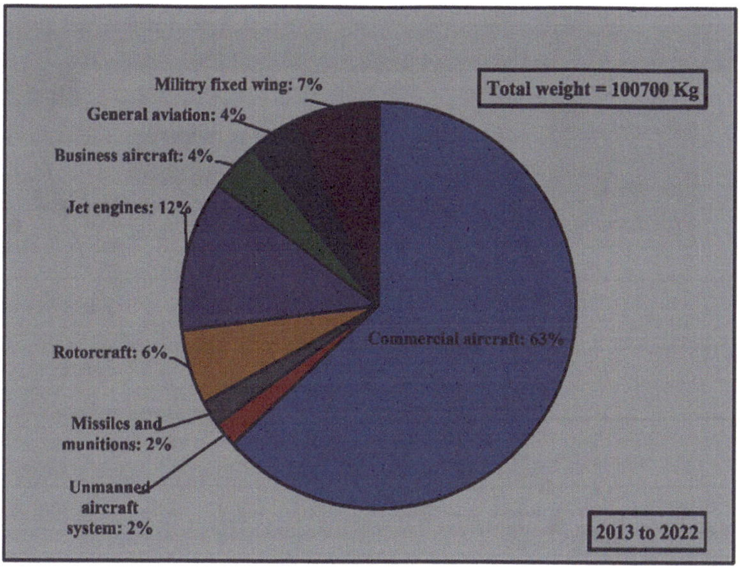

Fig. 5.21 The composite materials that were employed in the different aviation industry segments between 2013 and 2022 (Al-Furjan et al. 2022)

Airbus 320 has numerous composite components, including the fin and tail plane. In comparison to its aluminium alloy counterpart, this has enabled an 800 kg weight reduction. Composites make up more than 20% of the A380's airframe. Major uses for carbon- and glass-fibre reinforced polymers and composites include wings, fuselage sections (such the undercarriage and rear end of the fuselage), tail surfaces, and doors. A well-known graph depicting the rising use of carbon fibre composites in military and commercial aircraft throughout time is shown in Fig. 5.22 (Hiken 2017).

FRPs are utilised in a broad variety of aerospace applications due to their versatility, ranging from the creation of load-bearing structural elements to interior components. FRP composites are utilised in both military and commercial aircraft for a number of parts, including the fuselage and more substantial components like luggage bins, trolleys, radomes, and brackets. FRP composites are also helping to innovate lightweight satellites and rockets for space travel and exploration. The aerospace sector in North America is a significant user of FRP composites due to its requirement for lightweight materials that have the potential to decrease carbon emissions and increase fuel efficiency (Siengchin 2023). Since FRP composites are used in many aircraft parts, such as the wings, fuselage, and tail sections, their demand is expected to grow in the years to come. The Boeing Commercial Airplanes-developed and -built 787 Dreamliner has become the best-selling passenger aircraft in history. The various components of this aircraft are made of carbon fibre reinforced composites. The wings, the fuselage, the interiors, the doors, and the tail, among many others, are all fashioned from these composites. Boeing's most recent commercial aircraft project, Dreamliner 787 (Fig. 5.23), which has 50% CFRP structures, 20%

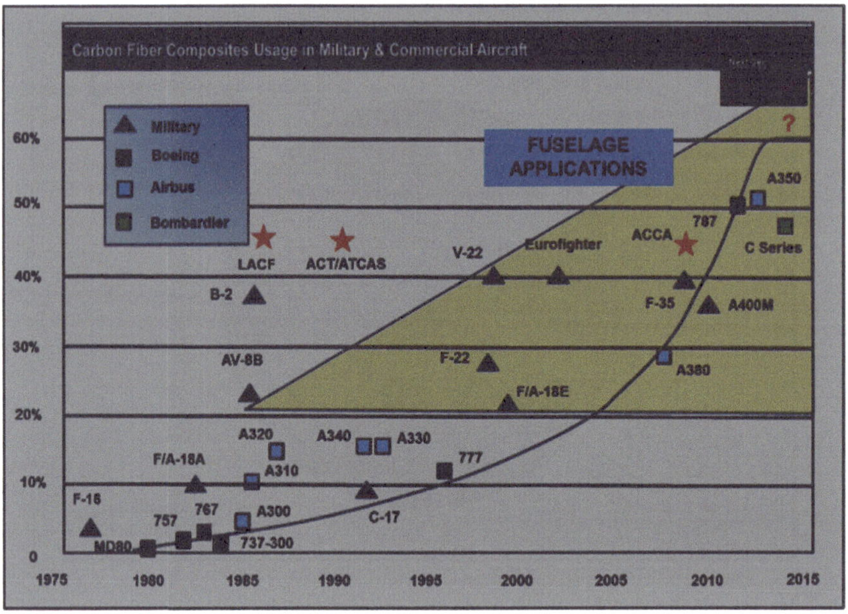

Fig. 5.22 The application of carbon fibre composites in both military and commercial aircraft over time (Hiken 2017)

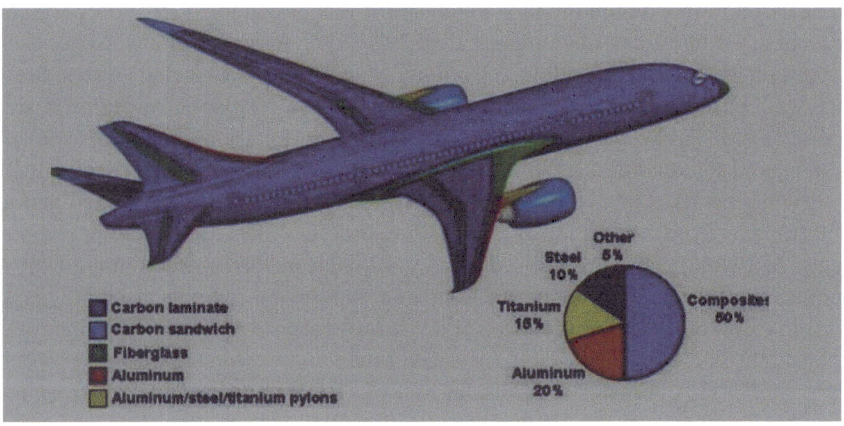

Fig. 5.23 The CFRP parts of dreamliner Boeing 787 (Boeing sets pace for composite usage in large civil aircraft 2005)

aluminium, 15% titanium, 10% steel and 5% other, demonstrates the progression of composite material use (Boeing sets pace for composite usage in large civil aircraft 2005).

Currently, over half of these fuel-efficient aircraft are built with carbon-reinforced plastic composites in compliance with safety regulations. Additionally, a number of businesses, including Airbus, Boeing, Bombardier, BAE equipment, Raytheon, GE Aviation, and Lockheed Martin, have begun incorporating thermoplastics and composites into their aviation and defence-related equipment. The aviation and propulsion industries benefit from additive manufacturing (3D printing) in a number of ways, including its suitability for many material groups, its ability to produce complex geometries without the need for tools or dies, and its ability to produce designs that are lighter and more functionally superior (Siengchin 2023). Additive manufacturing is very helpful for low-volume production, which is typical in the aviation sector. It shortens machining times, gets rid of precious alloy waste, and decreases buy-to-fly ratios.

In many types of aircraft, the principal substrate for the inner structure of the fuselage is made of glass textiles. Because of fibreglass's strength, dimensional stability, resistance to corrosion, and lightweight nature, it is used in many essential aircraft goods. Applications for S-2 and E-glass are especially important in the aviation sector. These two types of fibreglass provide the fibre toughness, impact resistance, and compressive and tensile modulus required for many aircraft applications. Around 1950, fibreglass was used for the first time on an aeroplane similar to a Boeing. The 2012 Boeing 787 Dreamliner is one of the aircraft models that employs fibreglass. According to them, fibreglass makes up 50% of aeroplanes (Zhang et al. 2023). Manufacturers are using composite materials in their aerospace goods more and more. Fibreglass has several benefits for aeroplanes, the first of which is its reduced weight. Because of the fibre-based matrix architecture, this material is stronger than aluminium. Additionally, fibreglass offers a smoother surface. Thus, another fantastic benefit of employing this fuel-efficient material is that no additional treatment is required to smooth the surface. Fibreglass is one of the composite materials or materials that resists corrosion well. This material is also resistant to cracking and can withstand bending conditions. The design of this material is more robust than that of aluminium so that, while in use, it requires no extra maintenance or repairs (Masuelli 2013). Applications for glass fibre-reinforced composite were common in aircraft and included bulkheads, ducting, antenna enclosures, luggage racks, engine cowlings, and instrument enclosures.

Carbon fibre is used to make several aircraft components, including the wings, fuselage, and engine cowlings. One of the primary benefits of utilising carbon fibre in aircraft construction is its potential to lower the aircraft's weight. Performance and fuel efficiency may both increase as a result. Additionally, strong, carbon fibre can contribute to increased aeroplane safety. The most well-known aircraft made of carbon fibre is the Boeing 787 Dreamliner (Parveez et al. 2022) It has carbon fibre for its tail, wings, and fuselage. This contributes to an aircraft's weight reduction of up to 20%, improving fuel economy. Another aeroplane that incorporates carbon fibre is the Airbus A350 XWB. Both the fuselage and the wings of the A350 XWB are constructed from carbon fibre. This contributes to an aircraft's weight reduction of up to 30%, improving fuel economy (Smock 2008). One business jet that uses carbon fibre is the Gulfstream G650. Carbon fibre is used to make the fuselage of the

G650, which helps the aircraft fly lighter and perform better. Composite laminates were employed more and more in Dreamliner, Boeing 767 and 787, Airbus A350, A380, and other vehicles, accounting for more than 50% of the total weight. Of the composites used in the Airbus A380, 25% are made up of the airframe, of which 22% are CRFP or GRFP (Breuer 2016). The Airbus structures' material breakdown is depicted in Fig. 5.24 (Aamir and Tolouei-Rad 2019).

Some experts predict that more than two-thirds of future military aircraft will be composed of composite materials. At least one-third of the structural components of contemporary military aircraft, such as the F-22, are made of carbon fibre composites. The multirole, twin-engine, SARAS re-engineered light transport aircraft is intended for both military and commercial use (Lafayette and Vista n.d.). Its wings are made of carbon fibre composite (CFC). Unlike previous aeroplanes, it will be equipped with a pressurised cabin divider made of composite material that will undergo its first testing specifically for this purpose. With the construction of the Airbus 380, a 280-ton aeroplane (average operational empty weight) composed of 25% composite materials, the use of composites saw a significant increase. Several important structural elements, such as the centre wing box, wing ribs, tail cone, vertical and horizontal tail planes, and tail cone are made of carbon-fibre composites (see Fig. 5.25) (Hamerton and Mooring 2012). The Boeing 787 and Airbus A350 are indeed notable examples of commercial aircraft that extensively use composite materials in their construction. The Boeing 787 is renowned for using carbon fibre-reinforced epoxy composite material for around half of its structural mass. Likewise, about half of the structural mass of the Airbus A350 is made of composite materials.

Aramid fibres are perfect for ballistic and aeronautical applications because of their remarkable strength and impact resistance. Aramid fibres are notable for being

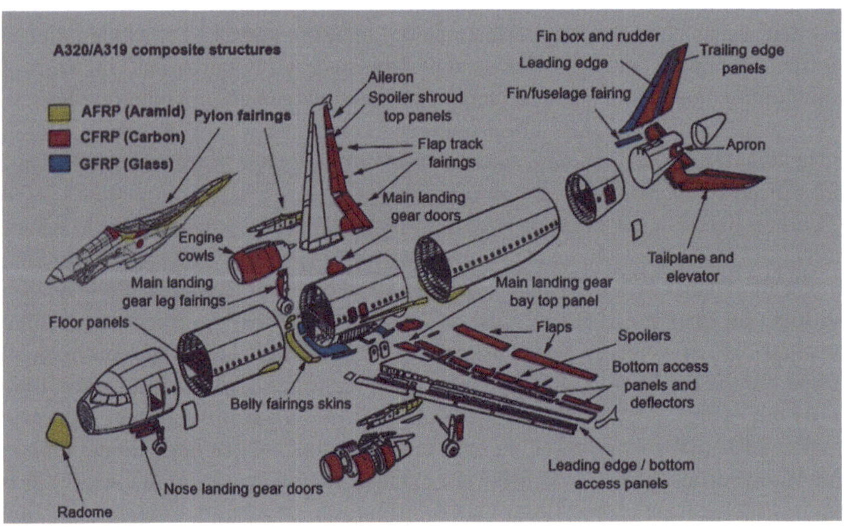

Fig. 5.24 Illustrates the Airbus constructions' material breakdown (Aamir and Tolouei-Rad 2019)

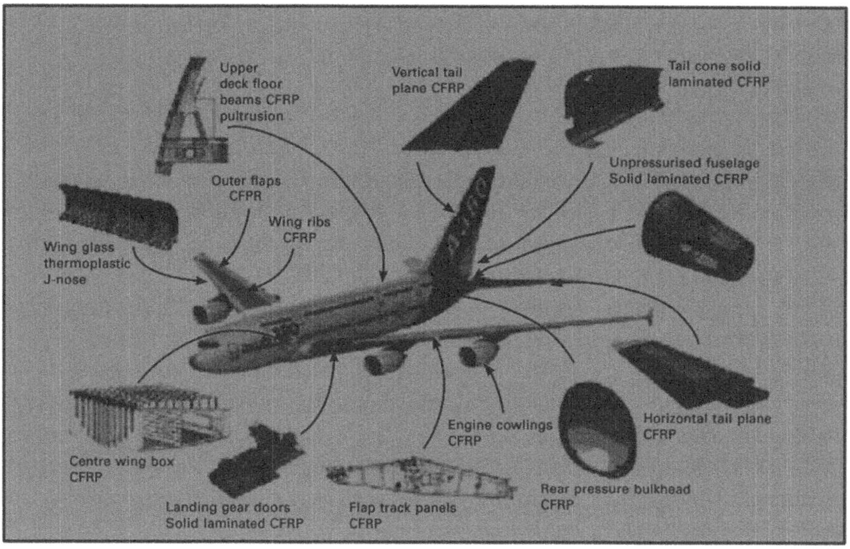

Fig. 5.25 The application of carbon-fibre composites in several important structural elements (Hamerton and Mooring 2012)

strong, lightweight, and having a high level of fatigue resistance. Arimid fibres are frequently utilised in honeycomb core constructions in the aerospace industry, where their resilience to fatigue and capacity to absorb vibrations extend the life and functionality of aircraft parts. Its benefits include flame retardancy, thermal insulation, sound absorption, high modulus, light weight, and good structural stability. It can be applied to the national defence, aerospace, rail transportation, and other fields as a lightweight, highly-strength structural material (mostly as a core material) significant area (Lowe 2005). Because of its exceptional performance, aramid honeycomb core material has emerged as a significant weight-reducing material in the aerospace. Along with the large rigid and secondary stress-bearing components of the aircraft, the principal application parts include flaps, ailerons, vertical tail leading edge, Rudder, canards, fuselage, blades, radome, wall panels, hatches, floors, and other components (Siengchin 2023).

For defence and military applications, fibre reinforced polymer (FRP) composites are a good choice because of their high strength, low weight, resistance to corrosion, ability to prototype complex designs, and other qualities. Their anti-corrosiveness, fatigue resilience, and light weight made them more attractive in the defence industry than traditional metals and steels. This was due to the fact that salty seawater causes increased corrosion in structures. Applications for FRP composites are numerous and include fighter aircraft, naval ships, undersea structures, shelters, and military vehicles. These materials were preferred because of their great strength and low weight, dependable operation, and ease of maintenance over the course of their useful lives. For many years, the wings and fuselage of fighter aircraft have been made of carbon fibre-epoxy composites to reduce weight and increase structural effectiveness. The percentage of the composite material used in fighter aircraft varies; it is as low as 4% in the F-14 Tomcat and as high as 5% in the F-16 Fighting Falcon (Sadraey 2012). Examples of aircraft with higher percentages of composite material are the F/A-18 E/F SuperHornet (20%), AV8B HarrierII (22%), Dassault Rafale B/C (25%), and F-22 Raptor (25%).The majority of structural materials used in late-generation aircraft, such the F-35 Lightning II (35%) and Eurofighter (40%) are composites. Figure 5.26 depicts the structural uses of composites in the F-35, primarily carbon fibre–epoxy and carbon fibre–bismaleimide (Alam 2022).

Even though synthetic composites, like carbon fibre and glass fibre composites, have structural advantages, they also come with a number of drawbacks, like high raw material costs and an environmental impact at the end of their useful lives because they are not recyclable or biodegradable. Additionally, the aircraft sector is emphasising sustainability. Natural fibres and resins are the basis for the development of bio-based composites, which aim to lessen the environmental impact of aircraft operation and production. When it comes to their service life, these materials are meant to be recyclable or biodegradable, with performance comparable to that of standard composites. The goal of the Flightpath 2050 report, released by the Advisory Council for Aviation Research and Innovation in Europe (ACARE), was to reduce the carbon footprint and dependency on petroleum-based products while protecting the environment and energy supply by integrating recyclable and environmentally friendly materials into aircraft technologies. (Mansor et al. 2019). Radar translucent, dome-shaped structures that protect aircraft radar antennas from aerodynamic stress, inclement weather, and bird strikes are referred to as aviation radomes. Figure 5.27

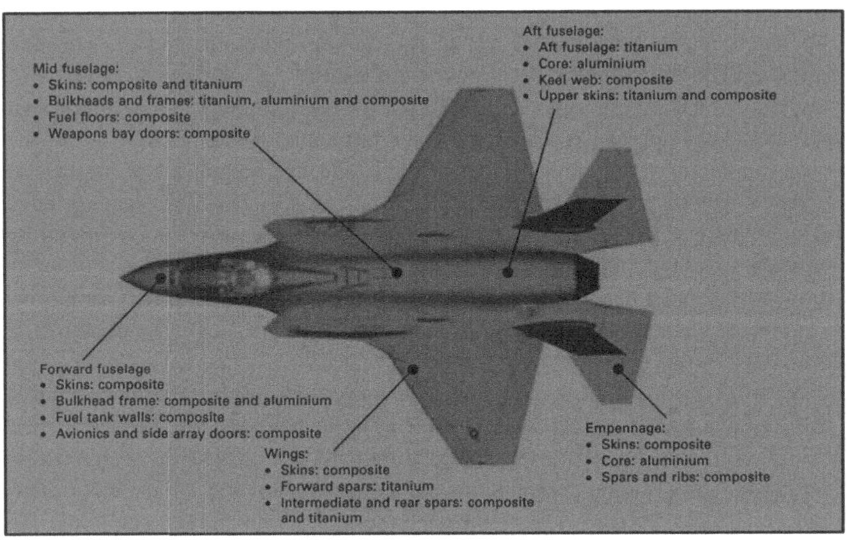

Fig. 5.26 Illustrates how composites are used in the F-35's structure (Alam 2022)

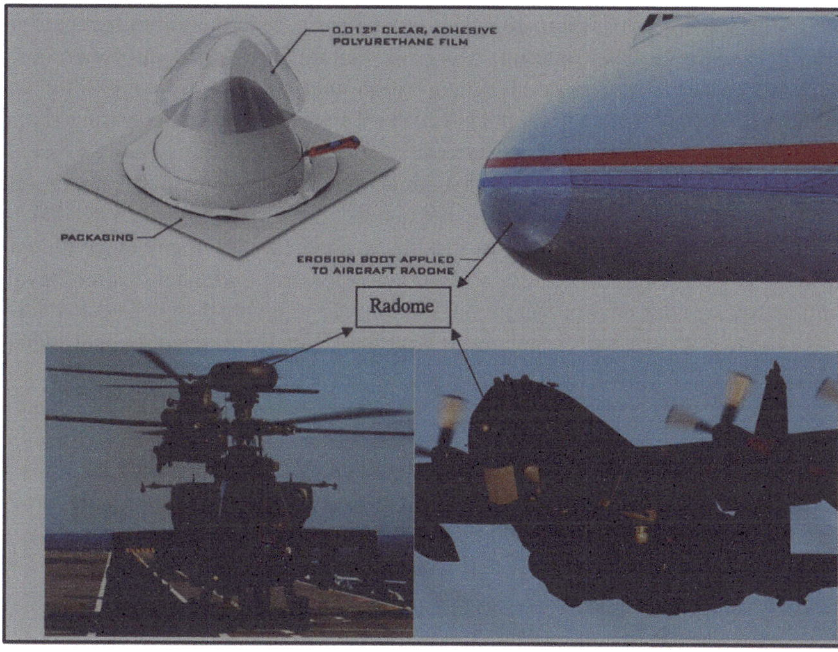

Fig. 5.27 Possible use of a polymer composite reinforced with natural fibre in radomes

shows the potential application of natural fibre reinforced polymer composite in radomes.

Natural fibre composites, or NFCs, got their start in the aviation sector with the launch of Gordon Aerolite, which was essentially flax roving that had been treated with phenolic resin. Gordon Aerolite's rigidity, high tensile strength, and low weight make it a good material for aviation. This was fitted to the Bristol Breinheim bomber as an experimental main spar. However, as more modern materials became available, the Gordon Aerolite's use decreased. The possibility for weight reduction with ramie fibre composites for aviation wing box components has been investigated. According to theoretical calculations, utilising ramie fibre composites instead of 7000-series aluminium alloy can reduce the weight of aircraft wing box components by 12% to 14% (Boegler et al. 2014). In a different study, Kling et al. (2015) also discussed the potential use of renewable materials in next-generation wing designs. The future wing design for a single-aisle aeroplane that they studied included an in-plane cantilevered thin wing made of certain sustainable materials. However, compared to wing boxes composed of aluminium and carbon fibre composite, their examination of the wing weight and other significant aircraft-related characteristics showed that the usage of NFPC results in heavier wing boxes and higher engine fuel consumption.

Through the European initiative known as "Cayley," Boeing Research and Technology Europe (Spain), INvent GmbH (Germany), Biomass, Biopolymer-Based Materials, and Bioenergy Aimplas (Spain), and Lineo (France) have currently joined forces to establish a consortium. The goal is to industrialise the solutions for environmentally friendly aeroplane interiors using flax, natural fibres, and reinforced renewable polymer matrix composites for the aviation industry (S B. Looking to lighten upaircraft interiors—with natural fibers? 2015). The composites are constructed of a recyclable thermoplastic foam core surrounded by skins consisting of a fire-resistant polymer matrix and natural fibres coated with flame-retardant non-halogenated chemicals. These sandwiched composites made of natural fibres are intended to be used as sides for cabins. Initiatives like this to green aircraft or any other alternative mode of transportation show how important it is to cope with new technology in addition to preserving environmental sustainability by reducing their negative effects.

In a similar vein, Airbus collaborates with the Council for Scientific and Industrial Research (CSIR) of South Africa to investigate the possibilities of Natural Fibre Polymer Composite (NFPC) uses on next-generation ecologically friendly aeroplanes. Their goal is to create an entirely reusable interior for an aircraft cabin. Currently, the interior components are entirely dependent on sophisticated, traditional composite materials like GFRP and CFRP, which are difficult to recycle. However, another example of potential uses for NCF is the X3 high-speed demonstrator built by Airbus Helicopters, which uses natural flax fibre in hybrid composite materials. The goal of the X3 high-speed hybrid demonstration is to combine speed and vertical take-off in a cost-effective and safe manner. Eloy et al. (2015) also looked at the potential use of NFPC for cotton fibre and polyurethane (PU) matrix interior components for aeroplanes. The use of PU resin generated from castor bean oil in this work was significant since it allowed for the complete formulation of composites utilising natural resources. Their investigation's findings demonstrated that, in comparison to other materials, cotton fibre-reinforced bio-based PU composites have poorer tensile properties; nonetheless, the difference is negligible, particularly when it comes to cotton fibre-reinforced epoxy composites. In order to improve product performance at the end of its useful life and achieve the goal of creating "green aircraft" in the future, this has shown that it is possible to replace synthetic composites in aerospace interior applications with completely bio-based composites.

5.6 Conclusion

The use of Fibre-Reinforced Polymers (FRP) composites in aerospace applications is a revolutionary advance in materials science, having a profound effect on the efficiency, performance, and design of aeroplanes and spacecraft. Because of their exceptional strength, light weight, and resistance to corrosion, fibre reinforced polymer (FRP) composites are now considered to be an essential material in the aerospace sector. The section on fibre-reinforced polymer properties explained how polymers and composites improve the overall performance of aerospace structures by delving into the composition and properties of these materials. To maximise the performance and design of aeronautical components, it is essential to comprehend the complex interaction between polymers and reinforcing fibres. Complex procedures are involved in the preparation of FRP composites, ranging from precision manufacturing techniques to the thoughtful selection of polymers and reinforcing fibres. As a reflection of the high standards required by the industry, the rigorous attention to detail in the preparation process is crucial to guaranteeing the optimal performance, durability, and reliability of aerospace components.

Because of their unique properties, FRP materials are perfect for a range of components, including structural elements, interior components, and external surfaces. The aerospace sector is still utilising FRP composites to lower weight, increase fuel economy, and enhance overall performance and safety of aeroplanes and spacecraft. An important development in materials science and engineering is the use of FRP composites in aeronautical applications. FRP composites will probably become even more important in the future as technology develops and shapes aerospace manufacturing and design. To ensure that FRP composites continue to be successful in aerospace applications and to unleash further advancements, this discipline has to continue with research and development.

References

Aamir M, Tolouei-Rad MG (2019) Recent advances in drilling of carbon fiber–reinforced polymers for aerospace applications: a review. Int J Adv Manuf Technol 105:2289–2308

Abbood IS, Odaa SA, Hasan KF, Jasim MA (2021) Properties evaluation of fiber reinforced polymers and their constituent materials used in structures—a review. Mater Today Proc 43:1003–1008. https://doi.org/10.1016/j.matpr.2020.07.636

Acquah C, Datskov I, Mawardi A, Zhang F, Achenie LEK, Pitchumani R et al (2006) Optimization under uncertainty of a composite fabrication process using a deterministic one-stage approach. Comput Chem Eng 30:947–960. https://doi.org/10.1016/j.compchemeng.2005.12.015

Agrawal AR, Pandelidis IO, Pecht M (1987) Injection-molding process control—a review. Polym Eng Sci 27:1345–1357. https://doi.org/10.1002/pen.760271802

Ahmad F, Choi HS, Park MK (2015) A review: natural fiber composites selection in view of mechanical, light weight, and economic properties. Macromol Mater Eng 300:10–24. https://doi.org/10.1002/mame.201400089

Alam M (20220 Recent advancements in advanced composites for aerospace applications: a review. Adv Compos Aerosp Eng Appl. https://doi.org/10.1007/978 3 030-88192-4_16

Al-Furjan MSH, Shan L, Shen X, Zarei MS, Hajmohammad MH, Kolahchi R (2022) A review on fabrication techniques and tensile properties of glass, carbon, and Kevlar fiber reinforced rolymer composites. J Mater Res Technol 19:2930–2959. https://doi.org/10.1016/j.jmrt.2022.06.008

Altin KM (2021) Investigation of friction performance and surface integrity of aramid fiber-reinforced polymer matrix composite. Polym Compos 42:6349–6361. https://doi.org/10.1002/pc.26302

Ansari A (2019) Aerospace composites market

Balasubramanian K, Sultan MTH, Rajeswari N (2018) Manufacturing techniques of composites for aerospace applications. Elsevier Ltd. https://doi.org/10.1016/B978-0-08-102131-6.00004-9

Balla VK, Kate KH, Satyavolu J, Singh P, Tadimeti JGD (2019) Additive manufacturing of natural fiber reinforced polymer composites: processing and prospects. Compos Part B Eng 174:106956. https://doi.org/10.1016/j.compositesb.2019.106956

Baran I, Cinar K, Ersoy N, Akkerman R, Hattel JH (2017) A review on the mechanical modeling of composite manufacturing processes. Arch Comput Methods Eng 24:365–395. https://doi.org/10.1007/s11831-016-9167-2

Blakey-Milner B, Gradl P, Snedden G, Brooks M, Pitot J, Lopez E et al (2021) Metal additive manufacturing in aerospace: a review. Mater des 209:110008. https://doi.org/10.1016/j.matdes.2021.110008

Boegler O, Kling U, Empl D (2014) ATI. Potential of sustainable materials in wing structural design

Boeing sets pace for composite usage in large civil aircraft (2005)

Breuer U (2016) Introduction. In: Commercial aircraft composite technology. Springer, Cham

Carbon Fiber Reinforced Plastic Market Size, Share and Trends Analysis Report By Raw Material, By Product (Thermosetting, Thermoplastic), By Application, By Region, And Segment Forecasts, 2023–2030

Chava S, Namilae S (2021) Continuous evolution of processing induced residual stresses in composites: an in-situ approach. Compos Part A Appl Sci Manuf 145:106368. https://doi.org/10.1016/j.compositesa.2021.106368

Chemical and Material (n.d.)

Chen Y, Zhang J, Li Z, Zhang H, Chen J, Yang W et al (2023) Manufacturing technology of lightweight fiber-reinforced composite structures in aerospace: current situation and toward intellectualization. Aerospace 10. https://doi.org/10.3390/aerospace10030206

Cheng X, Li X, Zhang J, Zhang Q, Cheng Y, Zeng B (2017) Evaluation of tensile properties of a composite-metal joint with a novel metal insert design by experimental and numerical methods. Chinese J Aeronaut 30:1004–1011. https://doi.org/10.1016/j.cja.2017.03.016

Composites Market Size (2022) Share and trends analysis report by product type (Carbon Fiber, Glass Fiber), By Manufacturing Process (Layup, Filament, Injection Molding, Pultrusion), By End-use, By Region and Segment Forecasts, 2023–2030

Compression (n.d.)

Diniță A, Ripeanu RG, Ilincă CN (2024) Advancements in fiber-reinforced polymer composites: a comprehensive analysis. Polymers (Basel) 16:1–49

El-Hassan H, El Maaddawy T, Šišková A (2019) Microstructure characteristics of GFRP reinforcing bars in harsh environment. Adv Mater Sci Eng 2019. https://doi.org/10.1155/2019/8053843

Eloy FS, Costa RRC, de Medeiros R, Ribeiro MLTV (2015) Comparison between mechanical properties of bio and synthetic composites for use in aircraft interior structures. In: Proceedings of meeting on aeronautical composite materials and structures. Sao Paulo

Ertekin M (2017) Aramid fibers. Elsevier Ltd. https://doi.org/10.1016/B978-0-08-101871-2.00007-2

Fiber Reinforced Polymer (FRP) Composites Market, By Fiber Type, By Application, By End-User, By Country, and By Region - Industry Analysis, Market Size, Market Share & Forecast from 2023–2030 2023

Galinska A, Cezary G (2020) Mechanical joining of fibre reinforced polymer. Polymers (Basel)

Gopu GN, Joseph SA (2022) Corrosion behavior of fiber-reinforced concrete—a review. Fibers 10. https://doi.org/10.3390/fib10050038

Gürgen S, Kuşhan MC, Diltemiz SF (2016) Fatigue failure in aircraft structural components. Handb Mater Fail Anal with Case Stud from Aerosp Automot Ind 261–77. https://doi.org/10.1016/B978-0-12-800950-5.00013-2

Hamerton I, Mooring L (2012) The use of thermosets in aerospace applications. Thermosets Struct Prop Appl 189–227. https://doi.org/10.1533/9780857097637.2.189

Hiken A (2017) The evolution of the composite fuselage—a manufacturing perspective. SAE Int J Aerosp 10:77–91. https://doi.org/10.4271/2017-01-2154

https://www.mordorintelligence.com/team ABS. Fiber Reinforced Polymer Market Size & Share Analysis—Growth Trends & Forecasts Source: 2023

https://www.mordorintelligence.com/team ABS. Natural Fiber Reinforced Composites Market Size & Share Analysis—Growth Trends & Forecasts (2024–2029) Source

Jayakrishna M, Vijay M, Khan B (2023) An overview of extensive analysis of 3d printing applications in the manufacturing sector. J Eng 2023:1–23. https://doi.org/10.1155/2023/7465737

Karim MA, Abdullah MZ, Deifalla AF, Azab M, Waqar A (2023) An assessment of the processing parameters and application of fibre-reinforced polymers (FRPs) in the petroleum and natural gas industries: a review. Results Eng 18:101091. https://doi.org/10.1016/j.rineng.2023.101091

Kling U, Empl D, Boegler OIA (2015) Future aircraft structures using renewable materials. In: Proceeding Dtsch. und Raumfahrtkongress. Rostock, Ger.

Leong YW, Thitithanasarn S, Yamada K, Hamada H (2013) Compression and injection molding techniques for natural fiber composites. Woodhead Publishing Limited. https://doi.org/10.1533/9780857099228.2.216

Lowe J (20050 Aerospace applications. Des Manuf Text Compos 405–23. https://doi.org/10.1533/9781845690823.405

Ma L, Liu F, Liu D, Liu Y (2021) Review of strain rate effects of fiber-reinforced polymer composites. Polymers (Basel) 13. https://doi.org/10.3390/polym13172839

Mangalgiri PD (1999) Composite materials for aerospace applications. Bull Mater Sci 22:657–664

Mansor MR, Nurfaizey AH, Tamaldin N, Nordin MNA (2019) Natural fiber polymer composites. Elsevier Ltd. https://doi.org/10.1016/b978-0-08-102426-3.00011-4

Masuelli M (2013) Introduction of fibre-reinforced polymers−polymers and composites: concepts, properties and processes. Fiber Reinf Polym Technol Appl Concr Repair. https://doi.org/10.5772/54629

Milosevic M, Dzunic D, Valasek P, Mitrovic S, Ruggiero A (20220 Effect of fiber orientation on the tribological performance of abaca-reinforced epoxy composite under dry contact conditions. J Compos Sci 6. https://doi.org/10.3390/jcs6070204

Muralidhara B, Kumaresh Babu SP, Suresha B (2019) Utilizing vacuum bagging process to prepare carbon fiber/epoxy composites with improved mechanical properties. Mater Today Proc 27:2022–2028. https://doi.org/10.1016/j.matpr.2019.09.051

Nagavally RR (2016) Composite materials—history, types, fabrication techniques, advantages, and applications. Int J Mech Prod Eng 25–30

Oku T (n.d.) Field assessment of concrete structures rehabilitated with FRP. Elsevier Ltd. https://doi.org/10.1016/B978-0-08-044163-4.50033-4

Parveez B (n.d.) Scientific advancements in composite materials for aircraft applications: a review. Polymers (Basel) 12

Parveez B, Kittur MI, Badruddin IA, Kamangar S, Hussien M, Umarfarooq MA (2022) Scientific advancements in composite materials for aircraft applications: a review. Polymers (Basel) 14. https://doi.org/10.3390/polym14225007

Prasad N, Dev S, Khande K, Chandra G, Prakash P, Sen K et al (2014) Study on aramid fibre and comparison with other composite materials. IJIRST-Int J Innov Res Sci Technol 1:303–306

Pultrusion Market. Trends, Opportunities and Competitive Analysis [2023 2028] (2023)

Rabbi MS, Islam T, Islam GMS (2021) Injection-molded natural fiber-reinforced polymer composites–a review. 1–21

Raj SSR, Dhas JER, Jesuthanam CP (2020) Challenges on machining characteristics of natural fiber-reinforced composites—a review. J Reinf Plast Compos. https://doi.org/10.1177/0731684420940773

Rajak DK, Pagar DD, Menezes PL, Linul E (2019) Fiber-reinforced polymer composites: manufacturing, properties, and applications. Polymers (Basel) 11

Rajak DK, Wagh PH, Linul E (2021) Manufacturing technologies of carbon/glass fiber-reinforced polymer composites and their properties: a review. Polymers (Basel) 13. https://doi.org/10.3390/polym13213721

Rajput MS, Industry S (2018) Damage tolerance of impacted composite sandwich structures damage tolerance of impacted composite sandwich structures Moeen ud din Saeed Rajput Doctoral Thesis Stockholm. Sweden. https://doi.org/10.13140/RG.2.2.27066.03525/1

Services A, Lafayette W, Vista B (n.d.) NASA composite materials development: lessons learned and future challenges
S B. Looking to lighten upaircraft interiors—with natural fibers? 2015
Sadraey MH (2012). Wing Design. https://doi.org/10.1002/9781118352700.ch5
Siengchin S (2023) A review on lightweight materials for defence applications: present and future developments. Def Technol 24:1–17. https://doi.org/10.1016/j.dt.2023.02.025
Smock D (2008) Boeing 787 dreamliner represents composites revolution
Sukiman NL, Zhou X, Birbilis N, Hughes AE, Mol JMC, Garcia SJ et al (2005) Aluminium alloys—new trends in fabrication and applications
Summerscales J, Searle TJ (2005) Low-pressure (vacuum infusion) techniques for moulding large composite structures. Proc Inst Mech Eng Part L J Mater des Appl 219:45–58. https://doi.org/10.1243/146442005X10238
Tao Y, Hadigheh SA, Wei Y (2023) Recycling of glass fibre reinforced polymer (GFRP) composite wastes in concrete: a critical review and cost benefit analysis. Structures 53:1540–1556. https://doi.org/10.1016/j.istruc.2023.05.018
Thyavihalli Girijappa YG, Mavinkere Rangappa S, Parameswaranpillai J, Siengchin S (2019) Natural fibers as sustainable and renewable resource for development of eco-friendly composites: a comprehensive review. Front Mater 6:1–14. https://doi.org/10.3389/fmats.2019.00226
Venugopal J, Dhanasakkaravarthi B, Surakasi R, Rinawa ML, Manjunatha LH, Alshgari RA et al (2022) Effect on compression molding parameters in mechanical properties of MWCNT/glass fiber/epoxy composites. Adv Polym Technol 2022. https://doi.org/10.1155/2022/9295407
Vijayan DS, Sivasuriyan A, Devarajan P, Stefańska A, Wodzyński Ł, Koda E (2023) Carbon Fibre-Reinforced Polymer (CFRP) composites in civil engineering application—a comprehensive review. Buildings 13. https://doi.org/10.3390/buildings13061509
Vinayagamoorthy R (2017) A review on the polymeric laminates reinforced with natural fibers. J Reinf Plast Compos 36:1577–1589. https://doi.org/10.1177/0731684417718385
Volk M, Yuksel O, Baran I, Hattel JH, Spangenberg J, Sandberg M (2022) Cost-efficient, automated, and sustainable composite profile manufacture: a review of the state of the art, innovations, and future of pultrusion technologies. Compos Part B Eng 246:110135. https://doi.org/10.1016/j.compositesb.2022.110135
Wazeer A, Das A, Abeykoon C, Sinha A, Karmakar A (2023) Composites for electric vehicles and automotive sector: a review. Green Energy Intell Transp 2:100043. https://doi.org/10.1016/j.geits.2022.100043
Yazdani N, Garcia EC, Riad M (2018) Field assessment of concrete structures rehabilitated with FRP. Elsevier Ltd. https://doi.org/10.1016/B978-0-08-102181-1.00008-3
Yunus S, Salleh Z, Aznan MA, Berhan MN, Kalam A, Rashid AA et al (2014) Impacted of vacuum bag woven kenaf/fiberglass hybrid composite. Appl Mech Mater 660:572–577. https://doi.org/10.4028/www.scientific.net/AMM.660.572
Zhang J, Lin G, Vaidya U, Wang H (2023) Past, present and future prospective of global carbon fibre composite developments and applications. Compos Part B Eng 250
Zhang M, Matinlinna JP (2012) E-glass fiber reinforced composites in dental applications. SILICON 4:73–78. https://doi.org/10.1007/s12633-011-9075-x

Chapter 6
Metal Nanoparticles for Aeronautical Engineering Applications

K. Aruna Prabha and L. Srinivasa Rao

Abstract By employing diverse techniques such as chemical reduction, sol–gel methods, electrochemical deposition, green synthesis, and template-assisted fabrication it is possible to precisely control the size, shape, composition, and surface features of nanoparticles to create materials with unique capabilities that meet the demands of the aerospace industry. The characterization techniques like Electron microscopy, X-ray diffraction, dynamic light scattering, spectroscopy (UV–Vis, FTIR, Raman), surface area analysis, and other characterization methods offer important insights into the size, composition, crystallinity, surface chemistry, stability, and dispersibility of metal nanoparticles.

Keywords Metal nanoparticles · Aeronautical · Nanocomposites · Aviation engineering · Coatings · Sensors · Propulsion systems and characterization

6.1 Introduction on Metal Nanoparticles

The area of aviation engineering is expanding its innovations and use of advanced materials and technology in this field improves performance, safety and efficiency. Metal nanoparticles exhibit tremendous potential for a variety of aerospace applications for having distinct physical, chemical and biological properties at the nanoscale level. Aeronautical systems function better, last longer and consume less energy by incorporating metal nanoparticles into a variety of parts and materials. They improve several elements such as aircraft performance, functionality and design. Because of their special qualities, they are useful in a variety of aerospace applications helping to develop new materials, coatings, sensors, and propulsion systems. Additionally, components for aviation and spacecraft are made robust, lighter and more resilient

K. Aruna Prabha (✉) · L. Srinivasa Rao
VNR Vignana Jyothi Institute of Engineering and Technology, Hyderabad, Telengana, India
e-mail: arunaprabha_k@vnrvjiet.in

L. Srinivasa Rao
e-mail: srinivasarao_l@vnrvjiet.in

Visakh P. M. (ed.), *Nanomaterials for Defense, Aeronautical and Aerospace Applications*, Advanced Structured Materials 236,
https://doi.org/10.1007/978-3-031-92723-2_6

by incorporating these metal nanomaterials in components of aircraft wings, fuselages, and engines. Presently used composites, coatings and electrical devices are other areas of applications (Dhinakaran et al. 2021). They have enhanced features such as light weightness, weatherproofing, thermal stability and different sensing and monitoring systems inside an aircraft (Wu et al. n.d.). Composites containing nanomaterials improve mechanical strength, fatigue resistance, and thermal stability in the field of structural materials. These improvements aid in the creation of robust yet lightweight components, hence meeting the vital requirements of aircraft performance and fuel efficiency with metal nanoparticles. Their application in catalysts, coatings, and fuel additives has shown promise in improving combustion efficiency, lowering emissions, and improving thermal management and prefaced advanced fuel propulsion systems. They have remarkable qualities for sensing applications, which make it possible to create extremely selective and sensitive sensors for use in aircrafts. Their application in environmental monitoring systems, gas sensors, and structural health monitoring helps to raise the bar for safety and dependability in aircraft operations (Wenwu et al. 2019). They are appealing for a variety of aeronautical applications because of their high surface area, size-dependent characteristics, and adaptability. An overview of the characteristics, production processes, and uses and applications of metal nanoparticles in the aerospace industry is given in this chapter.

6.1.1 Types of Metal Nanoparticles

The manufacture of nanoparticles commonly uses metals including aluminum (Al), gold (Au), cadmium (Cd), silver (Ag), titanium, platinum (Pt), copper (Cu), cobalt (Co), iron (Fe), lead (Pb) and zinc (Zn). These exhibit exceptional UV–visible sensitivity, mechanical, electrical, catalytic, thermal, and antibacterial characteristics due to quantum effects and outsized surface-to-volume ratio, the properties, benefits and types of nanoparticles are illustrated in Fig. 6.1. Metal oxide nanoparticles are frequently produced to benefit from their increased reactivity and efficiency. The electrostatic interactions between the positive metal ions and the oxygen ions give rise to strong and stable ionic connectionism to form metallic oxides such as aluminum oxide (Al_2O_3), zinc oxide (ZnO) and titanium oxide (TiO_2). Some aviation components where electrostatic discharge of electromagnetic interference shielding is necessary, nanopowders of copper, aluminum, and iron are utilized (Ealias and Saravanakumar 2017; Khan et al. 2022). The following are a few varieties of metal nanoparticles where the aerospace industry uses them:

1. **Gold nanoparticles**: Used in sensing, catalysis and medicine applications, gold nanoparticles are recognized for their stability, biocompatibility etc. These qualities of heat transfer can be improved by dispersing gold nanoparticles in lubricants or coolants. Because of their special optical qualities, like surface plasmon resonance they are useful in optical and electronic applications in the aerospace

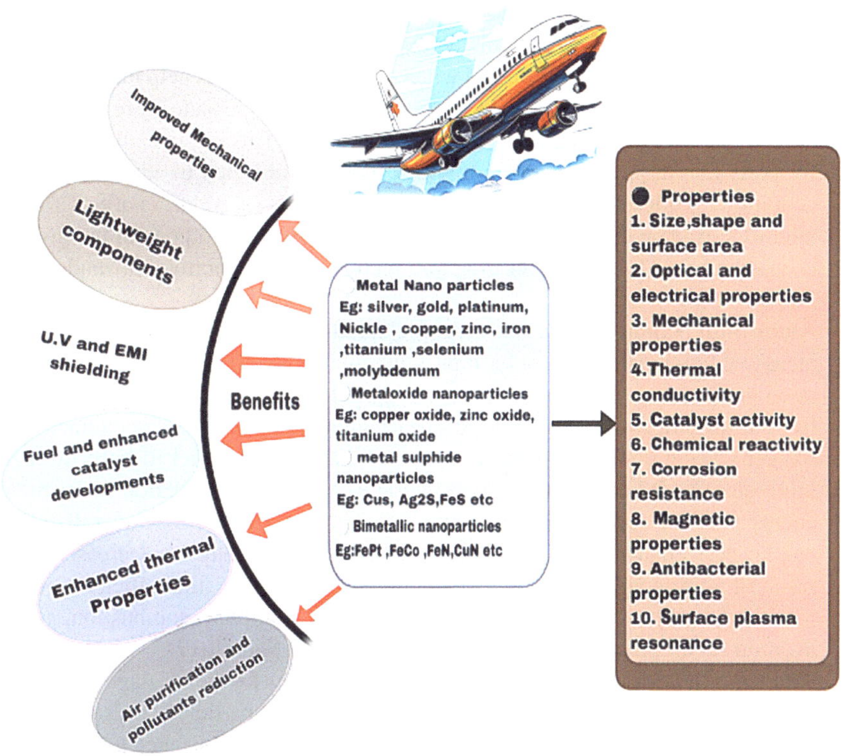

Fig. 6.1 Metal nanoparticles types, properties and benefits

industry. They are employed in the creation of cutting-edge optical sensors, displays and communication systems for aircrafts (Hassan et al. 2022).

2. **Silver Nanoparticles**: Strong antibacterial and antifungal effects are exhibited by silver nanoparticles. Antimicrobial coatings for airplane interiors minimizes the danger of microbiological contamination. Door handles, armrests, tray tables and other interior surfaces of airplanes can all be coated with silver nanoparticles contained materials. To improve cabin air purification, silver nanoparticles can be added to air filtration systems and water purification systems (Vela-Cano et al. 2021).
3. **Platinum nanoparticles**: These particles are useful due to their great activity and stability and are valuable as fuel cell catalysts that produce power on board and help make flying more environmentally friendly. Being catalytic is a well-known characteristic for platinum nanoparticles. They can be utilized in aerospace to improving fuel combustion efficiency in jet engines. Platinum catalysts have the potential to lower emissions and enhance propulsion systems overall efficiency (Stephen et al. 2019).

4. **Copper nanoparticles**: Copper nanoparticles have the ability to improve a material's electrical conductivity. This feature is useful in aircraft electrical components and wiring. Materials loaded with copper nanoparticles help make airplane electrical systems more effective. Aircraft surfaces can be made more resistant to corrosion and erosion by using copper nanoparticles. This is especially important for parts like the leading edges of wings that are subjected to adverse weather conditions while in flight. When flying over bodies of water, coatings for an aircraft's exterior can be made with copper nanoparticles to prevent the growth of marine life. This can lower drag and increase fuel economy (Gawande et al. 2016).
5. **Aluminum Nanoparticles**: These as fuel additives increase aviation fuels' efficiency of combustion, possibly improving engine performance. When incorporated into aviation fuels, aluminum nanoparticles can help promote more effective burning. Because of their large surface area, nanoparticles can interact with oxygen during combustion more effectively, causing the fuel to burn through more quickly. When aluminum nanoparticles are added to aviation fuels, more energy may be released during combustion. Higher thrust and overall better engine performance may result from this. Emissions may be decreased as a result of aluminum nanoparticles' which enhanced combustion efficiency. Less unburned hydrocarbons and other pollutants by complete combustion, make aviation sector greener and cleaner (Kannaiyan and Sadr 2017).
6. **Titanium dioxide nanoparticles**: Aerospace coatings for the fuselage and wings and other surfaces, can include titanium dioxide nanoparticles. These coatings have improved self-cleaning capabilities, durability, and resistance to corrosion. Furthermore, titanium dioxide's photocatalytic activity might aid in the breakdown of organic pollutants on the surfaces, thus maintaining the cleanliness of the aircraft. Because titanium dioxide has the capacity to block ultraviolet (UV) light, it is frequently found in sunscreens. Titanium dioxide nanoparticle contained coatings can be used on airplane windows or surfaces to reduce the effects of UV radiation and shield passengers and interior items from UV-induced deterioration (Shi et al. 2013).
7. **Zinc oxide nanoparticles**: have remarkable and unique properties that make them suitable for various applications, including coatings for sensors. Zinc oxide nanoparticles possess the ability to be used in the production of photo detectors that sense light. They are appropriate for uses in light-sensing applications, flame detection, UV monitoring and other situations requiring sensitivity to ultraviolet (UV) radiation. Temperature sensors can benefit from the addition of zinc oxide nanoparticles, which will increase their sensitivity in identifying temperature changes. Numerous industries, including aerospace, automotive, and electronics, depend on this application (Raha and Ahmaruzzaman 2022).
8. **Metal sulfides NPs**: can be added to coatings to prevent corrosion, such as zinc sulfide or molybdenum disulfide nanoparticles. In the aerospace business, protecting aircraft from environmental elements that can cause corrosion, like moisture and salt, is essential (Xing et al. 2022). Applications for heat management can make use of iron or copper sulfide. These nanoparticles can improve a

material's thermal conductivity, which qualifies it for use in engine and electronic components heat dissipation (Onwudiwe 2019).

9. **Bimetallic nanoparticles**: They are made up from two distinct metal elements and had gained prominence in a variety of industries, including aerospace and aircraft production. Pt–Pd and Ag–Cu based bimetallic nanoparticles possess the capacity to build fuel cell systems by escalating energy conversion efficiency and catalytic activities for use in aircraft propulsion systems. Fe-Pt bimetallic nanoparticles can be used to create magnetic materials that can be used in actuators and sensors in aircraft systems. The application of Cu-Ni bimetallic nanoparticles in conductive coatings can enhance electrical conductivity and corrosion resistance in electronic components and aircraft structures (Liu et al. 2023).

- **Properties of Metal nanoparticles**

The distinct physical, chemical, and electrical characteristics of metal nanoparticles make them an intriguing and adaptable class of nanomaterials that have attracted a lot of interest in a variety of sectors. Their high surface area-to-volume ratio, compact size and quantum effects, exhibit enhanced optical, optoelectrical, catalytic, antibacterial, antiviral properties. These characteristics render them appealing for a multitude of uses in fields like electronics, sensing, medicine, catalysis, and more. Key properties of metal nanoparticles include the following.

1. **Quantum Effects related to avionics**: Materials electrical and optical characteristics can alter at the nanoscale. When quantum effects become important, size-dependent phenomena result. The electrical characteristics of materials can be affected by quantum phenomena at the nanoscale in avionics and electronic systems. The performance of electronic components in aircraft systems may be impacted by the display of modified electronic behavior in nanoscale semiconductor materials. Semiconductor nanoparticles known as quantum dots have applications in aircraft imaging, sensors, and displays. The effectiveness and performance of these devices can be enhanced by the size-dependent electronic characteristics of these quantum dots (James Singh et al. 2021).
2. **Control of Size and Shape**: Changing the size and form of metal nanoparticles is crucial as it changes their properties. The smaller particle size means that there are lot more atoms on the surface. Controlling the size and shape of nanometals is an important part of nanotechnology and has a number of benefits for aviation. The features such as their enhanced strength-to-weight ratio, toughness, fatigue resistance and customized thermal conductivity, tuned plasmonic resonance with optimized catalytic activity and precision thus improving performance and efficiency in a variety of aviation applications (Liu et al. 2020).
3. **Surface Area**: The size and shape of the nanoparticles determine the surface area to volume ratio of the particles, which in turn affects features like the devices' ultraviolet visibility. The higher surface area of nanoparticles increases their reactivity and qualifies them for use in sensing and as catalyst (Joudeh and Linke 2022).

4. **Optical properties and optoelectricity:** Activating electromagnetic radiation and collective oscillation of free electrons in the metal causes surface plasmon resonance (SPR) of metal nanoparticles. Resonance at particular wavelengths causes changes in the optical characteristics of the nanoparticle as well as increased light absorption and scattering. Particularly in gold and silver nanoparticles, the electrons oscillate collectively in response to incoming light called as surface plasmon resonance. This feature of optical detection using plasmonic resonance response can detect structural stress or damage. It also provides real-time information about the integrity of crucial components and aids in applications of sensing and imaging (Li et al. 2023; Fernando et al. 2017).
5. **Catalytic Activity**: The inhibited properties such as high surface area and reactivity of metal nanoparticles, such as platinum and palladium, can be employed as catalysts in propulsion systems. By accelerating the oxidation of carbon monoxide and unburned hydrocarbons in exhaust streams, these nanoparticles help reduce emissions and increase combustion efficiency. As a result, the combustion process is optimized, improving fuel efficiency (Ramli et al. 2023).
6. **Electrically conductive**: Metal nanoparticles have the potential to have increased electrical conductivity, particularly when they are composed of components like gold and silver. The development of electronic devices and sensors depends heavily on metal nanoparticles. These distinct electrical characteristics can be used in conductive inks, printable electronics and sensors that identify different gasses or biomolecules. Aluminum NPs is one of the many conductive materials used to make the fuselage, wings, and tail of an aircraft. This conductivity is necessary as it reduces the chance of lightning strikes, dissipates electrical charges and stop static electricity from building up (Altammar 2023).
7. **Thermal aspects**: It is possible to incorporate metal nanoparticles, such as copper or silver nanoparticles, into materials to increase their heat conductivity. This characteristic can be helpful in controlling the distribution and dissipation of heat in a variety of aviation components, including engine and electronic systems. To improve the thermal characteristics of heat transfer fluids like coolants or lubricants, metal nanoparticles can be added. In aviation, this is important for controlling engine and other essential component's temperature. Better engine performance and fuel efficiency are a result of increased heat transfer efficiency (Coetzee et al. 2020).
8. **Mechanical Properties**: Metal nanoparticles can improve its suitability for aviation applications where frequent loading and unloading cycles occur. The ductility and toughness of materials can be affected by metal nanoparticles, depending on their size and kind. To sustain a variety of loads and stresses during operation, airplane components must maintain a balance between strength and ductility. The tensile strength of composite materials can be increased by incorporating metal nanoparticles into polymers or other matrix components. This may result in airplane construction components that are stronger and more resilient (Ozkan Ata et al. 2023).

9. **Chemical Reactivity**: Metal nanoparticles' small size, high surface area-to-volume ratio, and quantum effects give them a special chemical reactivity. The change are based on the kind of metal, its size, shape, and surroundings. Because of their large surface area and active sites, they are frequently employed as catalysts and take part in electron transfer activities to impact redox processes. Comprehending and managing the chemical reactivity of metal nanoparticles is crucial in customizing their characteristics in aircrafts. In order to increase the effectiveness, robustness, and performance of numerous aviation-related technologies, researchers are still investigating and developing these nanomaterials (Viñes et al. 2014).
10. **Environmental Attributes**: Metal nanoparticles have potential for cleaning up the environment. They can be applied for removal of contaminants from air and water by means of adsorption and catalytic degradation. Although these have many uses, there are still difficulties and worries related to their use, including possible toxicity and environmental effects during production, disposal and use. The proper development and use of these nanomaterials depend on comprehending and resolving issues (Guo et al. 2014).

- **Advantages of Metal NPs**

There are various benefits of incorporating metal nanoparticles into aircraft components and are described below.

1. **Enhanced mechanical characteristics**: Nanomaterials are more flexible, resilient, and harder than other materials. As a result, aircraft parts are made stronger and more resilient to harsh environments. These nanoparticles can be incorporated in, structural lubricant coatings, adhesives etc., as they have exceptional young modulus, stress and strain properties. Additionally, this feature can help to construct mechanically more robust nanodevices with a range of applications. Through the process of embedding NPs in the metal and polymer matrix, mechanical strengths, tribological properties can be boosted. The reason for this is that relatively little wear and friction may result from the NPs rolling in the lubricated contact region. Furthermore, NPs possess favorable sliding and delamination properties, which could lessen wear and friction, as a result they enhance the lubricating property of element. As the coating increases toughness and wear resistance, it can result in a variety of mechanically strong features. It has been effectively established that NPs based on carbon, titanium, and alumina may provide coatings to improve mechanical characteristics (Liu et al. 2023). The inclusion of nano-TiN can help improve the mechanical properties of ceramite. When comparing the fractional mechanical properties of nano TiN tailored cermet with standard cermet, the latter exhibits fracture toughness, bending strength and hardness. Adding micro WC and nano-ZrO_2 lowers the surface friction coefficient and wear rate of ceramite (Liu et al. n.d.). Adding nano-TiN to the matrix instead of nano-Al_2O_3 can improve a cermet composite's hardness, flexural strength and fracture toughness more successfully. Furthermore, the size, ratio and number of the nanoparticles all have an impact on

the mechanical properties of nanocomposites (Chmielewski and Pietrzak 2007). Mechanical properties of a composite are usually affected when nanoparticles are added to improve a material's electrical or thermal conductivity. Due to their high surface area to volume ratio, smaller nanoparticles are probably more reactive than bigger ones, which would be more useful for reinforcing. Thermal radiation and electrical current can flow through surfaces where adjacent nanoparticles come into contact with one another. The conductivity across these channels is determined by the particulate electrical and thermal conductivity. Increased electrical conductivity can be achieved by functionalizing nanoparticles, which create a molecular network inside the matrix with varying properties depending on the kind of functionalization (Rohatgi and Schultz 2007).

2. **Light weight components**: After stability and safety, one of a material's most important requirements is lightweighting. The aerospace industry is always looking for ways to lighten airplanes without sacrificing the mechanical integrity of their frames and components. When a plane consumes less fuel the lighter it becomes. In addition to lowering carbon emissions and increasing fuel efficiency, using less fuel also lowers the cost of each flight. Weightlessness had a major influence on mechanical, structural, and industrial aspects. The influence of nanocomposites are examined for low-earth orbit applications. Carbon nanotubes have clear advantages over polymeric composites due to the metal matrix's intrinsic high temperature stability, high strength, high modulus, wear resistance, and thermal and electrical conductivity. Aluminum nanocomposites are anticipated to surpass the weight reduction, attained through their use in aerospace applications since these metal matrices are stronger and more rigid than fiber composites and nanocomposites based on polymers. (Bhat et al. n.d.). It was discovered that adding ZrO_2 nanoparticles to polyimide decreased the coefficient of friction and lowered the wear rate. The lightweight nanoparticles also caused the structures overall mass to decrease. Nanocomposite system would provide a good tribological material for use in spacecraft. An analysis conducted showed that incorporating SiC and Al_2O_3 nanoparticles into metal matrix composites (MMCs) enhances the fatigue strength of aerospace components. The enhanced fatigue behavior was mostly determined by the size and distribution of the grains and it was seen that the fatigue strength of the composite increased as the percentage of nanoparticles increased (Kasgoz et al. 2015).
3. **Enhanced thermal aspects**: To improve heat dissipation, thermal interface materials can be treated with metal nanoparticles, such as copper or aluminum. This is essential for controlling the thermal loads in avionic systems and avoiding vital parts from overheating. TiC is noted for its great hardness and high melting point, making it a perfect material for different industrial applications, such as cutting tools, wear-resistant coatings, and abrasives. It can also be used in tough conditions due to its great chemical resistance and thermal stability. Furthermore, TiC is utilized as a high-temperature furnace lining because of its high melting point. It is a desirable material for a variety of industrial applications due to its unique blend of mechanical and thermal qualities. Zirconium is a silver-gray, multipurpose metal that is prized for its ductility, strength, malleability, and glossy look. Its

physical and chemical characteristics are comparable to those of titanium and also resists heat and corrosion. This metal has the same degree of toughness as copper and is lighter than steel (Xing et al. 2022). Al_2O_3 nanoparticle/nickel matrix nanocomposite showed improved fire resistance, reduced heat transference and strong mechanical and thermal stability. These composite materials can ward off lightning and can be regarded as an alternative to copper and aluminum, which are prone to galvanic corrosion under specific environmental circumstances (Ma et al. 2017). The suspended nanoparticles in nanofluids offer great promise for many industrial applications like heat exchangers, cooling appliances, reactors, solar collectors etc. since they increase the thermal conductivity of a typical fluid. Furthermore, the nanofluid may be considered a viable fuel and coolant for use in an aircraft power system. These fluids contain Pd@DSHPG or Pd@OSHPG remain stable even at reduced temperature. Thermal conductivity increased as temperature and the mass percentage of nanoparticles increased (Qin et al. 2020).

4. **Fuselage**: The fuselage structure can be strengthened by the application of metal nanoparticles, particularly the ones composed of lightweight metals like titanium or aluminum. By strengthening and extending the material's strength, this reinforcement helps the fuselage endure mechanical stress, pressure changes, and possible impacts. Researchers have found that strengthening an aircraft's fuselage with a cobalt nanoparticle can be beneficial. A defect-free fuselage can be up to 100 times stronger than the fuselages that are currently being manufactured, according to preliminary research. This would result in a lighter plane and a decrease in the amount of jet fuel used since less material would be required to give the fuselage the same strength. This further increases fuel efficiency and creates stronger, safer, and more affordable materials. (Siengchin 2023).
5. **Advanced Sheilding**: Surface engineering techniques are employed in the aerospace and defense industries to improve the performance and longevity of vital components that are exposed to harsh environments. The reliability and safety of engine components and aero-automotive parts are crucially dependent on wear-resistant coatings and corrosion protection technologies in the aero-automotive sector, while surface modifications in the electronics and semiconductor industry facilitate the creation of high-performance devices. Coatings containing metal nanoparticles can resist corrosion and shield aircraft surfaces from environmental deterioration. These coatings can lower maintenance costs and increase the longevity of aircraft parts. Nanoparticle TiO_2 self-cleaning coatings eliminate the need for expensive surface cleaning, as they are especially beneficial for skyscraper maintenance. Using micron-sized inorganic fillers can increase the coating's resistance to scratches, but by scattering visible light, they give the coating a matt or semi-matt appearance. However, by using nanoparticles, scattering of light can be minimized greatly. Nanoparticles like SiO_2, ZrO_2, and AlOOH incorporated into UV-curable lacquers, improves the resistance to abrasion (Gu et al. 2020). Polymers or composites can have their electrical conductivity enhanced by the addition of silver or copper nanoparticles. Applications like lightning strike protection in aeronautical structures and electromagnetic interference (EMI) shielding require this. Electrically conductive nanoparticles

are often produced from gold, aluminum, platinum, palladium, graphite, silver or copper. The capacity of silver nanoparticles (AgNP) to increase the phase transition materials' thermal conductivity has made them a promising choice (Bamogo et al. 2022).

6. **Improvised Fuel Catalysts**: By acting as catalysts in fuel cells, platinum or palladium nanoparticles can help produce cleaner, more effective electricity for airplanes. The environmental impact of fuel cells can be minimized by either supplementing or replacing conventional combustion engines. The rate-limiting Oxygen Reduction Reaction (ORR) of this process is typically catalyzed by platinum and other platinum group metal nanoparticles in the Polymer Electrolyte Fuel Cell (PEMFC), a fuel cell that is frequently studied (Hile et al. 2022).

- **Preparation of metal nanoparticles**

Metal nanoparticles can be produced by physical, biological or chemical mechanisms. Physical procedures include vapor condensation and laser ablation techniques, while common chemical approaches involve reducing metal ions in solution to produce nanoparticles as seen in Fig. 6.2. One element makes up metal nanoparticles, which are single-element nanomaterials. There can be single atoms or collections of several atoms. Among the most often manufactured nanoparticles are Au, Ag, Pt, Cu, Pd, Re, Zn, Ru, Co, Cd, Al, Ni, and Fe. There are two basic ways to conceptualize the processes involved in creating nanoparticles approaches that are top–bottom and bottom-up (Hile et al. 2022).

6.1.2 Top-Down Synthesis Method

Using physical, chemical or mechanical techniques, bulk materials are reduced into nanoparticles by means of the top-down and top–bottom approaches. Taking a bulk material, changing it into the desired nanostructure and then stabilizing the resultant nanosized metal nanoparticles by adding colloidal protective agents is the basic idea of top–bottom synthesis. This method begins with larger structures and works its way down to smaller units until the required size is reached. In order to produce nanoparticles using this technology, thermal, lithographic, high-energy lasers and mechanical energy is employed. Atomization, laser ablation, electron beam evaporation, arc discharge, focused ion beam lithography, vapor condensation, electron beam evaporation, and electron deposition are a few examples of these categories (Baig et al. 2021).

1. **Milling and mechanochemical processing-related mechanical attrition**: This technique is an efficient way to create metal nanoparticles as it controls and reduces size and characteristics of the particles. Using high-energy mills, milling is performed to create unique nanocrystalline powders. A planetary ball mill is one of the efficient mills used in the mechanical alloying synthesis process.

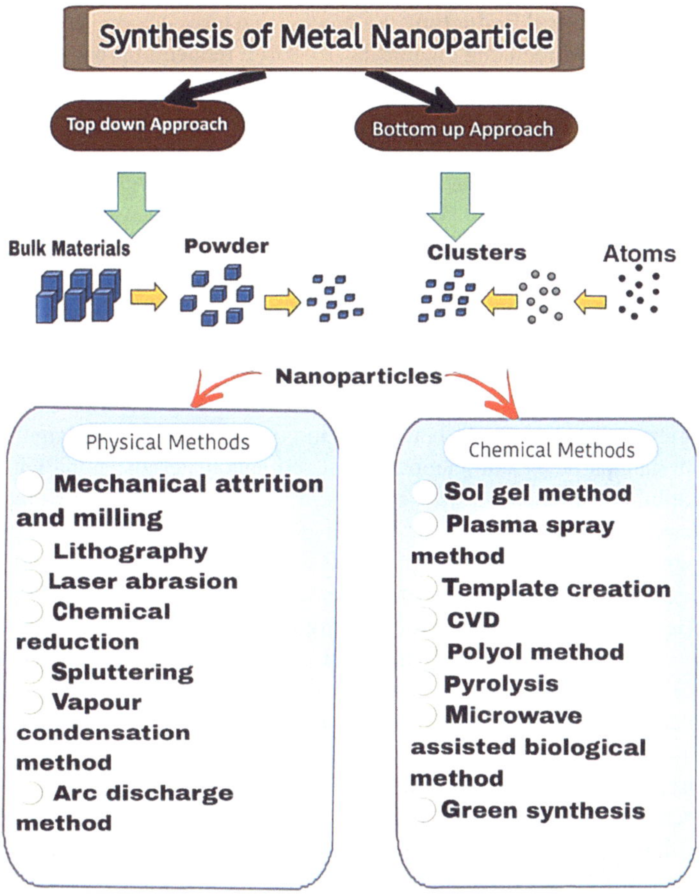

Fig. 6.2 Preparation methods of metal nanoparticles

This approach forms nanosized particles by having the vials revolve on disk as well as their own axis in which they are placed and mounted. A range of metal nanoparticles, including noble metals like gold and silver and transition metals like iron and copper, are synthesized using ball milling. The other techniques include cutting, grinding and etching (Shojaei et al. 2022).

2. **Lithography Methodologies:** Lithography transfers a pattern onto a substrate and creates exact features at nanoscale. Etching or deposition are used in later stages to define the nanostructures. Wet or dry etching is another technique used in nanolithography. Nanostructures or their arrays can be created via directed growth or limited growth from a few nanometers to several nanometers using nanolithography. These techniques have the benefit of enabling the fabrication of huge amounts of 1D nanostructures from a wide range of readily available materials (Colson 2013).

3. **Method of Chemical Reduction**: is a process that turns metal ions into nanoparticles by applying reducing chemicals. To start a reduction process, a reducing agent is usually applied after a metal salt has been dissolved in a liquid media and the reaction occurs at high temperatures. Citric acid, hydrazine sodium borohydride ($NaBH_3$) are frequently employed as reducing agents in the production of noble metal nanoparticles, such as silver and gold (Szczyglewska et al. 2023).
4. **Method of Laser Ablation (LAM)**: This technique makes use of a laser to irradiate the metal target, causing vaporization and condensation of the vapor into nanoparticles. Nanoparticles are formed when vaporized material in the plasma condenses. Anisotropic nanomaterials including carbon nanotubes, nanorods and nanowires can be synthesized on a large scale using laser ablation. High-purity, small-sized, well-crystallized catalysts are essential to the anisotropic growth of these one-dimensional nanomaterials in these methods. Furthermore, laser ablation offers the advantage of being able to consistently produce large amounts of both nanoparticles and nanoparticle-based nanomaterials, along with strong compatibility with aerosol flow systems. Metal and metal oxide nanoparticles synthesized by laser ablation uses helium gas to create gold nanoparticles. Using laser ablation on metallic bulk silver salts in solution phase, silver nanoparticles can be produced. One benefit of the laser ablation approach over other conventional techniques for producing metal colloids is the absence of chemical reagents in the solutions. Consequently, this approach produces pure colloidal AgNPs. Other materials that can be manufactured include oxides with good magnetic and electrical properties, such as titania nanoparticles, Ni/NiO nanoparticles, alloy nanoparticles of Co and Pt, and strontium ferrite (Kim et al. 2017).
5. **Spluttering**: The physical vapor deposition (PVD) method for synthesizing metal nanoparticles in top-down method is sputtering. It is mostly used for thin-film deposition, while it can also be modified to create nanoparticles in specific situations. The procedure is irradiating a target substance with high-energy ions, which causes atoms or molecules to be expelled and settle on a substrate to produce nanoparticles or a thin film. Sputtering onto iconic liquids for some amount of time can produce upscaled Cu NPs compared to mechanical and chemical processes of synthesis. Nevertheless, unlike other approaches, sputter synthesis just requires the actual sputter procedure to produce the stabilized NPs, no extra chemicals or synthesis steps are required (Meischein and Ludwig 2021).
6. **The Method of Vapor Condensation (VCM):** The process of creating metal nanoparticles by condensation of metal vapor into nanoscale particles is known as the Vapor Condensation Method (VCM). Using this technique, nanoparticles of many materials, including Ag, Au, and PbS, can be created. The vaporization process, has been extensively used to produce AgNPs, in which the target materials are vaporized by the heat source and then quickly condensed to provide the synthesized nanoparticles (Yusuf 2018).
7. **The efficient Arc Discharge Method (ADM)**: Metal nanoparticles such as, nanosilver, nano γ-Al_2O_3 can be synthesized using the Arc Discharge Method (ADM), Carbon-encapsulated metal nanoparticles, which are also referred to as metal-filled nanotubes or metal-filled nanocapsules can also be efficiently

produced. This technique involves an electric arc created under a regulated environment between two metal electrodes. The Ag wires containing surface layer evaporates and condenses the water layer during the arc-discharge. The clear solution with distinctive light yellow hue is converted into golden-brown colloidal suspension of silver nanoparticles (Sonanwane et al. 2022).

6.1.3 Bottom-Up Synthesis Method

The bottom-up, or constructive technique is the method of creating materials starting with atoms and working their way up to clusters and nanoparticles. For the production of nanoparticles, the most widely used bottom-up methods are pyrolysis, biosynthesis, chemical vapour deposition (CVD), spinning, and sol–gel. Sedimentation and reduction processes, green synthesis and biological synthesis, are instances of this approach. One effective illustration of bottom-up synthesis is the use of colloidal dispersion in the production of nanoparticles. There are four types of bottom-up synthesis of nanomaterials, gaseous phase, liquid phase, solid phase, and biological technique. The most widely used technique for preparation, particularly for the preparation of metal nanoparticles is wet chemical, liquid-phase reduction (Sonanwane et al. 2022).

1. **The sol–gel** procedure involves turning a colloidal solution (sol) into a gel, which is desiccated and heated to generate nanoparticles. The sol–gel method is a versatile technique used for the preparation of various functional materials, including photocatalysts, nonlinear optical materials, ferroelectrics, and superconductors. Functional materials like photocatalysts, ferroelectrics, nonlinear optical materials, and superconductors can be made using this technique. Simple and complicated nano metal oxides, including nano TiO_2, ZrO_2, Al_2O_3, ZnO, WO_3, Nb_2O_5, and other rare earth oxides, can be created using sol–gel synthesis (Marinescu et al. n.d.).
2. **Creation of templates**: Using nanometer-sized templates, template fabrication is one of the most widely used and possibly least expensive methods of nanolithography. This technique works well for growing nanowires by vapor phase, sol–gel, or electrodeposition. It is possible to disconnect nanowires from the templates and control them separately. Initially, organized nanopore templates need to be created. Metals are electrodeposited inside the nanopores using solutions containing metal salts that are either basic or acidic. It makes use of a three-terminal potentiostat. As the metal is deposited on the cathode, or working electrode, generally Ti and Au are metallized on one end of the porous membrane (Charitidis et al. 2014).
3. **Plasma spray method**: One technique for producing nanoparticles that can be employed even in an open environment is plasma spray synthesis. A high-temperature plasma is created throughout the process and directed onto a substrate or precursor material. The plasma spray method can be utilized in the synthesis of several kinds of nanoparticles, including metal oxides, ceramics, and composite

materials. The collection of the generated nanoparticles is difficult due to the extraordinarily high flow velocity of the particles. The method is simple, affordable and has potential for mass production (Jang et al. 2021).

4. **Chemical Vapor Deposition (CVD)**: Metal nanoparticle production is commonly achieved through the use of Chemical Vapor Deposition (CVD). Gaseous precursors react chemically in CVD to form a thin layer or deposit on a substrate. Controlled growth of nanoparticles on a substrate is made possible by CVD when it comes to nanoparticle synthesis. Nanoparticles are produced from metals such as platinum (Pt), silver (Ag), gold (Au), cadmium (Cd), cobalt (Co), iron (Fe), copper (Cu), and zinc (Zn) by this method (Escorcia-Díaz et al. 2023).
5. **Polyol Method**: In this approach, polyols like ethylene glycol serve as both a reducing agent and a solvent. Metal precursors are dissolved in a polyol, and high temperatures are used to reduce the metal. A reliable technique for producing metal NCs with precise dimensions, forms, compositions, and crystallinities is the polyol approach. The creation of metal, metal oxide, and semiconductor NCs in a variety of forms, including cubes, nanowires, nanorods, and nanobars, are also accomplished using this method. Copper, platinum, and palladium are just a few of the metal nanoparticles that are frequently created using the polyol process (Ruz and Sudarsan 2021).
6. **Pyrolysis**: Flame spray and spray pyrolysis produce the final metal or metallic oxide particles with aerosol decomposition and thermolysis. It involves atomizing metal salt solutions and heating the liquid droplets to a high temperature in a furnace or flame. Ultrasonic pyrolysis creates a variety of metallic and metal oxide particles in nano and micron sizes, starting with Au and Au salts. Au and Ag NPs, particles can be decked out from nanoparticles of Ag/TiO_2 and Au/Fe_2O_3 and can be obtained as a powder or as a suspension in solution. The choice of the precursor solution is important for all these procedures since the final particle's composition and structure will be affected by various metal salts. The properties of the produced nanoparticles are also greatly influenced by the process parameters, which include temperature, pressure, and residence time in the furnace or flame (Majerič and Rudolf 2020).
7. **Green synthesis**: is the application of sustainable and ecologically friendly techniques, frequently using microbial systems or plant extracts as reducing agents. Metal ions are reduced by plant extracts or microbes to create nanoparticles. Based on the intended nanoparticles, metal salts are selected as precursors. Copper sulfate ($CuSO_4$) or zinc acetate ($Zn(CH_3COO)_2$) for metal oxide nanoparticles, and gold chloride ($AuCl_4^-$) for gold nanoparticles are examples of common metal salts. A reaction mixture is made by mixing the metal precursor solution with the plant extract. The plant extract works as a reducing agent to create nanoparticles from metal ions. Utilizing natural plant extracts production of metal and metal oxide nanoparticles such as gold, silver, copper oxide, and zinc oxide (ZnO) are possible. The benefits are it is economical, ecologically friendly and frequently eliminates the need for dangerous chemicals (Álvarez-Chimal and Ángel Arenas-Alatorre 2023).

8. **Synthesis with Microwave Assistance**: The approach uses microwaves as a source of energy to synthesize nanoparticles. Microwaves are used to heat the reaction mixture, enabling more rapid and uniform heating. It speeds up the reduction process after metal salts have dissolved in a solvent. These include faster heating, more control over the distribution of particle sizes, and shorter reaction times. Lead salt and sulfur sources are typically used in the microwave irradiation method of producing PbS nanoparticles. With this method, the lead source is PbS derived from $Pb(NO_3)_2$ and $Pb(Ac)_2$ (Onwudiwe 2019).

6.2 Characterization of Metal Nanoparticles

Characterization plays a key role in understanding the characteristics of metal nanoparticles and optimizing their applications. A range of techniques are employed to characterize metal nanoparticles, providing information about their stability, surface charge, composition, size, and form. Typical methods for describing metal nanoparticles include the following and can be seen in Fig. 6.3.

6.2.1 Analysis of Shape and Size

1. **Transmission electron Microscopy (TEM)**: Characterizing metal nanoparticles is made possible by this powerful method which enables high-resolution imaging

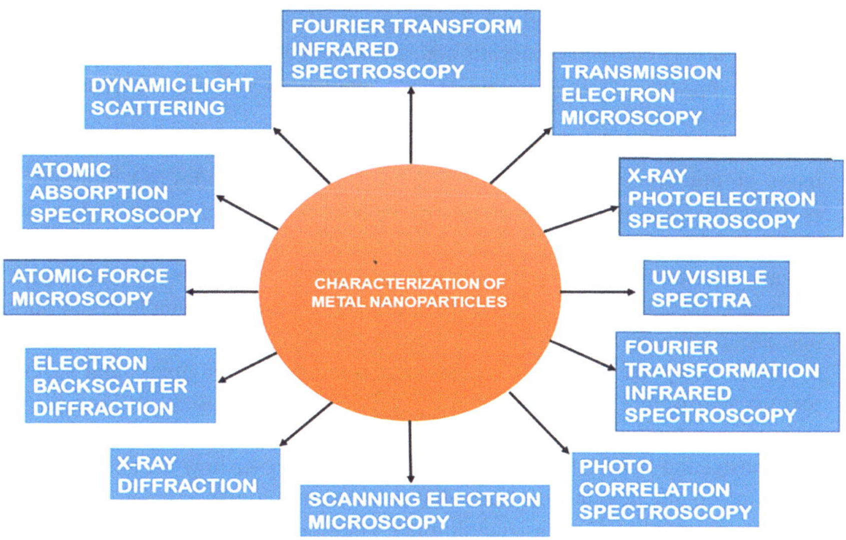

Fig. 6.3 Characterization and evaluation techniques of metal nanoparticles

of materials at the nanoscale, yielding details about the distribution, size, shape, and structural properties of the particles. Gold nanoparticles (Au–NPs), silver atom clusters (Ag–ACs), tantalum single atoms (Ta–SAs), carbon materials functionalized with iron cobalt (Fe–Co) and titania (TiO_2) NPs, and platinum loaded Ceria (Pt–CeO_2) nanocomposite are among the specific nanomaterials that are examined. Electron energy loss spectroscopy (EELS), aberration corrected bright field TEM (BF-TEM), high-angle dark-field scanning TEM (HAADF-STEM), and BF-TEM electron tomography (ET) are among the TEM techniques used to study nanomaterials (Anjum 2016).

Among the most effective techniques for examining a single nanomaterial's morphological properties is electron microscopy (EM), which makes it possible to take direct, high-resolution photographs of individual nanomaterials. Furthermore, TEM has a great deal of potential in statistical analysis because pictures of hundreds of nanomaterials can be obtained simultaneously with ease. Only qualitative descriptions are possible for the forms and structures of nanoparticles, and morphological assessments of nanomaterials are still mostly carried out using hundreds of nanomaterials that researchers have gathered (Lee et al. 2020).

2. **SEM Scanning Electron Microscopy**: Particle morphology analysis and surface imaging can be accomplished by means of scanning electron microscopy (SEM). It is common tool for studying size and morphology. SEM uses a concentrated electron beam to scan a material, interacting with the atoms to produce three-dimensional surface topography. The specimen & surface are scanned by high-energy electron beams coming from the SEM. It differs from conventional light microscopes where light waves produce an expanded image. When an electron beam is struck on a specimen surface, it reacts under a scanning electron microscope. Presence of gold nanoparticles on the cell surface could be obtained by imaging back scattered electrons. This method allows for the observation of the interaction between nanomaterials and macrophage cells that have been dosed with varying concentrations of metal oxide nanoparticles (CeO_2, TiO_2, and ZnO) (Goldstein et al. 2014).
3. **Atomic Force Microscopy**: Particle height, size, and shape can be measured using atomic force microscopy (AFM), which also provides 3D imaging. It is a versatile and efficient microscopic method that analyzes materials at the nanoscale. It meets the needs of researchers and engineers by providing various surface measurements and capturing a three-dimensional topographical image. Compared to other imaging techniques, AFM involves less sample preparation and produces images with atomic resolution and height information. It can be applied to polymer nanocomposites to evaluate surface roughness and surface texture on different kinds of materials. Additionally, because of its remarkable three-dimensional resolution, it can be categorized as a non-destructive approach (Baalousha and Lead 2012). This makes it possible to see how the nanoparticles are arranged and how their surfaces looks and allows to identify both individual and group particles. Synthesized alumina, zirconium oxide, and titanium oxide nanoparticles and/or agglomerates on various surfaces use AFM in the dynamic

mode to characterize them for determining size, shape, and distribution of the nanoparticles (Rao et al. 2007).

6.2.2 Examination of Structure

1. **XRD X-ray diffraction**: XRD usually provides information on the phase type, lattice parameters, and size of crystalline grains in materials. The idea of constructive interference of X-rays scattered by crystal lattice planes is the foundation of X-ray diffraction. Depending on the crystal lattice spacing, X-rays are diffracted at different angles when they impact a crystal. It is a very helpful characterisation tool for examining the chemical composition, physical structure, and crystallographic organization of materials and thin films non-destructively. Additionally, it can be used to quantify the strain, grain size, phase composition and defect structure, among other structural characteristics of these crystalline phases. Nanoparticles are analysed when they exhibit a diffraction peak as it indicates their crystalline nature within this size range. The broadening of this peak is attributed to the reduced particle size. In contrast to the bulk material random grain orientation, the preponderance of the line in XRD suggests that the nanoparticle grains have been selectively reoriented in one direction. These lines XRD positions are comparable to those found in bulk metals. This shows that the particles under investigation have a pure nanoparticle lattice with bulk-like periodicity (Mourdikoudis et al. 2018b).

6.2.3 Evaluation of Composition and Optical Properties

1. **UV–Vis spectroscopy**: This method measures the amount of light that a sample absorbs and scatters, the amount is called the extinction and it is calculated as the total of the light that is absorbed and scattered. In its most basic form, the intensity of a UV/visible light beam is measured both before and after it passes through a sample, which is positioned between a light source and a photodetector. The wavelength dependent extinction spectrum of the sample is quantified by comparing these observations at each wavelength. Usually, the data is displayed as extinction vs wavelength. In order to ensure that the sample extinction spectrum does not contain any spectral features from the buffer, each spectrum in the background is corrected using a buffer blank. SPR detects the oscillation of conduction electrons on the surface of nanoparticles. The position of the SPR (plasmon resonance of surfaces) peaks is highly sensitive to the size and shape of the nanoparticles. The SPR peak may move, expand or split into many peaks as the size or shape varies. These alterations may be tracked and information about the geometry of the nanoparticles can be deduced using UV spectroscopy. In this the metal nanoparticle features can be effectively characterized, including size, shape, and surface properties in the colloidal dispersion system. Moreover,

UV-Vi spectroscopy makes it possible to quantify the quantity of precursor metal ions needed to create metal nanoparticles such ad gold, palladium, manganese oxide etc. (Okitsu and Kumar 2013).

2. **X-ray photoelectron spectroscopy**: X-ray Electron Spectroscopy for Chemical Analysis (ESCA), commonly referred to as Photoelectron Spectroscopy (XPS), is a potent surface-sensitive analytical method for examining the chemical state and elemental composition of various materials. When used on metal nanoparticles, XPS offers insightful data regarding the electrical structure, oxidation states, and surface chemistry of the nanoparticles (Khan et al. 2022).
3. **AAS analysis**: Metal nanoparticles must be appropriately prepared. Simultaneous sizing and measurement of nanoparticles such as gold NPs can be done by AAS (Atomic Absorption Spectroscopy). This could entail breaking down the matrix and releasing the metal ions into solution by digesting the nanoparticles with the proper acids. AAS instruments are comprised of a sample chamber, a detector, a monochromator to isolate the metal absorption line, and a light source. The device calculates how much light the metal atoms in the sample are absorbed. This is an extensively used analytical method for figuring out how much of each element is present in a sample. It can be utilized to measure the metal content of metal nanoparticles (Friedland et al. 2021).

6.2.4 Surface and Geometrical Evaluation

1. **Dynamic Light Scattering**: establishes the zeta potential and hydrodynamic size of the nanoparticles in solutions Information on the physical properties of the sample can be obtained by detecting the light scattered by light interactions with matter. One method that is frequently used to determine the size of nanoparticles (NPs) in colloidal solutions in the nano- and submicrometer ranges is dynamic light scattering (DLS).. In most light scattering investigations, the sample is exposed to a monochromatic beam, and the scattered light is then recorded at a specific angle by a detector. DLS computes particle sizes assuming spherical particles. If the forms of the metal nanoparticles are non-spherical, DLS is a useful technique for determining the size distribution, polydispersity, and colloidal stability of metal nanoparticles. To get accurate and significant results, however, meticulous attention to sample preparation, concentration effects, and instrument parameters is necessary. They demonstrate how a number of variables, including MNP surface coating, suspension concentration, particle shape, and colloidal stability, affect the size value determined by DLS measurement (Carvalho et al. 2018).
2. **Electron Backscatter Diffraction (EBSD)**: A potent method for examining the crystallographic orientation of materials, particularly metal nanoparticles, is electron backscatter diffraction (EBSD). It is utilized to describe the nanoscale microstructure and crystallographic orientation of materials. EBSD can reveal

information on the orientation relationships, crystallographic phases and grain structure of gold (Au) and nickel (Ni) nanoparticles (Carneiro and Simões 2020).

3. **Fourier Transform Infrared Spectroscopy (FTIR)**: It is a flexible technique for analyzing vibrational modes, functional groups that are present on the surface of the particles. It is used in many disciplines, including material science, biology, and chemistry. The results of fourier transform infrared spectrum displays peaks at particular wavenumbers, each of which represents a vibrational mode of a particular chemical bond. For example TiO_2 NPs, PEG and 2ME have assigned vibrational modes and functionalized TiO_2-PEG and TiO_2-PEG–2ME. PEG–2ME, the TiO_2 intermolecular interactions can be observed as change in peak position by FTIR in composites (Dutta 2017; León et al. 2017).The functional groups contained in the nanoparticles can be identified by examining the locations and intensities of these peaks. FTIR reveals the bond's stretch as well as the structure and corresponding bands of the produced nanoparticles. It is possible to identify the reactive surface sites that cause the surface reactivity as well as the chemical composition of the NPs' surface under particular circumstances. Researchers can better understand the nature of interactions between nanoparticles and their surroundings by using FTIR to uncover information on the surface chemistry of the nanoparticles (Eid 2022).

6.3 Applications of Metal Nanoparticles in Aerospace Engineering

Metal nanoparticles' special qualities and attributes make them useful in a variety of aerospace engineering applications as seen in Fig. 6.4. Among the noteworthy applications are:

1. **Strength and Durability**: Materials used in the construction of airplanes can have their mechanical qualities improved by the addition of metal nanoparticles. Strength, toughness, and durability must all be increased in order to resist the challenging circumstances encountered during flight, such as turbulence, high altitudes, and fluctuating temperatures. This is important because the aircraft industry wants lightweight, highly durable materials to increase overall performance and fuel efficiency. Findings proved that individual nanoparticles or the nanoparticles operating on a surface possess adhesion, friction, hardness, and elastic modulus, as well as movement laws with enhancing mechanical characteristics (Guo et al. 2014). Subsequently, lubrication additives, nanoparticle reinforced composite coating are among the primary uses of nanoparticles due to their unique mechanical qualities. Ceramics made of alumina and nano-Al_2O_3 offer greater flexural strength. Nano-Fe_2O_3 exhibited greater compressive and flexural strengths compared to the blank group. Nano-Al_2O_3 ceramics are more robust than microscale monolithic alumina ceramics. Its structural dimensions are in nanoscale when contrasted with ordinary materials. Nanomaterials offer

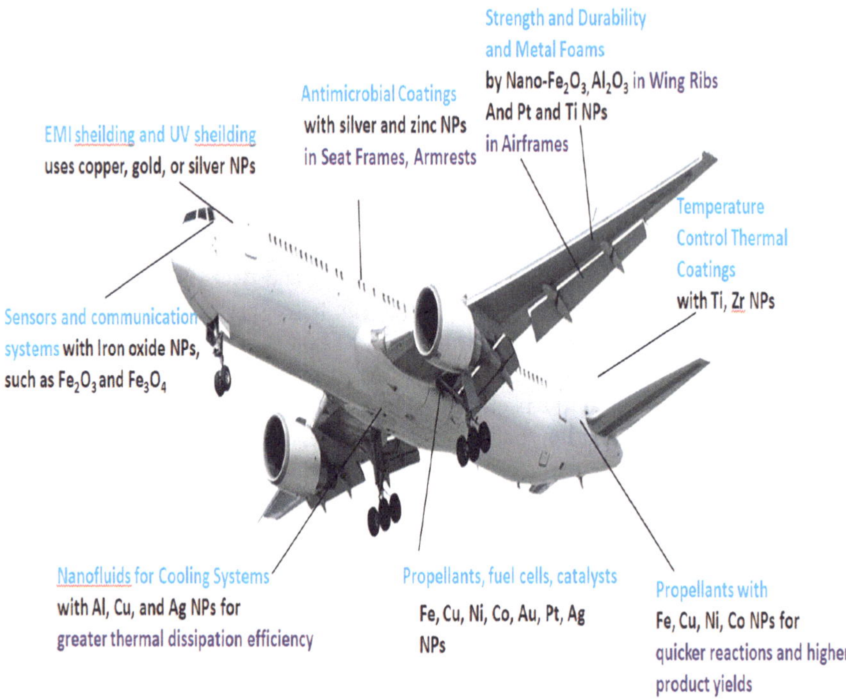

Fig. 6.4 Applications of metal nanoparticles in Aircrafts

special qualities as they can be distributed throughout the composite matrix, which improve the material's tensile and brittle properties. Composite materials, which blend of various materials having performance characteristics, are frequently used in aircraft wings. Metal nanoparticles can be added to composite materials to improve its strength, stiffness, and durability. Examples of metal nanoparticles include titanium, nickel and aluminum (Wu et al. n.d.).

2. **Propellants**: More active sites and a large surface area on metal nanoparticles specifically encourage quicker reactions and higher product yields. The two primary categories of these particles are non-noble-metal Fe, Cu, Ni, Co, etc. based nanoparticles and noble-metal such as Au, Pt, Ag, etc. supported metal nanoparticles. Solid rocket propellants employ metal nanoparticles like aluminum nanoparticles, as additions. By adding these nanoparticles, the energy released during combustion can be increased, improving the efficiency of the propulsion systems. Palladium and platinum metal nanoparticles can be employed as catalysts in propulsion systems. By increasing the efficiency of combustion reactions, they improve the performance of jet propulsion systems and rocket engines. In the field of catalysis, palladium nanoparticles have become especially well-known. They have been used to develop various catalysts for a number of organic transformations, including the production of C–C

bonds. In addition, fuel cells, sensors, and hydrogen storage have all made substantial use of palladium-based nanostructures (Narayan et al. 2019).

3. **Fuel cells**: Due to higher surface area-to-volume ratio than the bulk material, the metal component of the nanoparticles play a prominent role in the application as green catalysts. Due to their great dispersibility and catalytic activity, platinum-supported nanoparticles, in addition to gold, have been widely used in many fields, including electrocatalytic oxidations for fuel cells (Alaqarbeh et al. 2023).
4. **Propulsion System Catalysis**: Iron oxides and other magnetic nanoparticles have been used to support metallic Pt (Platinum) in order to give the oxidation of alcohols, CO and vitamin precursors to build a stable and efficient catalyst. Palladium (Pd) is another fascinating material that has been extensively researched in order to create Pd-supported nanoparticles. In order to create active catalysts for a variety of catalytic applications, metallic Pd was supported on a variety of substrates, including porous carbons, nanoparticles, polymers, etc. In general, at mild circumstances, metal nanoparticles including Pt, Pd, and Au showed high activity and selectivity towards the selective oxidation of alcohols. The temperature of the catalyst, in which the metallic nanoparticles are dispersed in solution, is lower than the solvent's boiling point.

   Because metallic nanoparticles are transparent to light, they can be used as photocatalysts when they are dissolved in liquids. The preparation makes it simple to regulate the metallic nanoparticles' size and shape. Even for reactions occurring in a gaseous phase, metal nanoparticles can function as catalysts when they are immobilized on solid substrates. The catalyst in question has a temperature below the solvent's boiling point where the metallic nanoparticles are dispersed throughout the solution. Metallic nanoparticles distributed in fluids can be used as photocatalysts since they are light-transparent. Metallic nanoparticle size and form may be readily controlled during preparation. They can function as catalysts for processes even in the gaseous phase when they are immobilized on solid substrates (Khan et al. 2023).
5. **Temperature Control Thermal Coatings**: Materials used in aircraft constructions must be extremely strong and stiff to retain mechanical integrity due to the high phase temperatures qualities (Parveez et al. 2022). To enhance heat dissipation and thermal stability, metal nanoparticles can be added to thermal coatings. Coatings offer defense against inclement weather, radiation, and corrosive substances in severe environmental conditions. As a result, aluminum alloy maintenance and replacement costs are decreased and the lifespan of aeronautical components are increased. In order to shield aircraft components from extremely high or low temperatures during atmospheric re-entry or high-speed flight, this is essential. Anodic and cathodic fortification, passivations, and protective coatings are often employed techniques to reduce corrosion rates and extend the life of certain metals. Furthermore, fresh techniques to defense that are connected to nanotechnology have just been created. Examples of these include surface treatments, coatings applied layer-by-layer, coatings that self-clean, coatings made of sol–gel, ceramic coatings, coatings made of nanoscale

alloys, and coatings applied on top of the layers. Nanotechnology-related coatings greatly improve aluminum alloy corrosion resistance as compared to traditional coating materials (Nhiem et al. 2023).Titanium alloys are among the lightest materials utilized in contemporary airplanes, especially for producing parts for gas turbine engines (Aruna Prabha et al. 2023). Enhancing titanium alloys' resistance to heat and wear will enable the substitution of less costly, lighter substitutes for pricey nickel alloys, greatly lowering the weight of the airplane. A multilayer metal-ceramic nanocoating is intended to shield the turbine and compressor blades of contemporary jet engines constructed with titanium alloys (Chaisaenrith and Taksakulvith 2021). Electron-beam physical vapor deposition is used to deposit yttria-stabilized zirconia nanocoatings. This is the next generation of thermal barrier coatings (TBCs) for gas turbine engines in industrial and aviation applications (Kumar et al. 2021).

6. **Metal Foams**: By utilizing nanoparticles, lightweight metal foams with enhanced strength and thermal characteristics can be produced. These foams can be used in structural parts to lighten their overall mass without sacrificing functionality. Examples of these include surface treatments, layer-by-layer coatings that self-clean, coatings made of sol–gel, ceramic coatings, coatings made of nanoscale alloys, and coatings applied on top of the layers. Through a range of methods, with chemical deposition and impregnation, metal nanoparticles can enter the porous structure of metal foams. These include carbon nanotube-reinforced aluminium with functional qualities such as low density necessary for light-weight applications, refractoriness, useful for fireproof, acoustic and thermal-insulating qualities for noise and heat management. They also possess mechanical strength for demanding constructions under load situations, and their particular energy absorption capacity, which is required for crash safety applications (Duarte and Ferreira 2016).
7. **Sensors and communication systems**: The application of metal nanoparticles is useful in cutting-edge sensing technologies. Sensors that can track changes in pressure, temperature, and other environmental variables can be made using nanomaterials. They can improve a sensor's sensitivity and selectivity, which makes it possible to utilize it to identify different gases, chemicals, and even structural damage in aircraft parts. Metal nanoparticles are incorporated in sensors and NEMS devices for aeronautical applications. These tools can detect a wide range of environmental parameters with extreme sensitivity and selectivity, improving performance and safety. Magnetic nanoparticles (Iron oxide nanoparticles, such as Fe_2O_3 and Fe_3O_4) integrated polymer films and composites can be used in a range of data storage medium. Ceramic nanoparticles like barium titanate and barium strontium titanate are used to make super capacitors. It is possible to create a standardized fuel management division that regulates the fuel in aero-engines using MEMS (Micro Electro Mechanical) (Arepalli and Moloney 2015).
8. **EMI Sheilding And UV Sheilding**: In order to create conductive plastics for use in a variety of aviation components where electrostatic discharge of electromagnetic interference shielding is necessary, nanopowders of copper, aluminum

and iron are utilized. High electrical conductivity, large specific surface area, and minimum cost and diameter ranging from 20–200 nm metallic nanomaterial have significant potential as an affordable and effective substitute for costly and high-density conventional metallic materials used in electromagnetic interference shielding. Metal nanoparticles can be added to coatings or polymers that are used on aircraft parts. This is especially true with conductive metals like copper, gold or silver. These substances contribute to the formation of a conductive shield that either reflects or absorbs electromagnetic energy (Nan et al. 2023). The sun's UV light can deteriorate materials and have an impact on how well electronic components function. When included into coatings or other materials, metal nanoparticles can either absorb or reflect UV light, so averting harm and prolonging the life of vital components. Metal nanoparticles can be employed to give materials UV protection while preserving optical clarity. This is notably true of materials composed of titanium dioxide or zinc oxide. For uses like cockpit windows and other transparent surfaces in aircraft, this is essential. UV shielding is crucial for preventing the degradation of composite parts and structural materials in addition to electronic systems. Metal nanoparticles protect aircraft components from UV-induced deterioration, extending their overall lifespan and lowering the frequency of maintenance and replacement (Yousefi et al. 2023; Silva et al. 2023).

9. **Antimicrobial Coatings**: To prevent the growth of germs and other microbes, aircraft surfaces can be coated with metal nanoparticles that possess antimicrobial qualities, such as silver nanoparticles (Bruna et al. 2021). One of the most intriguing solutions to stop the spread of disease-causing germs and eliminate them is the antimicrobial coating. It provides an efficient way to prevent nosocomial infection that is non-invasive, inexpensive, simple to use, free of adverse effects, and ecologically friendly. Zinc oxide (ZnO) is one of the best materials for antimicrobial coatings due to its low toxicity, excellent biocompatibility to human organs, outstanding stability, high abundance, affordability, and high photocatalytic efficacy to destroy a range of infectious microbes. A variety of aircraft parts, such as tray tables, armrests, door handles, and other high-touch surfaces, can have the antimicrobial coating applied to them (Puspasari et al. 2022).
10. **Nanofluids for Cooling Systems**: The thermal conductivity of coolants used to cool aeronautical components can be improved by the dispersion of metal nanoparticles in the fluids. Critical system overheating may be avoided as a result of increased heat transfer efficiency. During the thermal study of the electronic components, there has been an issue with the thermal dissipation resistance. The nanofluid and flow characteristics of the coolant passing through the thermal cooling system are a solution to the coolant heat removal capacity constraint. Ag/Fe_3O_4 nanofluid mixture as a coolant has a greater thermal dissipation efficiency than using Ag nanofluids or de-ionized water as a coolant (Siricharoenpanich et al. 2020). As high-power fuel cell are developed, more and more demands are placed on heat dissipation. Conventional cooling methods increase heat dissipation capacity, however, their application is hindered by

high energy consumption and restricted engine compartment space. Promising substitutes that significantly increase the base liquid's capability for heat transfer are nanofluids made of nanoparticles and base liquids. Al, Cu, and Ag nanoparticles added to the base fluid may significantly enhance heat transfer efficiency (Ali and Salam 2020).

6.4 Conclusion

In conclusion it is inferred that the incorporation of metal nanoparticles into aeronautical engineering exhibits great potential for improving a number of elements of aircraft performance and design. Nanoparticles have special qualities that can improve the functioning, durability and structural integrity of aerospace materials and components.

A promising area of aerospace engineering is the synthesis of metal nanoparticles, which presents special chances to transform the functionality, performance, and design of aircraft. By employing diverse techniques such as chemical reduction, sol–gel methods, electrochemical deposition, green synthesis, and template-assisted fabrication it is possible to precisely control the size, shape, composition, and surface features of nanoparticles to create materials with unique capabilities that meet the demands of the aerospace industry.

The characterization techniques like Electron microscopy, X-ray diffraction, dynamic light scattering, spectroscopy (UV–Vis, FTIR, Raman), surface area analysis, and other characterization methods offer important insights into the size, composition, crystallinity, surface chemistry, stability, and dispersibility of metal nanoparticles.

Enhanced mechanical properties such as strength, stiffness, and fracture toughness are observed by distributing these nanoparticles throughout matrix composites. As a result, aircraft structures become stronger and lighter, allowing for improved safety margins, better payload capacity, and increased fuel efficiency. Furthermore, metal nanoparticles exhibit exceptional thermal and electrical conductivity, rendering them ideal for applications in heat management and electromagnetic shielding.

References

Alaqarbeh M, Adil SF, Ghrear T, Khan M, Bouachrine M, Al-Warthan A (2023) Recent progress in the application of palladium nanoparticles: a review. Catalysts 13:1343. https://doi.org/10.3390/catal13101343

Ali ARI, Salam B (2020) A review on nanofluid: preparation, stability, thermophysical properties, heat transfer characteristics and application. SN Appl Sci 2:1636. https://doi.org/10.1007/s42452-020-03427-1

Altammar KA (2023) A review on nanoparticles: characteristics, synthesis, applications, and challenges. Front Microbiol 17(14):1155622. https://doi.org/10.3389/fmicb.2023.1155622.PMID: 37180257;PMCID:PMC10168541

Álvarez-Chimal R, Ángel Arenas-Alatorre J (2023) Green synthesis of nanoparticles. A biological approach. In: Advances in green chemistry. IntechOpen. https://doi.org/10.5772/intechopen.1002203

Anjum DH (2016) Characterization of nanomaterials with transmission electron microscopy. IOP Conf Ser Mater Sci Eng 146:012001. https://doi.org/10.1088/1757-899X/146/1/012001

Arepalli S, Moloney P (2015) Engineered nanomaterials in aerospace. MRS Bull 40:804–811. https://doi.org/10.1557/mrs.2015.231

Aruna Prabha K, Naveen Reddy Ch, Srinivasa Rao L, Sreekesh Reddy S, Yeshwanth Kumar G (2023) Structural and thermal analysis on high-pressure steam turbine blade to determine the optimum material for its manufacturing. Mater Today Proc 92, Part 2:1193–1199. ISSN 2214-7853. https://doi.org/10.1016/j.matpr.2023.05.309

Baalousha M, Lead JR (2012) Rationalizing nanomaterial sizes measured by atomic force microscopy, flow field-flow fractionation, and dynamic light scattering: sample preparation, polydispersity, and particle structure. Sci Technol 46(11):6134–6142. Publication Date: May 17, 2012. https://doi.org/10.1021/es301167x

Baig N, Kammakakam I, Falath W (2021) Nanomaterials: a review of synthesis methods, properties, recent progress, and challenges. Mater Adv 2:1821–1871. https://doi.org/10.1039/D0MA00807A

Bamogo H, Ouedraogo M, Sanou I, Aubert J-E, Millogo Y (2022) Physical, hydric, thermal and mechanical properties of earth renders amended with dolomitic lime. Materials 15:4014. https://doi.org/10.3390/ma15114014

Bhat A, Budholiya S, Raj SA, Sultan MTH, Hui D, AUM Shah, Safri SNA (n.d.) Review on nanocomposites based on aerospace applications. From J Nanotechnol Rev https://doi.org/10.1515/ntrev-2021-0018

Bruna T, Maldonado-Bravo F, Jara P, Caro N (2021) Silver nanoparticles and their antibacterial applications. Int J Mol Sci 22:7202. https://doi.org/10.3390/ijms22137202

Carneiro Í, Simões S (2020) Recent advances in EBSD characterization of metals. Metals 10:1097. https://doi.org/10.3390/met10081097

Carvalho Patrícia M, Felício Mário R, Santos Nuno C, Sónia G, Domingues Marco M (2018) Application of light scattering techniques to nanoparticle characterization and development. Front Chem 6. ISSN 2296-2646. https://doi.org/10.3389/fchem.2018.00237

Chaisaenrith P, Taksakulvith P, Pavasupree S (2021) Effect of nano titanium dioxide in intumescent fireproof coating on thermal performance and char morphology. Mater Today Proc 47, Part 12:3462–3467. ISSN 2214-7853. https://doi.org/10.1016/j.matpr.2021.03.461

Charitidis CA, Georgiou P, Koklioti MA, Trompeta A-F, Markakis V (20140 Manufacturing nanomaterials: from research to industry. Manuf Rev 1:11. https://doi.org/10.1051/mfreview/2014009

Chmielewski M, Pietrzak K (2007) Processing, microstructure and mechanical properties of Al_2O_3–Cr nanocomposites. J Euro Ceram Soc 27.1273–1279. https://doi.org/10.1016/j.jeurceramsoc.2006.05.093

Coetzee D, Venkataraman M, Militky J, Petru M (2020) Influence of nanoparticles on thermal and electrical conductivity of composites. Polymers 12:742. https://doi.org/10.3390/polym12040742

Colson P, Henrist C, Cloots R (2013) Nanosphere lithography: a powerful method for the controlled manufacturing of nanomaterials. Article ID 948510. https://doi.org/10.1155/2013/948510

Dhinakaran V, Swapna Sai M, Varsha Shree M (2016) Impact of nanomaterials in aerospace engineering, book nanomaterials and nanocomposites, 1st edn.t Published 2021 Imprint CRC Press Pages 13 eBook. ISBN9781003160946, Conference 2016, LNCS, vol 9999. Springer, Heidelberg, pp 1–13

Duarte I, Ferreira JMF (2016) Composite and nanocomposite metal foams. Materials 9:79. https://doi.org/10.3390/ma9020079

Dutta A (2007) Chapter 4—Fourier transform infrared spectroscopy. In: Thomas S, Thomas R, Zachariah AK, Mishra RK (eds) Micro and nano technologies, spectroscopic methods for nanomaterials characterization. Elsevier, pp 73–93. ISBN 9780323461405. https://doi.org/10.1016/B978-0-323-46140-5.00004-2

Ealias AM, Saravanakumar MP (2017) A review on the classification, characterisation, synthesis of nanoparticles and their application. IOP Conf Ser Mater Sci Eng 263:032019. https://doi.org/10.1088/1757-899X/263/3/032019..

Eid MM (2022) Characterization of nanoparticles by FTIR and FTIR-microscopy. In: Handbook of consumer nanoproducts. Springer, Singapore. https://doi.org/10.1007/978-981-16-8698-6_89

Escorcia-Díaz D, García-Mora S, Rendón-Castrillón L, Ramírez-Carmona M, Ocampo-López C (2023) Advancements in nanoparticle deposition techniques for diverse substrates: a review. Nanomaterials 13:2586

Fernando JFS, Zhang C, Firestein KL, Golberg D (2017) Optical and optoelectronic property analysis of nanomaterials inside transmission electron microscope. Small 13:1701564. https://doi.org/10.1002/smll.201701564

Friedland J, Brandt A, Leopold K, Güttel R (2021) Atomization of gold nanoparticles in graphite furnace AAS: modelling and simulative exploration of experimental results. Spectrochimica Acta Part B: Atomic Spectrosc 182:106249. ISSN 0584-8547. https://doi.org/10.1016/j.sab.2021.106249

Gawande MB, Goswami A, Felpin F-X, Asefa T, Huang X, Silva R, Zou X, Zboril R, Varma, RS (2016) Cu and Cu-based nanoparticles: synthesis and applications in catalysis. Chem Rev 116(6):3722–3811. https://doi.org/10.1021/acs.chemrev.5b00482

Goldstein A, Soroka Y, Frušić-Zlotkin M, Popov I, Kohen R (2014) High resolution SEM imaging of gold nanoparticles in cells and tissues. J Microsc 256:237–247. https://doi.org/10.1111/jmi.12179

Gu Y, Xia K, Wu D, Mou J, Zheng S (2020) Technical characteristics and wear-resistant mechanism of nano coatings: a review. Coatings 10:233. https://doi.org/10.3390/coatings10030233

Guo D, Xie G, Luo J (2014) Mechanical properties of nanoparticles: basics and applications. J Phys D Appl Phys 47:13001. https://doi.org/10.1088/0022-3727/47/1/013001

Hassan H, Sharma P, Hasan MR, Singh S, Thakur D, Narang J (2022) Gold nanomaterials—the golden approach from synthesis to applications. Mater Sci Energy Technol 5:375–390. https://doi.org/10.1016/j.mset.2022.09.004

Hile DD, Swart HC, Motloung SV, Koao LF (2022) 3—Zinc selenide semiconductor: synthesis, properties and applications. In: Pawade VB, Dhoble SJ, Swart HC (eds) Woodhead publishing series in electronic and optical materials, nanoscale compound semiconductors and their optoelectronics applications. Woodhead Publishing, 2022, pp 67–84. ISBN 9780128240625. https://doi.org/10.1016/B978-0-12-824062-5.00001-4

James Singh K, Ahmed T, Gautam P, Sadhu AS, Lien D-H, Chen S-C, Chueh Y-L, Kuo H-C (2021) Recent advances in two-dimensional quantum dots and their applications. Nanomaterials 11:1549. https://doi.org/10.3390/nano11061549

Jang HJ, Jung EY, Parsons T, Tae H-S, Park C-S (2021) A review of plasma synthesis methods for polymer films and nanoparticles under atmospheric pressure conditions. Polymers 13:2267. https://doi.org/10.3390/polym13142267

Joudeh N, Linke D (2022) Nanoparticle classification, physicochemical properties, characterization, and applications: a comprehensive review for biologists. J Nanobiotechnol 20:262. https://doi.org/10.1186/s12951-022-01477-8

Kannaiyan K, Sadr R (2017) The effects of alumina nanoparticles as fuel additives on the spray characteristics of gas-to-liquid jet fuels. Exp Thermal Fluid Sci 87:93–103. ISSN 0894-1777

Kasgoz A, Akin D, Durmus A (2015) Effects of size and shape originated synergism of carbon nano fillers on the electrical and mechanical properties of conductive polymer composites. J Appl Polym Sci 132:1–11. https://doi.org/10.1002/app.42313

Khan Y, Sadia H, Ali Shah SZ, Khan MN, Shah AA, Ullah N, Ullah MF, Bibi H, Bafakeeh OT, Khedher NB et al (2022) Classification, synthetic, and characterization approaches to nanoparticles, and their applications in various fields of nanotechnology: a review. Catalysts 12:1386. https://doi.org/10.3390/catal12111386

Khan Y, Sadia H, Shah SZA, Khan MN, Shah AA, Ullah N, Ullah MF, Bibi H, Bafakeeh OT, Khedher NB, Eldin SM, Fadhl BM, Khan MI (2022) Classification, synthetic, and characterization approaches to nanoparticles, and their applications in various fields of nanotechnology: a review. Catalysts 12(11):1386. https://doi.org/10.3390/catal12111386

Khan ME, Mohammad A, Ali W, Imran M, Bashiri AH, Zakri W (2023) Properties of metal and metal oxides nanocomposites. In: Khan ME, Aslam J, Verma C (eds) Woodhead publishing series in composites science and engineering, nanocomposites-advanced materials for energy and environmental aspects. Woodhead Publishing, pp. 23–39. ISBN 9780323997041, https://doi.org/10.1016/B978-0-323-99704-1.00027-8

Kim M, Osone S, Kim T, Higashi H, Seto T (2017) Synthesis of nanoparticles by laser ablation: a review. KONA Powder Particle J 34:80–90. Released on J-STAGE February 28, 2017, Advance online publication April 30, 2016, Online ISSN 2187-5537, Print ISSN 0288-4534. https://doi.org/10.14356/kona.2017009

Kumar A, Moledina J, Liu Y, Chen K, Patnaik PC (2021) Nano-micro-structured 6–8% YSZ thermal barrier coatings: a comprehensive review of comparative performance analysis. Coatings 11:1474. https://doi.org/10.3390/coatings11121474

Lee B, Yoon S, Lee JW, Kim Y, Chang J, Yun J, Ro JC, Lee J-S, Lee JH (2020) Statistical characterization of the morphologies of nanoparticles through machine learning based electron microscopy image analysis. ACS Nano 14(12):17125–17133. Publication Date: November 24, 2020. https://doi.org/10.1021/acsnano.0c06809

León A, Reuquen P, Garín C, Segura R, Vargas P, Zapata P, Orihuela PA (2017) FTIR and raman characterization of TiO_2 nanoparticles coated with polyethylene glycol as carrier for 2-methoxyestradiol. Appl Sci 7:49. https://doi.org/10.3390/app7010049

Li Y, Liao Q, Hou W, Qin L (2023) Silver-based surface plasmon sensors: fabrication and applications. Int J Mol Sci 24:4142. https://doi.org/10.3390/ijms24044142

Liu M, Sun Z, Liu P, Hai W, Chen Y (2023) Microstructure and Properties of Ti(C, N)-based cermets with AlxCoCrFeNiTi binder. Materials (Basel). 16(7):2894. https://doi.org/10.3390/ma16072894.PMID:37049188;PMCID:PMC10096151

Liu Y, Pan B, Li H, Lang D, Zhao Q, Zhang D, Wu M, Steinberg CEW, Xing B (2020) Can the properties of engineered nanoparticles be indicative of their functions and effects in plants. Ecotoxicol Environ Saf 205:111128. ISSN 0147-6513. https://doi.org/10.1016/j.ecoenv.2020.111128

Liu N, Xu YD, Li H, Li GH (2022) Effect of nano-micro TiN addition on the microstructure and mechanical properties of TiC based cermets. J Euro Ceram Soc 22(13):2409–2414. https://doi.org/10.1016/S0955-2219(02)00010-9

Ma P, Jia Y, Konda Gokuldoss P, Yu Z, Yang S, Zhao J, Li C (2017) Effect of Al_2O_3 nanoparticles as reinforcement on the tensile behavior of Al-12Si composites. Metals 7:359. https://doi.org/10.3390/met7090359

Majerič P, Rudolf R (2020) Advances in ultrasonic spray pyrolysis processing of noble metal nanoparticles-review. Materials (Basel) 13(16):3485. https://doi.org/10.3390/ma13163485. PMID:32784637;PMCID:PMC7476056

Marinescu L, Ficai D, Oprea O, Marin A, Ficai A, Andronescu E, Holban A-M (n.d.) Optimized synthesis approaches of metal nanoparticles with antimicrobial applications. https://doi.org/10.1155/2020/6651207

Meischein M, Ludwig A (2021) Upscaling nanoparticle synthesis by sputter deposition in ionic liquids. J Nanopart Res 23:129. https://doi.org/10.1007/s11051-021-05248-8

Mourdikoudis S, Pallares RM, Thanh NTK (2018a) Characterization techniques for nanoparticles: comparison and complementarity upon studying nanoparticle properties. Nanoscale 10:12871. https://doi.org/10.1039/C8NR02278J

Mourdikoudis S, Pallares RM, Thanh NTK (2018b) Characterization techniques for nanoparticles: comparison and complementarity upon studying nanoparticle properties. Nanoscale 10:12871–12934. https://doi.org/10.1039/C8NR02278J

Nan Z, Wei W, Lin Z et al (2023) Flexible nanocomposite conductors for electromagnetic interference shielding. Nano-Micro Lett 15:172. https://doi.org/10.1007/s40820-023-01122-5

Narayan N, Meiyazhagan A, Vajtai R (2019) Metal nanoparticles as green catalysts.Materials 12:3602. https://doi.org/10.3390/ma12213602

Naseem T, Durrani T (2021) The role of some important metal oxide nanoparticles for wastewater and antibacterial applications: a review. Environ Chem Ecotoxicol 3:59–75. ISSN 2590-1826

Nhiem LT, Oanh DTY, Hieu NH (2023) Nanocoating toward anti-corrosion: a review. VJCH 61:284–293. https://doi.org/10.1002/vjch.202300025

Okitsu K, Kumar C (ed) (2013) Springer materials UV-Vis spectroscopy for characterization of metal nanoparticles formed from reduction of metal ions during ultrasonic irradiation (UV-VIS and Photoluminescence Spectroscopy for Nanomaterials Characterization). https://materials.springer.com/lb/docs/sm_smf_978-3-642-27594-4_4. https://doi.org/10.1007/978-3-642-27594-4_4

Onwudiwe DC (2019) Microwave-assisted synthesis of PbS nanostructures. Heliyon 5(3):e01413. https://doi.org/10.1016/j.heliyon.2019.e01413. PMID: 30976689; PMCID: PMC6439236

Ozkan Ata S, Akay C, Mumcu E (2023) The effects of metal nanoparticles incorporation on the mechanical properties of denture base acrylic resin. Eur Oral Res 57(1):36–40. https://doi.org/10.26650/eor.20231079531.PMID:37020640;PMCID:PMC10069800

Parveez B, Kittur MI, Badruddin IA, Kamangar S, Hussien M, Umarfarooq MA (2022) Scientific advancements in composite materials for aircraft applications: a review. Polymers 14:5007. https://doi.org/10.3390/polym14225007

Puspasari V, Ridhova A, Hermawan A et al (2022) ZnO-based antimicrobial coatings for biomedical applications. Bioprocess Biosyst Eng 45:1421–1445. https://doi.org/10.1007/s00449-022-02733-9

Qin X, Yang S, Chen Y, Zhao XJ, Fang W, Luo D (2020) Thermal conductivity and stability of hydrocarbon-based nanofluids with palladium na noparticles dispersed by modified hyper-branched polyglycerol. ACS Omega 5(48):31156–31163. https://doi.org/10.1021/acsomega.0c04315

Raha S, Ahmaruzzaman M (2022) ZnO nanostructured materials and their potential applications: progress, challenges and perspectives. Nanoscale Adv 4:1868–1925. https://doi.org/10.1039/D1NA00880C

Ramli ZA, Pasupuleti J, Saharuddin TS, Yusoff YN, Isahak WN, Baharudin L, Yaw CT, Koh SP, Kiong ST (2023) Electrocatalytic activities of platinum and palladium catalysts for enhancement of direct formic acid fuel cells: an updated progress. Alexandria Eng J 76:701–733. ISSN 1110-0168. https://doi.org/10.1016/j.aej.2023.06.069

Rao A, Schoenenberger M, Gneccol E, Glatzel Th, Meyer E, Brändlin D, Scandella L (2007) Characterization of nanoparticles using atomic force microscopy. J Phys Conf Ser 61:971. https://doi.org/10.1088/1742-6596/61/1/192

Rohatgi K, Schultz B (2007) Lightweight metal matrix nanocomposites—stretching the boundaries of metals Prof. P. Univ Wisconsin–Milwaukee Mater Matters 2(4):16

Ruz P, Sudarsan V (2021) Polyol method for synthesis of nanomaterials. In: Tyagi AK, Ningthoujam RS (eds) Handbook on synthesis strategies for advanced materials. Indian Institute of Metals Series, Springer, Singapore. https://doi.org/10.1007/978-981-16-1807

Shi H, Magaye R, Castranova V et al (2013) Titanium dioxide nanoparticles: a review of current toxicological data. Part Fibre Toxicol 10:15. https://doi.org/10.1186/1743-8977-10-15

Shojaei TR, Soltani S, Derakhshani M (2022) Chapter 6—Synthesis, properties, and biomedical applications of inorganic bionanomaterials. In: Barhoum A, Jeevanandam J, Danquah MK (eds) Micro and nano technologies, fundamentals of bionanomaterials. Elsevier, pp 139–174. ISBN 9780128241479. https://doi.org/10.1016/B978-0-12-824147-9.00006-6

Siengchin S (2023) A review on lightweight materials for defence applications: present and future developments. Defence Technol 24:1–17. https://doi.org/10.1016/j.dt.2023.02.025

Silva MR, Alves MF, Cunha JP, Costa JL, Silva CA, Fernandes MH, Vilarinho PM, Ferreira P (2023) Nanostructured transparent solutions for UV-shielding: Recent developments and future challenges. Mater Today Phys 35:101131. ISSN 2542-5293. https://doi.org/10.1016/j.mtphys.2023.101131

Siricharoenpanich A, Wiriyasart S, Srichat A, Naphon P (2020) Thermal cooling system with Ag/Fe_3O_4 nanofluids mixture as coolant for electronic devices cooling. Case Stud Thermal Eng 20:100641. ISSN 2214-157X. https://doi.org/10.1016/j.csite.2020.100641

Sonanwane DB, Shah AM, Jaiswal N (2022) Review on application of nanoparticles and classification, synthesis. Res J Pharmacol Pharmacodyn 14(2):117–4. https://doi.org/10.52711/2321-5836.2022.00020

Stephen AJ, Rees NV, Mikheenko I, Macaskie LE (2019) Platinum and palladium bio-synthesized nanoparticles as sustainable fuel cell catalysts. Front. Energy Res 7:66. https://doi.org/10.3389/fenrg.2019.00066

Szczyglewska P, Feliczak-Guzik A, Nowak I (2023) Nanotechnology-general aspects: a chemical reduction approach to the synthesis of nanoparticles. Molecules 28:4932. https://doi.org/10.3390/molecules28134932

Vela-Cano M, Garcia-Fontana C, Osorio F et al (2021) Silver-derived antimicrobial coatings for the prevention of microbial biofilms in metal pipes. Water Air Soil Pollut 232:313. https://doi.org/10.1007/s11270-021-05264-3

Viñes F, Gomesb JRB, Illas F (2014) Understanding the reactivity of metallic nanoparticles: beyond the extended surface model for catalysis. Chem Soc Rev 43:4922–4939. https://doi.org/10.1039/C3CS60421G

Wu Q, Miao W-S, Zhang Y-D, Gao H-J, Hui D (n.d.) Mechanical properties of nanomaterials: a review. J Nanotechnol Rev. https://doi.org/10.1515/ntrev-2020-0021

Xing W, Xu Y, Song C, Deng T (2022) Recent advances in thermal interface materials for thermal management of high-power electronics. Nanomaterials 12:3365. https://doi.org/10.3390/nano12193365

Xu W, Kasar AK, Menezes PL, Dieringa H, Varma RS, Gupta M (2019) Metals 9(3):330. https://doi.org/10.3390/met9030330

Yousefi F, Mousavi SB, Heris SZ et al (2023) UV-shielding properties of a cost-effective hybrid PMMA-based thin film coatings using TiO_2 and ZnO nanoparticles: a comprehensive evaluation. Sci Rep 13:7116. https://doi.org/10.1038/s41598-023-34120-z

Yusuf M (2018) Silver nanoparticles: synthesis and applications. Handb Ecomater 30:2343–2356. https://doi.org/10.1007/978-3-319-68255-6_16.PMCID:PMC7122521

Chapter 7
Impact Simulation of Polycarbonates in Aeronautical and Aerospace Applications

Ritu Parna Roy and Adib Bin Rashid

Abstract This chapter examines the crucial role of impact simulation in evaluating the performance of polycarbonate in aeronautical and aerospace applications. Polycarbonates are known for their lightweight and durable properties, making them suitable for various aerospace components. The discussion covers simulation techniques like Finite Element Analysis (FEA), Discrete Element Method (DEM), and Computational Fluid Dynamics (CFD) specifically tailored for impact analysis. Case studies are used to explore the application of these simulations in scenarios such as bird strikes, hailstone impacts, and collisions on aerodynamic surfaces. The chapter also emphasizes the significance of experimental correlation and validation to ensure the accuracy of simulation results. Furthermore, it delves into optimization strategies for improving impact resistance through design modifications and material advancements. The challenges and limitations of impact simulation are addressed, and potential future trends such as the integration of machine learning and collaborative research initiatives are discussed. The chapter concludes by highlighting the regulatory considerations for impact resistance standards in aerospace applications, underscoring the importance of compliance and certification.

Keywords Polycarbonates · Impact simulation · Aerospace applications · Finite Element Analysis (FEA) · Discrete Element Method (DEM) · Computational Fluid Dynamics (CFD) · Lightweight materials

R. P. Roy
Department of Aeronautical Engineering, Military Institute of Science and Technology (MIST), Dhaka, Bangladesh

A. B. Rashid (✉)
Department of Industrial Production Engineering, Military Institute of Science and Technology (MIST), Dhaka, Bangladesh
e-mail: adib@me.mist.ac.bd

Visakh P. M. (ed.), *Nanomaterials for Defense, Aeronautical and Aerospace Applications*, Advanced Structured Materials 236,
https://doi.org/10.1007/978-3-031-92723-2_7

7.1 Introduction

In the aviation industry, precision, durability, and unwavering reliability are paramount for all components and systems. Aerospace engineers face the significant challenge of combining lightweight materials with high impact strength to meet strict safety standards. Polycarbonates have emerged as a leading solution due to their ability to withstand drastic atmospheric changes and resist harsh environmental conditions, making them a common choice in aviation and aerospace applications. Polycarbonate (PC) is an engineering plastic widely utilized in the aviation, automotive, and aerospace industries because of its high impact resistance, thermal stability, fatigue resistance, light weight, and strength (Kausar 2017; Costa et al. 2021; Ibrahim et al. 2011). For instance, bird strikes frequently cause windshields and canopies to crack or perforate. In space exploration, astronauts face serious dangers from space debris impacts; even debris as small as 1 mm can carry impact energies up to 71 J, posing significant safety threats. Understanding and accurately predicting the impact responses and potential failures of advanced materials like polycarbonate is crucial for ensuring safety and reliability.

In the 1930s, polymers like PVC, polystyrene, polyethylene, and acrylics were rapidly developed and commercialized. Enhancements to polystyrene led to materials such as SAN and ABS, improving properties but sometimes compromising transparency. Acrylics gained popularity for their resistance to harsh weather conditions and UV radiation. The introduction of polycarbonate in the early 1950s shifted the landscape of transparent materials, balancing softness and brittleness. First synthesized in 1898 by German chemist Alfred Einhorn ("Tracing the History of Polymeric Materials: How polycarbonate came about, virtually simultaneously, through the efforts of two industry giants 2024). polycarbonates later gained prominence due to their durability, transparency, and versatility. They can withstand significant impact and stress and be molded into complex shapes. While some polycarbonate items may contain bisphenol A (BPA), raising health concerns (Vandenberg et al. 2007) polycarbonates have become essential across various industries.

Composite structures have become integral in aerospace research for component design at multiple levels. Validating constitutive models or novel materials typically requires extensive high-quality test data, including low and high-velocity impact tests under different scenarios. However, experimental validation demands significant resources and specialized equipment. Numerical and computational simulation methods, such as Finite Element (FE) and Discrete Element (DE) analyses, offer more efficient alternatives, allowing for quicker and more abundant analyses. This chapter explores the use of polycarbonate in aerospace engineering through computational simulation approaches and experimental methods focused on predicting behavior during low and high-velocity impacts.

7.2 Properties and Characteristics of Polycarbonates Relevant to Impact

Polycarbonates are a class of thermoplastic polymers characterized by the presence of carbonate groups in their chemical structure. Renowned for their unique combination of mechanical, thermal, and physical properties, polycarbonates have become indispensable in applications requiring high impact resistance, transparency, and thermal stability. Following are the common mechanical properties of polycarbonate:

- **High Impact Resistance**

One of the most significant attributes of polycarbonate is its exceptional toughness. Compared to other engineering plastics, polycarbonate exhibits outstanding impact strength, enabling it to absorb substantial energy without fracturing. This property is particularly beneficial for critical aerospace components like cockpit canopies and windows, which must endure sudden impacts without compromising structural integrity (Table 7.1).

- **Strain-Rate Sensitivity**

Polycarbonate's mechanical properties are influenced by the rate at which load is applied. Under high strain rates, such as those experienced during high-velocity impacts, the material exhibits increased strength and stiffness. This strain-rate sensitivity enhances its ability to absorb impact energy effectively.

- **Thermal Properties**

– ***Glass Transition Temperature (T_g)***

Polycarbonate possesses a high glass transition temperature of approximately 147 °C. This property ensures that the material maintains its structural integrity over a wide temperature range, which is essential in aerospace environments where components may be exposed to extreme thermal fluctuations.

– ***Thermal Stability***

Table 7.1 Mechanical properties of polycarbonate

Property	Typical value	Notes
Density	~1.2 g/cm^3	Contributes to lightweight structures
Tensile Strength	60–70 MPa	Balances strength and flexibility
Elastic Modulus	2,000–2,400 MPa	Provides necessary stiffness
Impact Strength	High	Exceptional toughness against impacts
Elongation at Break	100–120%	Indicates high ductility
Glass Transition Temp (Tg)	~147 °C	Ensures thermal stability

The material's resistance to thermal degradation allows it to withstand repeated thermal cycling without significant loss of properties. This stability is crucial for aerospace applications involving high-altitude flight or space missions, where temperatures can vary drastically (Table 7.2).

- **Physical Properties**
  - ***Lightweight and Transparency***

    With a density of approximately 1.2 g/cm^3, polycarbonate contributes to the overall weight reduction in aerospace structures, enhancing fuel efficiency and payload capacity. Its inherent transparency and optical clarity make it suitable for applications such as aircraft windows, lenses, and light covers, where visibility and light transmission are critical.
  - ***Dimensional Stability***

    Polycarbonate exhibits low shrinkage and warpage, maintaining its dimensions under various environmental conditions. This dimensional stability is vital for components that require tight tolerances and consistent performance.

- **Fracture Behavior**
  - ***Energy Absorption Mechanisms***

    Under impact loading, polycarbonate can deform plastically, absorbing energy that might otherwise cause brittle fracture in other materials. The material's ability to undergo significant plastic deformation before failure allows it to absorb and dissipate impact energy effectively.
  - ***Failure Modes***

    Generally, polycarbonate exhibits ductile fracture with noticeable necking. However, at sub-zero temperatures, the material may become more brittle, necessitating careful consideration in design for low-temperature environments.

- **Environmental Resistance** See Table 7.3.

Table 7.2 Thermal properties of polycarbonate

Property	Value	Notes
Glass Transition Temp (Tg)	~147 °C	Above this material becomes more rubbery
Melting Temperature	225–230 °C	Important for processing considerations
Thermal Conductivity	0.19 W/m·K	Low; affects thermal insulation properties
Coefficient of Thermal Expansion	65–70 × 10^{-6}/K	Impacts dimensional stability under temperature changes

Table 7.3 Environmental resistance of polycarbonate

Environmental factor	Resistance level	Improvement factor
UV Radiation	Moderate	UV stabilizers recommended for prolonged exposure
Chemical Resistance	Moderate	Addition of acids/alcohols resistant and avoiding strong alkalis
Moisture Absorption	Low	Coating can be used
Weathering	Good	Durable with appropriate additives

- **Behavior Under Impact Conditions**
  - ***Temperature Effects***

    Polycarbonate retains its toughness and strength at elevated temperatures but may become less ductile at very low temperatures. This temperature dependence affects its impact resistance and must be considered when designing components for extreme environments.
  - ***Ballistic Performance***

    Due to its high impact strength, polycarbonate is used in applications requiring ballistic protection, such as bullet-resistant glass and protective gear. Its ability to resist penetration from high-velocity projectiles makes it valuable for aerospace components that may encounter debris impacts.

7.3 Additives and Modifications

7.3.1 Enhancing Material Properties with Material Modifications

The modifications typically involve altering the polymer structure through additives or blending with other materials, which can significantly enhance the impact resistance by improving energy absorption and distribution during an impact event (Huang and Liu 2017).

These modifications may include the incorporation of rubbery materials, which can provide a toughening effect, or the use of nanofillers that reinforce the polymer matrix, leading to improved mechanical properties and durability. The processing conditions, such as temperature and shear rate, play a crucial role in determining the final properties of the modified polycarbonate, influencing factors like crystallinity and phase separation. The way it will be changed from the requirement of a specific property. In Table 7.4 few overviews are given for polycarbonate performance changes with certain additives.

Table 7.4 Effects of additives on polycarbonate properties

Modification	Additives	Effects on properties	Applications	References
Composites	Metal Particles	Improved impact resistance, toughness, and stability	General-purpose structures	Xie et al. (2022)
	Wear-Resistant and UV Agents	Improved impact resistance, stability, and environmental durability	Wear-resistant outdoor applications	Method for Modifying PolycarbonatelTrea (2024)
	Nanosilica, hydrophobic coating and Polysiloxane Particles	Enhanced impact strength, toughness, and durability	High-durability for industrial use	Ramakrishnan et al. (2021)
Flame Retardants	Phosphorus-Nitrogen Flame Retardants	Enhanced flame retardancy while maintaining mechanical properties and safety regulations	Flame-resistant polycarbonate for electrical and aviation components	Zhao et al. (2024)
Nanomaterial Additions	Graphene Oxide (GO)	Increased tensile strength by 26%, improved impact strength and thermal stability	Advanced polycarbonate materials in aerospace and automotive sectors	The impact of graphene oxide on the mechanical and thermal strength properties of polycarbonate—Maziyar Sabet (2023)
	Carbon Fiber	Improved interfacial adhesion, tensile and flexural strengths, and resin impregnation	Reinforced composite materials in aerospace and automotive	Yao et al. (2021)
High-Impact-Resistant Compositions	Elastomer and CNT Additives	Enhanced fracture toughness and impact strength through network structures	Reinforced polycarbonate blends in automotive and aerospace sectors	Li et al. (2016)

7.3.2 *Structural and Surface Modifications*

Surface modifications, such as, the application of amorphous hydrogenated carbon (a-C:H) layers on PC can improve its hardness and reduce scratch susceptibility, with the interlayer region playing a crucial role in bonding stability (The Growth Behavior of Amorphous Hydrogenated Carbon a-C:H Layers on Industrial Polycarbonates—A Weak Interlayer and a Distinct Dehydrogenation Zone 2024). The effect of top surface lubrication, specifically using Paraffin wax, during friction stir welding

(FSW) of polycarbonate sheets results in defect-free joints with improved surface quality (Ahmed et al. 2021). Chemical polishing using ammonia can reduce surface roughness and render the PC surface hydrophobic, which is beneficial for microfluidic applications by increasing optical transparency and preventing cross-contamination in two-phase systems (Ogończyk et al. 2020).

Surface treatments such as plasma exposure in O2/Ar atmospheres and light-activated chlorine dioxide radical oxidation improve the adhesion and wettability of PC surfaces. Plasma treatment achieves superhydrophilicity, enhancing the adhesion of coatings like germanium films, which is crucial for aerospace applications requiring layered materials (Tailoring polycarbonate surfaces for improved Ge film adhesion: The role of plasma treatments in O2/Ar atmospheres—ScienceDirect2024). Similarly, chlorine dioxide radical treatment adds polar groups to PC surfaces, improving metal adhesion without compromising transparency (Jia et al. 2020).

7.3.3 Processing Techniques

Polycarbonate (PC) can be treated with numerous effective processes, each providing specific advantages. Ultrasonic vibration improves the dispersion of carbon nanotubes (CNT) in PC, enhancing mechanical properties at lower concentrations (Gao et al. Feb. 2016). The use of low-pressure plasma with monomers like acrylic acid and allylamine enhances membrane permeability without affecting bulk properties(Enhancement of polycarbonate membrane permeability due to plasma polymerization precursors—ScienceDirect 2024). Light-activated chlorine dioxide radical (ClO_2•) efficiently oxidizes PC surfaces, enhancing wettability and adhesion without damaging the material (Jia et al. 2020). The solid-state reaction with primary amines can functionalize PC but requires careful control to avoid degradation (Zimmerer et al. 2018).

7.3.4 Recycling and Environmental Considerations

The recycling and modification of polycarbonate (PC) have become increasingly important in addressing both environmental impact and material performance. Various innovative processing techniques aim to enhance sustainability while maintaining the mechanical properties of PC. Supercritical Fluid (SCF) technology has emerged as a promising approach, offering reduced emissions and energy consumption during PC production, with the added benefit of minimizing impurities (Alternative Separation Process of Thermoplastic Polymer for Environmental Assessment | Recent Progress in Science and Technology 2024). For electronic waste recycling, density separation and solvent-based processes can effectively recover polycarbonate using NMP and DCM, though careful management is required to prevent material degradation(Separation and Solvent Based Material Recycling of Polycarbonate from Electronic Waste I ACS Sustainable Chemistry and Engineering 2024). Additionally, metal-free synthesis of polycarbonate polyols through CO_2 copolymerization presents an environmentally sustainable method, employing recyclable initiators such as triethylborane (Patil et al. 2020).

While these modifications significantly enhance the impact performance of polycarbonate materials, it is crucial to consider the trade-offs involved. For instance, while certain additives improve impact resistance, they may affect other properties like transparency or chemical resistance. Therefore, the choice of modification should align with the specific application requirements and environmental considerations.

7.4 Structural Composition

The molecular structure of polycarbonate distinguishes it from other engineering plastics due to its unique characteristics. Polycarbonate is composed of repeating units that contain two carbocyclic six membered aromatic rings, which are the main segments forming its molecular structure (Nishi et al. 2018) shown in Fig. 7.1. Adopted from Rodriguez et al. (2003) (Fig. 7.2).

Polycarbonate's molecular structure features a carbonate-based backbone linked to aromatic rings, providing exceptional mechanical strength, thermal stability, and

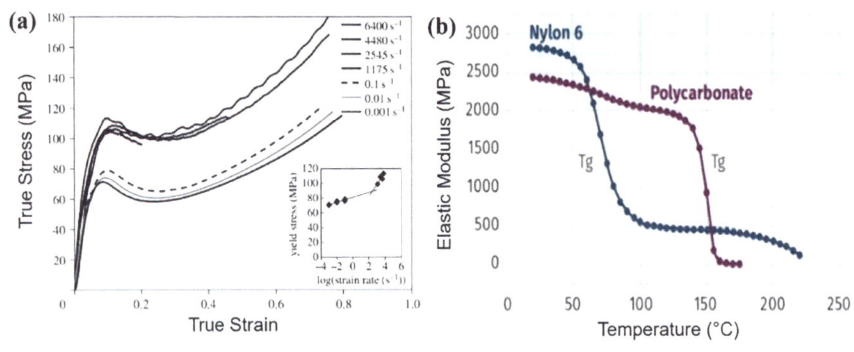

Fig. 7.1 **a** illustrates the stress–strain behavior of polycarbonate in room temperature at different strain rates (Kendall and Siviour May 2014). **b** Illustrates Elastic Modulus versus Temperature Behavior for amorphous polycarbonate

Fig. 7.2 Molecular structure of polycarbonate

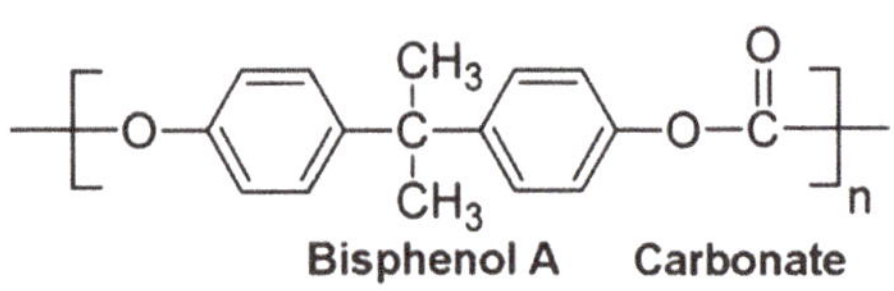

optical clarity. The carbonate groups (–O–C(O)O–) contribute to hardness and heat resistance, while the aromatic benzene rings enhance transparency and impact resistance. As an amorphous polymer, polycarbonate is exceptionally tough and easily moldable, and its hydrolytically stable carbonate linkages ensure greater durability compared to polyesters. The breakdown of Polycarbonate's Molecular Structure:

- PC has a chemical formula of $C_{16}H_{14}O_3$, consisting of repeating units of carbonate groups (–O–C(O)O–) linked by aromatic benzene rings (Brydson 1999). The aromatic benzene rings give PC its transparency and impact resistance (Nishi et al. 2018).
- The carbonate groups provide polycarbonate with its characteristic hardness, rigidity, and heat resistance. Polycarbonates contain more hydrolytically stable carbonate linkages compared to polyesters, which predominantly have ester linkages (Chemical recycling to monomers: Industrial Bisphenol-A-Polycarbonates to novel aliphatic polycarbonate materials 2022; Brunelle 2005). This indicates that the carbonate groups in polycarbonate contribute to its characteristic hardness, rigidity, and heat resistance.
- PC is an amorphous polymer, meaning it lacks a crystalline structure, which contributes to its toughness and moldability (Kausar 2017).
- The molecular architecture of polycarbonate, characterized by linear structures and the presence of end groups, can impact the interactions between the polymer and other components in composite materials (Huang et al. 2000). The linear structure of polycarbonate, as opposed to cyclic structures found in some materials, affects the relative interactions within the polymer matrix, influencing properties like free energy of formation and overall material performance.

Polycarbonates (PCs) were synthesized via the reaction of Resorcinol and Hydroquinone with phosgene in pyridine, resulting in distinct polymer characteristics; the hydroquinone variant was crystalline and insoluble, while the resorcinol variant exhibited amorphous properties with a melting point around 200 °C (Bischoff and Hedenström 1902). Subsequent research by Bischoff and von Hedenström, as well as Carothers and Natta, explored various synthesis methods, ultimately leading to the pivotal discovery of poly(ethylene terephthalate) by Whinfield and Dickson in 1941, which catalyzed further developments in aromatic ring polymers. The commercial success of bisphenol A-derived PCs emerged by 1958, with contemporary advancements encompassing thermoplastic alloys, polyester carbonates, and siliconeePC block copolymers, prominently represented by SABIC's Lexan and Bayer's Makrolon.

There are various methods for synthesizing polycarbonate, each with its unique approach and advantages (Balaji et al. 2018). Here are different methods of synthesizing polycarbonate:

- **Phosgene process**

The primary commercial method for producing polycarbonate involves reacting phosgene with dihydroxy compounds like bisphenol A. While this process efficiently yields high-quality polycarbonate, it presents significant drawbacks due to the use of toxic phosgene gas (Brydson 1999; Huang et al. 2000). However, recently there are more sustainable and controlled approaches. For example (Table 7.5).

- **Non-phosgene methods**

These methods aim to produce polycarbonate without the need for toxic phosgene, solvents, and with minimal waste generation most common techniques are (Table 7.6).

- **Interfacial/solution process**

The interfacial polymerization process is widely used for industrial polycarbonate production. It involves reacting phosgene (carbonyl chloride) with bisphenol A at the interface of an organic solvent and an aqueous phase. The polycarbonate resin is then isolated using methods like concentration, spray drying, steam distillation, antisolvent precipitation, or mechanical devolatilization, enhancing its industrial application value. A catalyzed interfacial phosgenation process, with controlled feed ratios and pH levels, produces high-heat polycarbonate copolymers with improved hydrolytic stability, showcasing the advantages of this method (Successful polycarbonate synthesis using the photo-on-demand interfacial polymerization method | Kobe University News site, Kobe University 2024; Ebert et al. 2009; Mahood and Gautam 2015).

- **Aliphatic polycarbonates**

Aliphatic polycarbonates are emerging as sustainable alternatives to fossil fuel-derived plastics, despite being less commercially prevalent than aromatic polycarbonates from bisphenol A. Their synthesis involves the copolymerization of

Table 7.5 Synthesis of polycarbonate via phosgene method

Method	Description
Organocatalyzed Ring-Opening Polymerization	This method uses glucose as a natural starting material to create polycarbonates (Polycarbonates Derived from Glucose via an Organocatalytic Approach | Journal of the American Chemical Society 2024)
Phase Boundary Process with Specific Energy Input	By carefully controlling the energy input during the mixing of aqueous and organic phases, this process reduces the excess of phosgene, resulting in polycarbonates with lower oligomer content and specific di-chain terminator carbonate levels (Bachmann et al. 2022; 바흐만롤프, 미헬레폴커, 하이일얀, 코크펠릭스 2021)
Continuous Two-Phase Interface Phosgene Process	This approach improves the conversion rates of branching agents, lowers costs, and enhances product quality by preventing crosslinking reactions (Zhang et al. 2016)
Novel Photochemical Reaction for Phosgene Synthesis	This innovative method synthesizes phosgene from chloroform, leading to safer and more efficient phosgenation reactions and systems in organic synthesis (Tsuda 2023)

Table 7.6 Syntheses of polycarbonate via non-phosgene method

Method	Description
Using Carbon Monoxide or Carbon Dioxide	Utilizes carbon monoxide or carbon dioxide instead of phosgene in the production process (Gao et al. 2016)
Transesterification of Bisphenol A with Diphenyl Carbonate	Replaces phosgene with diphenyl carbonate in the reaction with bisphenol A (Photo-on-Demand In Situ Phosgenation Reactions That Cross Three Phases of a Heterogeneous Solution of Chloroform and Aqueous NaOH I ACS Omega 2024)
Photo-on-Demand Polymerization	Uses light to initiate the polymerization reaction, eliminating the need for phosgene (Photo-on-Demand In Situ Phosgenation Reactions That Cross Three Phases of a Heterogeneous Solution of Chloroform and Aqueous NaOH I ACS Omega 2024)
Utilization of Ionic Liquids as Catalysts	Employs ionic liquids as catalysts, providing a cleaner production process for polycarbonates (孔立明, 陈志伟, 张艳艳, and 孔琳仪 2014)

carbon dioxide and epoxides, enhancing molecular weight and lowering production costs. These polycarbonates feature low glass transition temperatures (–74.9 °C to –73.2 °C), exhibit high reactivity under mild conditions, and possess over 99% carbonate linkages. Furthermore, they display inherent viscosities of 0.19–0.43 dL/g, reflecting low to moderate molecular weights. Such properties render aliphatic polycarbonates suitable for eco-friendly applications, providing performance advantages and less dependence on conventional petroleum-based materials. (Thorat et al. 2003; Xu et al. 2014).

7.5 Mechanical Properties Under Impact Loading

One of the most significant attributes of polycarbonate is its exceptional toughness. Compared to other engineering plastics, polycarbonate exhibits outstanding impact strength, enabling it to absorb substantial energy without fracturing.

7.5.1 Viscoelastic Behavior

The viscoelastic behavior of polycarbonate is characterized by its ability to absorb energy upon impact, as it internally softens in response to the propagation of stress waves within its structure (Mercier et al. 1965). This viscoelastic response allows polycarbonates to withstand impact forces by dissipating energy, thereby enhancing its impact resistance (Ibrahim and Bhowmik 2024). The black and red dots in Fig. 7.3 represent the viscoplastic strain–stress curve of PC at varying rates. Viscoplastic strain is minimal at low stress levels but becomes significant as stress increases. Furthermore, viscoplasticity exhibits a strong dependence on stress rate, with strain decreasing at higher rates (Yu et al. 2016). There are several factors to be noted while considering the viscoelastic response of PC under load 3 mentioned in the following Table 7.7 (Fig. 7.4).

7.5.2 High Ductility and Strain Hardening

These properties enhance its resistance to impact and perforation. Its capacity for plastic deformation under stress allows for energy absorption and fracture resistance. (Ballistic performance of Polycarbonate and Polymethyl methacrylate under normal and inclined dynamic impacts—ScienceDirect 2024).

7.5.3 Impact Fatigue and Fracture Behavior

The fatigue performance of polycarbonate shows that cracks begin on the bottom surface when exposed to biaxial tensile stresses from impacts. The behavior of impact fractures is shaped by notch geometry as well as molecular weight. Studies show that the presence of notches and differences in molecular weight considerably alter the impact fracture behavior of polycarbonate, thereby accentuating the critical nature of these parameters in the assessment of material reactions to impacts (Xu et al. 2024; Roy et al. 2001).

7.6 The Effect on the Molecular Structure

Coarse-grained molecular dynamics simulations have been applied to investigate the deformation and fracture characteristics of polycarbonate under different loading conditions. Studies demonstrate that the mechanical characteristics of polycarbonate

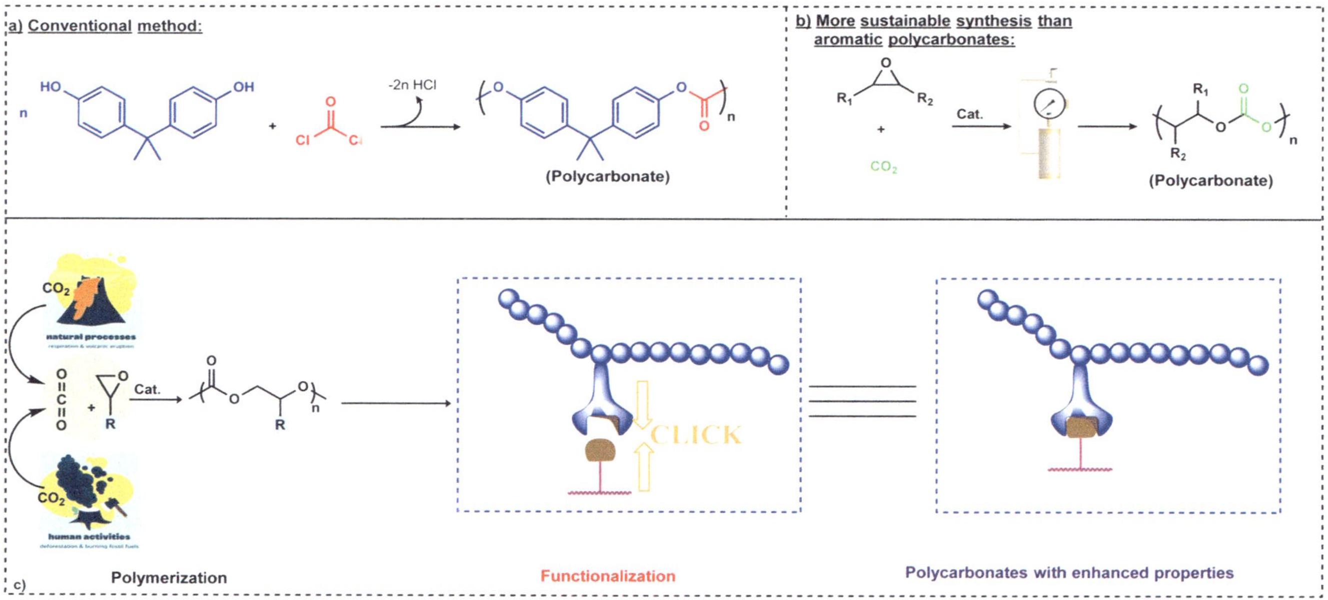

Fig. 7.3 Comparison of conventional and sustainable approaches for polycarbonate synthesis. **a** Conventional method of synthesis from bisphenol A (BPA) and phosgene (Ono and Erhard 2011) **b** Sustainable synthesis of polycarbonates(Copolymerization of carbon dioxide and epoxide—Inoue 1969) **c** Brief description about the synthesis and functionalization of polycarbonates synthesized via the copolymerization of CO_2 and epoxides. Figure adapted from Hassan et al. (2024)

Fig. 7.4 Stress versus viscoplastic strain (Yu et al. 2016)

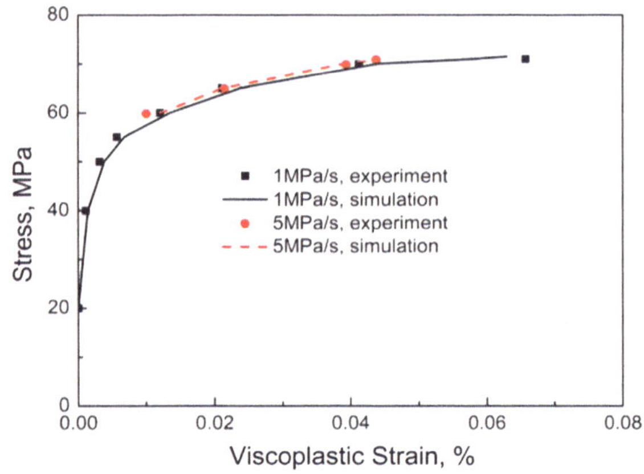

Table 7.7 Factors related to viscoelastic behavior of polycarbonate and its composites

Factor	Impact on viscoelastic behavior	Effect
Temperature	• At low temperatures, PC becomes stiffer and more brittle • At high temperatures, it becomes more ductile and compliant	Nonlinear viscoelastic behavior observed under shear loading from 0 to 140 °C Thermal history (annealing, quenching) affects viscoelasticity (Knauss and Zhu 2002; Torres and Frontini 2016)
Stress and Strain Rate	• Stress state (uniaxial, biaxial, or shear) significantly affects deformation and failure modes • High strain rates result in elastic, brittle behavior, while low rates lead to viscous, ductile behavior	Biaxial impact loading causes different deformation modes compared to uniaxial (Torres and Frontini 2016)
Volumetric Strain	Increased volumetric strain affects the material's stiffness and ductility, influencing its overall viscoelastic response	Volumetric (dilatational) strain alters stiffness, affecting performance under different loading conditions (Knauss and Zhu 2002)

are heavily dependent on its molecular design (Leelaprachakul et al. 2024; Umeno et al. 2019) (fig shown). Cyclic loading can lead to molecular degradation in polycarbonate, consequently resulting in a reduction of residual strength. The simulations elucidate the degradation process, offering a deeper understanding of fatigue damage mechanisms in polycarbonate (Leelaprachakul et al. 2024).

Understanding the mechanical behavior of polycarbonate under impact loading is crucial for designing products with improved performance. Research indicates that dynamic tension behavior is critical for assessing polycarbonate's mechanical quality, particularly regarding failure and fracture under impact (Macedo et al. 2019) (Albinmousa et al. 2021). Additionally, polycarbonate's mechanical properties can be affected by processing conditions, recycling, and reinforcement with materials like

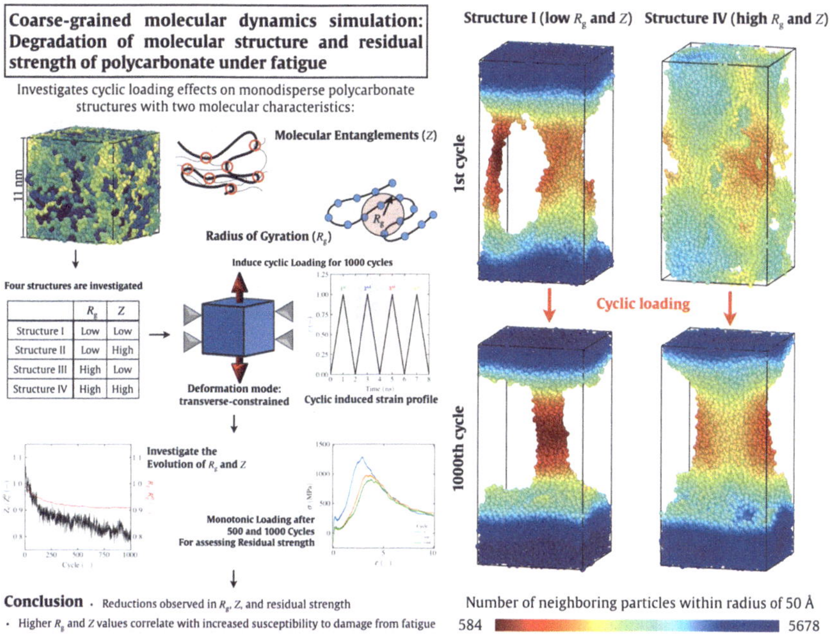

	R_g	Z
Structure I	Low	Low
Structure II	Low	High
Structure III	High	Low
Structure IV	High	High

Fig. 7.5 Coarse-grained molecular structure under cyclic loading (Leelaprachakul et al. 2024)

graphene (Pour et al. 2016). The mechanical response of polycarbonates is significantly influenced by loading rate and temperature. Static load properties are inadequate for analyzing dynamic situations. The mechanical behavior of polymers, such as polycarbonates, is notably influenced by the rate of loading. Transitioning from isothermal to adiabatic conditions alters the strain rate. The yield stress at strain rates exceeding 10 s^{-1} appears consistent across various polycarbonate grades. High-rate loading shows that strain softening and strain hardening behaviors of polycarbonates are largely unaffected by temperature increases (Kendall and Siviour 2013) (Figs. 7.5 and 7.6).

7.7 Previous Applications and Performance in Impact Scenarios with Polycarbonate

Polycarbonate has been employed in various high-impact applications across multiple industries. Aromatic polycarbonates derived from bisphenol-A are particularly important for their rigid molecular structure, optical clarity, exceptional impact resistance, toughness, and chemical stability (Hamouz et al. 2006). Polycarbonate, derived from Bisphenol-A, is known for being partially crystalline, transparent, and

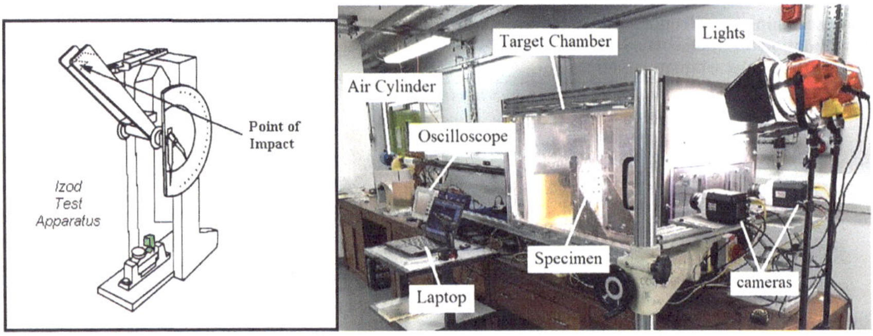

Fig. 7.6 Experimental set-ups for the high-velocity Izod and gas gun impact test (Ballistic performance of Polycarbonate and Polymethyl methacrylate under normal and inclined dynamic impacts—ScienceDirect 2024) (Liu et al. 2019)

possessing a high impact strength of 9.1 Ncm-1 and a high glass transition temperature of 149 °C (Kember et al. 2011). For various applications polycarbonates have been used in Dentistry for its high impact resistant and mixing with anti-bacterial agents (Rokaya et al. 2018).

– ***Safety and security glazing***

PC is used in bullet-resistant windows, riot shields, and safety goggles due to its high impact resistance. Advanced glazing technologies combine PC with reinforced films and bulletproof glass to increase strength while reducing thickness, enhancing compact safety doors and windows (Safety glass with improved impact resistance and safety strength and method for manufacturing the same 2020). Additives like dicarboxyl-terminated nylon 1212 improve wear resistance and low-temperature impact resistance, making PC ideal for machine tool windows (Polycarbonate composition with excellent wear resistance and low-temperature impact resistance and preparation method of polycarbonate composition 2019). Incorporating photosensitive and acid-producing groups enhances photoresist properties, while UV absorbers and flexibilizers improve light yellowing resistance and long-term UV stability (Polycarbonate-based composite material and preparation method and application thereof 2019; Application of polycarbonate as ultraviolet photoresist material 2020). Block copolymers of polycarbonate linked to hydrophilic polymers offer improved thermal stability and high hydroxyl content, enhancing their utility in safety glazing (Polycarbonate block copolymers and methods thereof 2020). These innovations have made PC crucial in protective equipment and structures like glass-polycarbonate panels in modern architecture, electric components' shields, aeronautics and astronautics (Research on Thin Glass-Polycarbonate Composite Panels 2018; Polycarbonate material selectively shielding near-infrared ray, preparation method, and application thereof 2019).

– ***Automotive industry***

PC is widely used for headlamp lenses, motorcycle windshields, and bumpers due to its shatter resistance and energy absorption properties (Balasubramanian Kandasubramanian 2024). The glazing technology in PC with hard-coating liquid, and plasma chemical vapor deposition provide high scratch and weather resistance, low optical distortion, and improved driver safety (Industrialization of automotive glazing by polycarbonate and hard-coating 2019; Emerging Applications of Polymers for Automobile Industries 2022). PC also replaces metallic parts in vehicles, providing superior impact resistance and mechanical properties. Carbon fiber-reinforced polycarbonate (CF/PC) is explored for use in automotive structures to improve mechanical performance and reduce molding times (Novel polycarbonate block amide copolymer as well as preparation method and application of polycarbonate block amide copolymer 2017). Polycarbonate's role in vacuum casting for helmets, vehicle parts, and bulletproof glasses shows its versatility in production techniques. Metal-coated plastic parts in the automotive industry are produced using sputtering technology to achieve thicker chromium coatings without defects, such as pinholes or cracking (Decorative Chromium Coatings on Polycarbonate Substrate for the Automotive Industry 2023). Aliphatic polycarbonate resins have been developed for use as thermally decomposable binders, enabling low-temperature degreasing and reducing residual carbon after sintering which can make it withstand varying temperatures (Novel aliphatic polycarbonate 2020). Furthermore, polycarbonate block amide copolymers provide enhanced thermal stability and stress cracking resistance (Automotive plastic panel 2015).

– ***Electronics***

Polycarbonate is used in phone cases, laptop housings, and other consumer electronics, offering protection from drops and impacts. It also replaces scratch-resistant metals, offering self-healing properties and alleviating static electricity and signal blocking issues (Leelaprachakul et al. 2024; Umeno et al. 2019). It can replace metal scratch-resistant materials, providing self-repairing properties and heat resistance for protective shells in electronic devices (Polycarbonate as well as preparation method and application thereof 2019). Flame-retardant PC, containing organosilicon flame retardants, also finds use in electrical equipment, offering enhanced flame resistance and lower environmental impact (Application of polycarbonate serving as electron beam photoresist material 2018). Moreover, polycarbonate is used in manufacturing polycarbonate films for speaker networks, incorporating inorganic filling agents for improved functionality and ease of processing (Polycarbonate film applied to speaker network for electronic appliances and manufacturing method thereof 2010).

– ***Sports equipment***

Polycarbonate helmets, visors and protective guards offer high impact resistance for player safety. Polycarbonate-siloxane block copolymers are employed in manufacturing baseball and softball bats for their exceptional impact fatigue resistance (Degradable polycarbonate sport fishing materials 2017; (Open Access) Study by Thermal Methods of Physico-Mechanical Properties of Polycarbonate used for High

Performance Sport Products 2008). With antioxidants and toughening agents, polycarbonate polymer can be formulated with high-impact-resistance aging-resistant anti-static and transparent which is suitable for sports equipment due to its durability and aging resistance (High-impact-resistance aging-resistant anti-static transparent polycarbonate polymer and preparation method and application thereof 2015).

7.8 Importance of Impact Resistance in Aerospace Components

In the aerospace sector, impact resistance is essential. Aerospace materials must withstand high-impact forces and temperature variations while preserving performance. The durability and efficacy of aircraft and spacecraft are contingent upon the materials selected. Aluminum is mostly used in the making of an aircraft (Maamoun et al. 2018; Batool et al. 2018). Impact properties are vital in aerospace structures to mitigate damage from various sources like maintenance incidents, collisions, and environmental factors. Composite materials play a pivotal role in aerospace technologies, focusing on improving thermo-mechanical properties for unmanned aerial vehicles, underscoring the significance of lightweight and high-strength materials in aerospace applications (Khan et al. 2021). High-impact-resistant materials are crucial for various aerospace components to ensure durability and safety in demanding operational conditions. Components that require such materials include.

7.8.1 Aircraft Skins and Fuselage

The outer skin of the fuselage is susceptible to impact from various sources, such as bird strikes, ground handling, emergency landings, hail, and debris. Ensuring the impact resistance of aircraft skins is crucial for maintaining structural integrity and safety (Zai et al. 2020).

7.8.2 Aircraft Structural Components

Lightweight materials like carbon fiber-reinforced polymers (CFRPs) are extensively used in aircraft structural components to reduce weight and improve fuel efficiency.

7.8.3 Aircraft Composite Structures

Aircraft composite structures are highly vulnerable to impact damage, whether from low-velocity sources during maintenance or high-velocity sources during flight events ([citation report] A model of low-velocity impact damage assessment of laminated composite structures 2024). Composite materials, such as carbon fiber composites, are widely employed in aircraft structures for their high specific stiffness and resistance to fatigue and Citation corrosion. Enhancing the impact resistance of composite structures is essential for ensuring the durability of aircraft components.

7.8.4 Aircraft Interiors

Lightweight materials with outstanding impact resistance are essential for aircraft interiors to ensure passenger safety and comfort. Polymethyl methacrylate (PMMA) is commonly used in aircraft windshields due to its impact-resistant characteristics (Fidan 2014).

7.8.5 Aircraft Engine Components

Engine parts, including fan blades, compressor blades, and casings, are vulnerable to impact from foreign object debris (FOD) or bird strikes. Impact-resistant materials and protective measures are critical to prevent engine damage (Liu and Han 2018). Components like fan containment structures in aircraft engines require high impact resistance to withstand potential foreign object damage. Lightweight hybrid structures are developed to provide impact resistance in critical engine components (Yang et al. 2022).

7.8.6 Aircraft Wings

Runway debris may impact the underside of wing structures, leading to hydrodynamic ram effects in fuel tanks and significant aircraft vulnerability (Artero-Guerrero et al. 2013). The wing leading edge is identified as one of the most vulnerable zones to bird strikes. Impact on the leading edge can result in structural damage and aerodynamic disturbances, affecting the overall performance of the aircraft (Yu et al. 2019). Unmanned aircraft wings made from laminar and sandwich composite materials require impact-resistant strength to withstand various loads and stresses. These materials are designed to be resistant to impact and lightweight for efficient flight performance.

7.8.7 Aircraft Cockpit Windows

Cockpit windows are exposed to potential impacts from bird strikes, hail, or debris. Using impact-resistant materials like strengthened glass or laminates can enhance the safety of cockpit windows (Doubrava et al. 2018).

7.9 Experimental Methods

Experimental methods used to test the properties of polycarbonate under impact loading encompass a range of techniques aimed at evaluating its mechanical behavior. Experimental tests on PC plates have revealed several common failure modes under ballistic impact, including cracking cone, elastic dishing, plugging, deep penetration, and petaling (Wright et al. 1993). The specific failure mode depends on factors such as the projectile shape, impact velocity, and plate thickness. For example, thin PC plates tend to exhibit elastic dishing and cracking cone failures, while thicker plates show plugging and deep penetration. There are certain properties of polycarbonate which is must to be considered for use in aeronautical and aerospace applications.

The key experiments and characterization techniques covered include:

- Strain rate testing over a large range to study the mechanical behavior of polycarbonate under different loading conditions.
- Detailed descriptions of the experimental facilities and methods used to measure the mechanical properties of polycarbonate.

Studies have employed various approaches such as **tensile testing**, **Izod impact testing**, **falling dart tests**, and **Charpy impact tests** to assess the impact resistance and fracture toughness of polycarbonate. These tests involve subjecting the material to controlled impact forces to measure its response to sudden loading conditions, providing insights into its ability to withstand impact without fracturing. Moreover, the effect of processing conditions on the mechanical properties of polycarbonate has been studied through capillary rheometry, thermogravimetric analysis, and atomic force microscopy (AFM) to understand the influence of processing parameters on the material's morphology and mechanical performance. These experimental methods provide valuable data on the structure–property relationships of polycarbonate blends and composites, aiding in the development of materials with enhanced impact resistance and mechanical strength (Table 7.8).

Table 7.8 Summary of experimental methods for testing polycarbonate impact resistance

Test method	Instruments	Standardized specimen	Purpose	Advantages	Disadvantages	Applications	References
Tensile Testing	UTM with tensile grips and extensometers	Dog-bone shaped specimens, according to ASTM D638	Measures tensile strength, yield strength, etc	Provides detailed mechanical data	Requires precise preparation, time-consuming	Aerospace structural parts	Zhang et al. (2023), Malek-Mohammadi et al. (2019), Goryl et al. (2024), Jaya Christiyan et al. (2018), Choi et al. May 2024)
Impact Testing	Charpy and Izod impact testers	Notched specimens according to ASTM D256 (Izod) or ASTM D6110 (Charpy)	Evaluates energy absorption during fracture	Quick, provides toughness data	Sensitive to notch quality	High-impact missile and aircraft components	Ballistic performance of Polycarbonate and Polymethyl methacrylate under normal and inclined dynamic impacts - ScienceDirect (2024)
Taylor Impact Test	High-speed projectile launcher	Cylindrical specimens, typically standardized according to specific research protocols	Simulates high-strain-rate impacts	Provides dynamic failure modes	Requires specialized equipment and high safety	Missiles, high G-load components	
Flexural Testing	UTM with flexural fixtures	Rectangular bars according to ASTM D790	Measures flexural strength and modulus	Provides stiffness and load-bearing data	Influenced by specimen dimensions and loading rate	Load-bearing aerospace components	Jaya Christiyan et al. (2018), Dimitrellou et al. (2024), Juracka et al. (2023)

(continued)

Table 7.8 (continued)

Test method	Instruments	Standardized specimen	Purpose	Advantages	Disadvantages	Applications	References
Falling Dart Test	Falling dart tester	Flat, rigid plastic specimens or films,according to ASTM D1709	Simulates real-world impacts	Useful for assessing real-world impact resistance	Results influenced by specimen thickness and dart shape	Protective applications (panels, shields)	Rafique (2015)
Gas Gun Test	Gas gun, high-speed cameras, sensors	Flat panels or specific shapes of materials required	Simulates high-velocity impacts	High precision, controlled impact simulation	High setup costs, requires specialized expertise	Aerospace debris and ballistic threats	Ghavanini et al. (2023), Rahner et al. (2014)

7.10 Simulation Techniques for Impact Analysis

In aerospace engineering, simulation techniques provide an efficient and cost-effective means to predict how materials behave under stress, impact, and environmental conditions. By using advanced software tools, engineers can simulate real-world scenarios like bird strikes or debris impacts without the need for costly and risky physical testing. These simulations help optimize materials and designs for safety and performance.

Software Tools

Several powerful software tools are commonly used in aerospace simulations:

- **ANSYS**: For finite element analysis (FEA) and computational fluid dynamics (CFD).
- **LS-DYNA**: Specialized in dynamic impact and crash simulations.
- **ABAQUS**: Widely used for complex stress analysis.
- **MATLAB and Simulink**: Ideal for numerical simulations and control system analysis.
- **COMSOL Multiphysics**: Allows for coupled simulations involving multiple physical phenomena.

Types of Impact Threats and Simulations

Aerospace vehicles face several threats that can lead to structural damage or material failure. Simulating these impacts is crucial for designing components that can withstand real-world conditions.

- **Bird Strikes**: High-speed collisions with birds, especially during takeoff and landing, can damage engines and windshields. Using **LS-DYNA** and **ABAQUS**, engineers simulate the bird's impact velocity and angle to assess damage and ensure the integrity of critical components.
- **Debris Impact**: Foreign object debris (FOD) on runways or in-flight can cause severe damage. Simulations with **ANSYS** help identify failure points and improve the resilience of the fuselage and wings.
- **Hailstorms**: Aircraft can encounter hail during flight, damaging the fuselage and wings. **LS-DYNA** simulations model the hail's size and velocity to determine the extent of cumulative damage on the aircraft's exterior.
- **Space Debris and Micro-Meteoroids**: Spacecraft face the risk of collisions with high-speed particles. Using **LS-DYNA** and **SPH** (Smoothed Particle Hydrodynamics), simulations predict the penetration depth and damage to spacecraft exteriors.
- **Drone Collisions**: The increasing use of drones has introduced a new risk for aircraft. Simulations in **ABAQUS** model the impact of drone collisions to evaluate structural damage and potential failure.

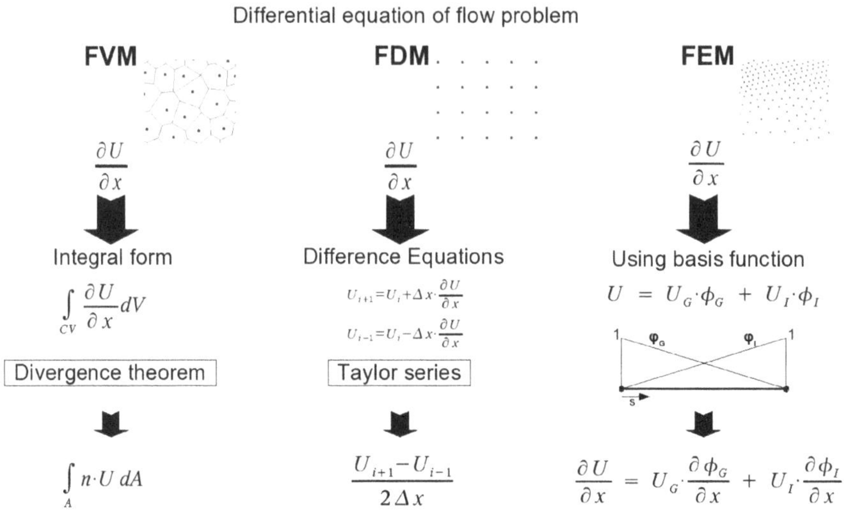

Fig. 7.7 Most commonly used numerical methods for simulation (Figure adapted from Milbradt and Abed (2008)

- **Lightning Strikes**: Direct or induced lightning strikes can damage structures and electronics. **COMSOL Multiphysics** simulates the effects of electromagnetic fields and current on the aircraft, assessing both structural and electronic vulnerabilities.
- **Turbulence and Wind Shear**: Sudden changes in wind direction and speed create stress on the aircraft's structure. Simulations with **MATLAB** and **ANSYS Fluent** help evaluate how aerodynamic forces impact the aircraft's stability and stress points (Fig. 7.7).

Several simulation techniques are employed to study the behavior of materials and structures under impact:

- **Finite Element Analysis (FEA)**: This method divides structures into small, finite elements and predicts how they will respond to forces, heat, and vibration. Widely used for impact analysis.
- **Material Modeling**: Involves creating detailed models of materials to simulate how they react to stress, temperature changes, or impacts.
- **Dynamic Analysis**: Focuses on understanding how materials and structures behave over time when subjected to varying loads, such as impacts.
- **Agent-Based Simulation**: Models the actions and interactions of independent agents (e.g., drones or debris) within a system to predict the overall effects.
- **Monte Carlo Simulations**: Uses statistical techniques to predict outcomes and evaluate risk, often used in decision-making scenarios.
- **Computational Fluid Dynamics (CFD)**: Simulates fluid flow and heat transfer to evaluate how air interacts with structures, critical for aerodynamic and thermal analysis.

7.10.1 Finite Element Analysis (FEA) for Impact Simulation

Finite Element Analysis (FEA) is essential in aerospace for evaluating aircraft component safety during impact events. It allows for the simulation of impacts on subjects and adjacent anatomical elements, aiding in predicting material behavior without the need for destructive tests (Merema et al. 2020). This method simulates material and structural responses to predict the behavior of materials such as metals, composites, and polymers when subjected to impacts, helping to optimize material designs for safety and performance(What Is FEA I Finite Element Analysis Documentation 2024). FEA has been extensively used in the study of impact resistance, such as in motorcycle helmet testing, where it predicts head-form accelerations considering friction effects (Mills et al. 2009). These FEA models have been able to reasonably predict the stress–strain behavior, energy absorption, and penetration depth compared to experimental results (Sarva et al. 2007; Stecconi and Landi 2020).

Tools like **SolidWorks** and **ANSYS** allow engineers to simulate impacts with high precision, helping in the design and optimization of mechanical components.

In impact simulations, **Explicit Time-stepping algorithms** are commonly employed to model dynamic events like aircraft or car crashes, offering detailed predictions of structural responses. It uses complex calculations to analyze how structures react to different loading scenarios, allowing for the evaluation of impact forces and internal damage, improving accuracy in crash simulations.

FEA models break structures into smaller elements, using numerical methods to solve for dynamic loading conditions. This process is especially valuable in aerospace engineering, where FEA is essential for studying how structures respond to impact loads(Improvement of Finite Element Solutions with Deep Learning 2022; Wellkamp and Reduction of epistemic uncertainty of a crash box model—experimental and numerical investigations 2019).

FEA predicts the load analysis by creating 3D models, defining material properties, and applying boundary conditions. The software then solves equations to predict deformation and stress concentrations, and results are analyzed to assess failure points and stress distribution.

Finite Element Analysis (FEA) uses two primary methods for solving problems i.e. explicit and implicit (Implicit Vs Explicit Finite Element Analysis: When to

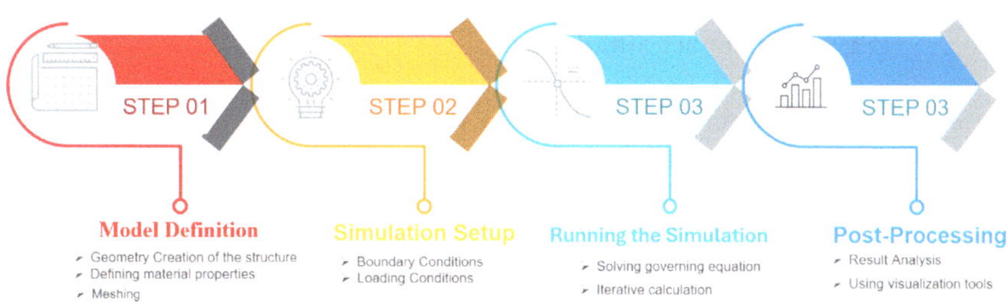

Fig. 7.8 Flow diagram of FEA method

Table 7.9 Explicit versus implicit solving method

Explicit method	Implicit method
The explicit method calculates the state of the system at a future time directly from the current state	The implicit method involves solving a system of equations that include both the current and future states
It uses an extrapolation approach, where the new state is determined based on the known data from the current state	This often requires iterative techniques, such as the Newton–Raphson method, to find a solution
Ideal for dynamic and highly non-linear problems, such as impact simulations, crash tests, and explosions	Implicit methods are well-suited for static, quasi-static, and slowly varying dynamic problems. They are commonly used in structural analysis, thermal analysis, and steady-state simulations

Use Which 2024). Among these two methods explicit method is mainly suitable for impact simulation of any materials (Table 7.9).

7.10.1.1 FEA in Aerospace Impact Simulation

In the aerospace sector, the application of FEA is widespread, helping simulate everything from low-velocity impacts during ground operations to high-velocity strikes from space debris. One standout application is in the simulation of composite materials, which are extensively used due to their excellent strength-to-weight ratio. Composite panels are particularly susceptible to damage from impacts, but FEA provides the tools to predict such damage and optimize the materials accordingly. Following are few impact scenarios and simulation techniques that presents high prospect of FEA in impact simulation (Table 7.10).

However, despite its efficacy, FEA's limitations and assumptions must be acknowledged. Simulation accuracy relies on high-quality material data and model realism. Furthermore, the complexity and cost of these simulations require continuous improvements in modeling techniques and computational capabilities (Table 7.11).

7.10.1.2 Machine Learning Integration with Finite Element Analysis (FEA) for Impact Simulation

The use of Finite Element Analysis (FEA) with Machine Learning (ML) is now redefining computational modeling. It is augmenting productivity, precision, and flexibility, with FEA's elaborate physical modeling features alongside ML's forecasting capabilities, promoting rapid and precise simulations of material responses during impact scenarios. The subsequent sections examine critical methodologies, applications, and challenges associated with the integration of ML and FEA for impact simulations.

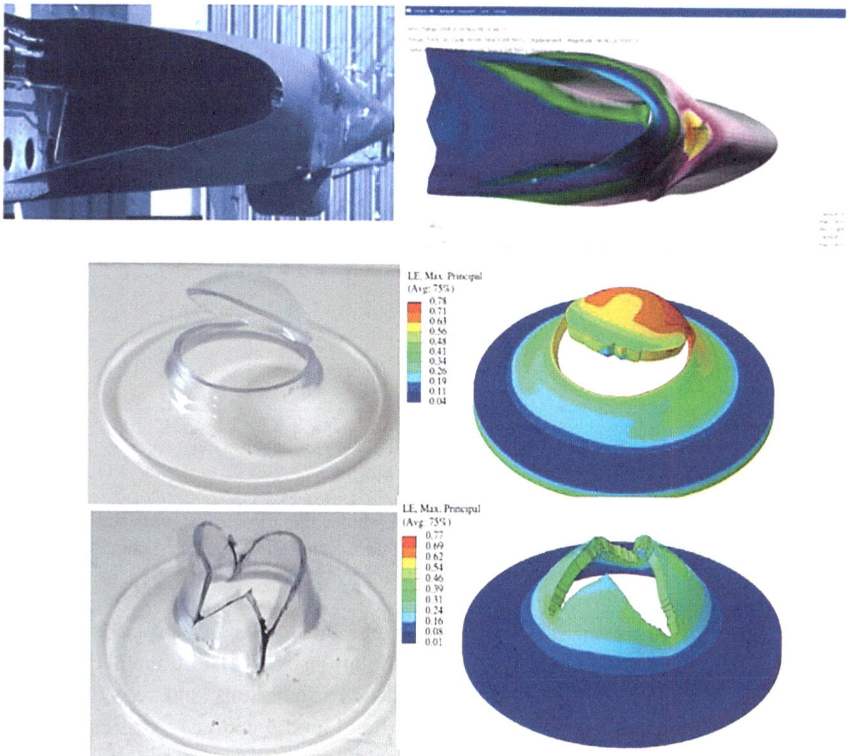

Fig. 7.9 Experiment versus FEM prediction for fracture aspect of specimen (Torres and Frontini 2016)

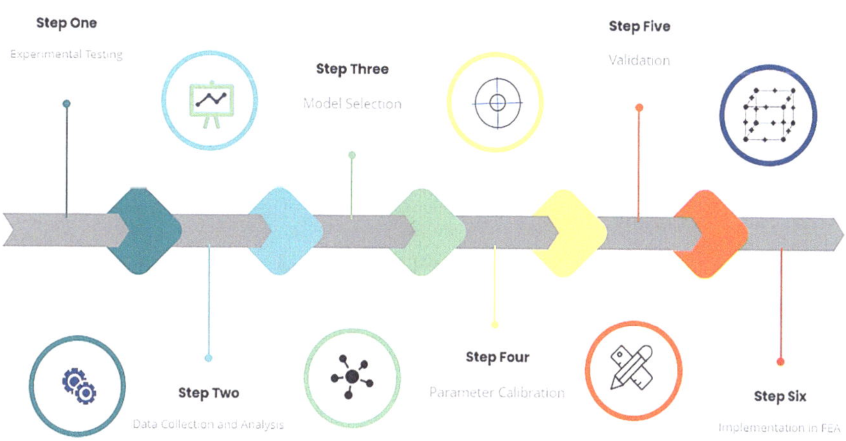

Fig. 7.10 Process of Material modeling

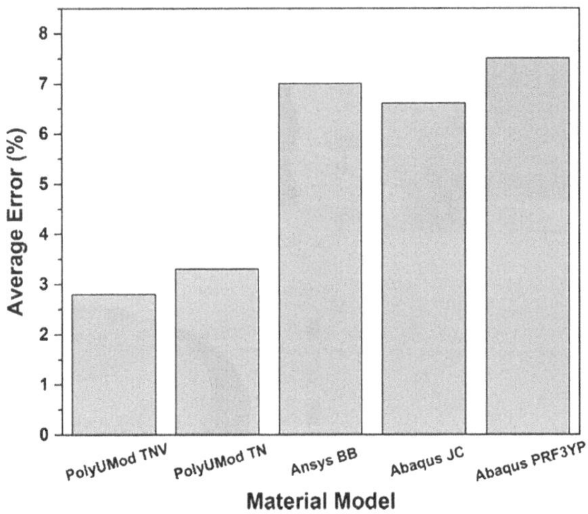

Fig. 7.11 Percentage of Error in different material models comparing to experimental data (Best Material Model for Polycarbonate at High and Low Strain Rates—PolymerFEM.com 2024)

Optimization of Material Model Parameters: Recent studies propose utilizing artificial neural networks to correlate FEA simulations with experimental data. This method effectively learns discrepancies between predictions and real outcomes, improving simulation fidelity beyond conventional model updates. Artificial Neural Networks (ANNs) can be employed to determine optimal material model parameters by analyzing high-speed impact experiments. This approach allows for the rapid emulation of models, facilitating the Bayesian calibration of parameters. ANNs can also directly infer material model parameters from impact data, offering a novel solution for inverse problems in engineering applications (Pogorelko et al. 2024). A hybrid framework combining ML and FEA can autonomously calibrate numerical models, reducing trial-and-error in material characterization. This framework uses Gaussian Process Regression to iteratively refine model inputs, enhancing the accuracy of simulations and minimizing errors in data reduction (Hybrid Physics and Machine Learning Modeling for Material Characterization and Failure Analysis | AIAA SciTech Forum 2024).

Acceleration of Multiscale Simulations: Unsupervised ML techniques, such as k-means clustering, can significantly speed up multiscale simulations by reducing the number of micro-scale problems that need to be solved. This method groups similar integration points, solving only one nonlinear problem per cluster, thus achieving speed-up factors of up to 85 without compromising accuracy (Yvonnet and Chaouch 2024).

ML models, including linear regression and ANNs, can predict stress–strain responses in porous hyperelastic materials, allowing for faster multi-scale analyses. These models can identify critical points for high-fidelity analysis, optimizing computational resources (Couto Carneiro et al. 2023).

Table 7.10 Impact scenarios and simulation techniques

FEA application	Material used	Method used	Field of study	Results	References
Impact on composite panels	Composite Materials (T800-F3900 carbon fiber composite)	LS-DYNA with rate-dependent damage model & Gas gun Impact Experimental Test	Aerospace structures	Accurate prediction of low and high-velocity impact damage in structure	Shyamsunder et al. (2022)
CFRP laminate impact	Carbon Fiber Reinforced Polymer	3-D Elastic–Plastic (EP) FEA model	Aerospace and transportation industry	Low-velocity impact tests showed close match to experimental results by drop-weight impact tests	Liu et al. (2020)
Low velocity impact	Poltcarbonate	3D thermo-mechanical constitutive FE model calibrated with experimental data	Automotive, aerospace, and safety equipment	The effective plastic stretch is a suitable measure for modeling ductile failure of PC in multi-axial impact scenarios	Torres and Frontini (2016)
Hypervelocity impact on spacecraft	Debris Shields (Aluminium alloy AA6061-T6)	Coupled FEM/DEM in LS-DYNA	Spacecraft shielding	Promising prediction of debris cloud formation and ballistic limits 1 to 14 km/s in Piekutowski's target	Færgestad et al. (2023)
Bird-strike simulation	Composite Laminates	SPH-FEM via Continuum Damage Mechanics by ABAQUS/Explicit method	Aircraft structures	Damage spreads mainly along fibers and boundaries due to stress. Strain energy rises quickly after contact, then slowly, showing high-frequency vibrations	Zhou et al. (2019)
Engine failure events	Aircraft Components	Open-source FEA framework	Aerospace safety	Realistic modeling of damage scenarios to improve aircraft survivability during engine failure	Cid et al. (2022)

(continued)

Table 7.10 (continued)

FEA application	Material used	Method used	Field of study	Results	References
Crash load simulation	Carbon fiber reinforced composite laminates 24-ply and 32-ply laminates	Explicit Time-stepping algorithm by ABAQUS(r) explicit solver	Automotive safety	Improved accuracy in predicting crash load cases for vehicle body design	Joglekar et al. (2021)
High overload impact simulation	Various Aerospace Materials	FEM with ANSYS/ LS-DYNA	High overload devices	Simulated impacts with overloads of 80,000–90,000 g for optimizing impact test devices	Li et al. (2024)
Ballistic impact simulation	Aluminum alloy 2024-T351 as target and rigid steel at projectile	FEA with Modified Johnson–Cook (JC) & Mohr–Coulomb damage model vs. SPH comparison	Aerospace defense	FE method was more consistent than the SPH method in predicting residual velocity after ballistic impact. While MMC damage model showed better result compared to the JC damage model when analyzing the damage response of the plate target under different ballistic impact parameters	Numerical analysis of ballistic impact through FE and SPH methods | AVESİS (2024)

Table 7.11 Actual and simulated gun test results including head angle, launch pressure, target collision velocity, target penetration depth, and maximum deceleration

Gas gun test	Head angle (Deg)	Launch pressure (Bar)	Target penetration depth (mm)	Target collision velocity (m/s)	Maximum deceleration (G)
FEM Simulation	60	–	142	95	7664
	60	–	218	145	9667
	90	–	104	95	10,495
	90	–	167	145	20,222
Actual	60	10–11	130	94	7560
	60	35–36	230	151	10,386
	90	10	95	93	13,332
	90	31–32	150	145	25,036

Enhanced Material Behavior Prediction: ML-based surrogate models, such as fully connected neural networks and long-short-term memory networks, can predict stress distributions and sequences during impact events with high accuracy. These models serve as efficient alternatives to traditional FEA, providing rapid insights into material stress analysis (Accelerating Analysis for Structure Design via Deep Learning Surrogate Models—Shao 2023). Recurrent Neural Networks (RNNs) are particularly effective in capturing path-dependent material behaviors, such as those seen in geological materials. New model architectures and training methods have been developed to address the limitations of RNNs, ensuring robust performance in FEA simulations under varying conditions (Machine learning based modeling of path-dependent materials for finite element analysis—ScienceDirect 2024).

7.10.2 Material Modeling for Polycarbonates

Material modeling for polycarbonates involves creating mathematical representations to predict how the material responds to stress, strain, temperature, and other environmental factors. For predicting its performance under various conditions, accurate modeling is mandatory (Table 7.12).

Table 7.12 Key factors in polycarbonate material modeling

Factor	Modeling approach	Application
Stress–strain	Elastic–plastic models	Yield, Hardening behavior
Viscoelasticity	Viscoelastic-plastic models	Creep, Stress relaxation
Temperature effects	Temperature-dependent models	Thermal fluctuations in aerospace
Strain rate	Strain-rate dependent models	High-velocity impact simulations

Constitutive Models:

Accurate constitutive modeling of PC is crucial for predicting its material behavior. Constitutive models were studied to understand the strain rate sensitivity and deformation mechanisms and to enhance simulation accuracy (Barbagallo et al. 2024) (Mirone et al. 2024). The development of a true stress–strain curve through experimental analysis allows for better predictions of PC performance under dynamic loading conditions. In the following table some of the common numerical models are shown that can be used for predicting polycarbonate behavior under different parameters (Table 7.13).

There are some common material models for evaluating material properties of polycarbonate amongst them MCalibration software was used to define the errors. The following figure plots the average error in the model predictions from the different material models examined in this study. The Abaqus PRF model has an error that is 160% larger than the error from the PolyUMod TNV model (Best Material Model for Polycarbonate at High and Low Strain Rates—PolymerFEM.com 2024).

7.10.3 Discrete Element Method (DEM) Simulations

The Discrete Element Method, known as DEM, was first generated by Cundall as a numerical method for discrete particle analysis in rock mechanics (Cundall and Strack 1979). The Discrete Element Method has developed to simulate various systems, specially granular materials. This gives an intricate look at the movement of particles, something that is hard to examine in a lab setting. Particle interactions are modeled via contact springs, with applications spanning various materials, such as polycarbonate composites (Balamonica and Goh 2019).

The Discrete Element Method (DEM) is utilized to resolve the motion of particles, while the Computational Fluid Dynamics (CFD) method is employed to calculate the interstitial fluid flow (Cleary 2004; Kloss et al. 2012). It provides a robust approach for simulating particulate systems (Pasha et al. 2016). With advancements in computing technology, it allows for obtaining detailed information on particle movement, forces, and energy consumption during bulk material crushing (Zhang et al. 2020). DEM is largely used as a valuable numerical tool for simulating the complex dynamic particle motion and interactions within tumbling mills (McBride et al. 2004) (Fig. 7.13).

The key difference between DEM and FEM lies in their underlying principles: FEM discretizes the domain into elements to solve continuous problems, while DEM models materials as discrete particles to analyze their interactions and behavior. The Finite Element Method (FEM) is used for solving problems in solid mechanics, including those involving damage, plasticity, and multifracture situations (Oñate et al. 2017) and it is suitable for continuum problems. DEM is specifically designed to handle discontinuous media and simulate the behavior of discrete particles.

Discrete Element Method (DEM) can be utilized to predict the mechanical properties and failure behavior of polycarbonate materials under various loading conditions.

Table 7.13 Common constitutive model for polycarbonate

Constitutive model	Description	Application	References
Linear Elastic Models	Models elastic behavior based on Hooke's Law for small deformations	Suitable for 2D elastic behavior of polycarbonate, influenced by physical aging and molecular weight distribution	Soloukhin et al. (2003), Mindlin (1964), Mirone et al. (2024)
Nonlinear Elastic Models	Captures nonlinear stress–strain behavior during larger deformations	Used in cases where polycarbonate exhibits significant nonlinear behavior under stress	
Plasticity Models	Describes permanent deformation post-yield using models like Johnson–Cook and Bergström-Boyce	Predicts the plastic deformation of polycarbonate under various loading conditions	Bischoff and Hedenström (1902), Torres and Frontini (2016), Improvement of Finite Element Solutions with Deep Learning (2022), Færgestad et al. (2023)
Viscoelastic Models	Models time-dependent behavior using Prony series and generalized Maxwell models	Captures creep, stress relaxation, and time-dependent deformation in polycarbonates under cyclic loading	Yu et al. (2016), A viscoelastic–plastic constitutive model for uniaxial ratcheting behaviors of polycarbonate (2015)
Von Mises Model	Predicts yielding under complex loading but doesn't consider hydrostatic pressure effects	Suitable for general yield behavior but limited for pressure-sensitive materials like polycarbonate	A two-surface damage-plasticity model based on a Drucker—Prager yield criterion—Zhang (2023)
Drucker-Prager Model	Extends von Mises to account for pressure-dependent yielding	Useful for polycarbonate under complex loading conditions with hydrostatic pressure effects	Drucker-Prager plasticity model in the framework of OSB-PD theory with shear deformation | Engineering with Computers (2024)
Generalized Maxwell Model	Describes the viscoelastic response of polymers under uniaxial conditions	Studies stress relaxation in polycarbonate, with accurate approximation of experimental data	Material modeling and solid phase forming of polycarbonate sheet - Lee (1986), The relationship between the effect of thermal pre-treatment and the viscoelastic behaviour of polycarbonate in the glassy state (2024)

(continued)

Table 7.13 (continued)

Constitutive model	Description	Application	References
Bergström-Boyce Model	Integrates viscoelastic and viscoplastic components for modeling wide strain rates and temperature ranges	Provides high accuracy for predicting polycarbonate behavior under dynamic loading conditions	Narijauskaitė et al. (2013), Wang et al. (2017)
Elasto-Viscoplastic Models	Models nonlinear, rate-dependent deformations considering viscoelasticity and temperature sensitivity	Captures complex mechanical behavior of polycarbonates under various loading and temperature conditions	
Bodner-Partom Model	Effective for rate-dependent plasticity by incorporating viscoplastic effects across strain rates	Simulates polycarbonate behavior under cyclic loading and extrusion processes	High fidelity computational characterization of the mechanical response of thermally aged polycarbonate—IOPscience (2024), Bocko et al. (2015), Wójcik and Skrzat (2022)
Arruda-Boyce Model	Incorporates temperature effects into the hyperelastic behavior of polycarbonates	Used for simulating the strain-rate-dependent viscoplastic behavior in polycarbonate materials	

By incorporating data from different research papers, DEM can simulate the behavior of polycarbonate composites with nanoparticles (Andriß et al. 2022), investigate the rate- and temperature-dependent deformation of polycarbonate analyze the mechanical response of polycarbonate under uniaxial tests at different temperatures and stress levels (Wang et al. 2017), and predict the intrinsic mechanical response of polymers filled with particles. Studies have shown that DEM simulations can effectively capture dynamic fracture instability characteristics in brittle solids like PMMA, reproducing crack propagation and branching under dynamic loading conditions (Krop et al. 2016). Additionally, DEM simulations have been employed to investigate the effective stiffness of assemblies of frictional, elastic spheres under anisotropic loading, showcasing the ability to analyze stress responses under different strain perturbations (Kou et al. 2019). DEM simulations have been used to predict the instability behavior of granular materials during constant shear-drained conditions, highlighting the importance of initial stress states on the evolution of anisotropic coefficients and macroscopic responses (Recchia et al. 2020). Extending the discrete element method to simulate nonlinear material behavior in contact zones enables a more comprehensive representation of the mechanical response in microscopic conditions (Shilko

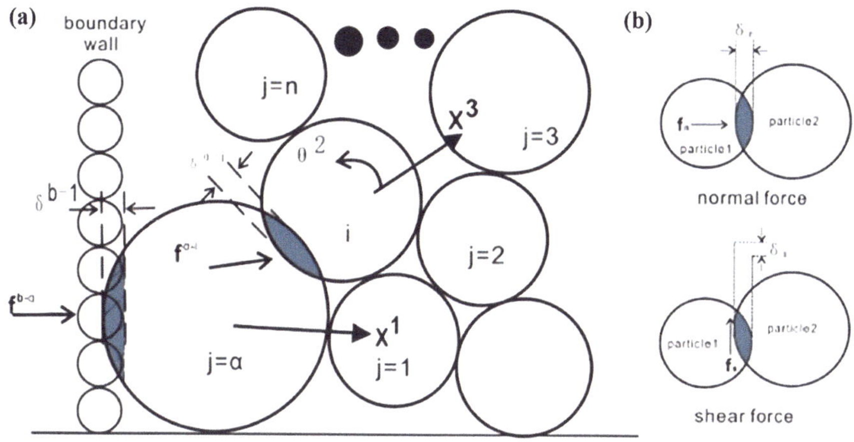

Fig. 7.12 Schematic Diagram of DEM: A. Boundary wall interacting with strata particles, causing movement and rotation due to self-gravity and external stress and B. Two forces (normal and shear) between particles (Wu 2021)

et al. 2021). This extension allows for the representation of materials with diverse rheological properties and scales, from microscopic to macroscopic, broadening the applicability of DEM to various material types like polycarbonate. DEM can offer a more precise representation of the geometry of the polyhedral particles (Nassauer et al. 2013) which can effectively capture the intricate shapes and structures found in polycarbonate materials.

- *DEM in Aerospace Impact Scenarios*

  The Discrete Element Method (DEM) is well-suited for simulating discontinuities in materials but lacks the capability to accurately represent the continuous response of materials like polycarbonates under impact loading conditions. While DEM is adapted to simulate discontinuities, it is not suitable for simulating continuous behavior. However, it is particularly effective for impact scenarios due to its ability to accurately model the behavior of materials under dynamic loading conditions as DEM is a numerical technique that represents materials as an assembly of discrete particles, allowing for the simulation of interactions, deformations, and fractures at the particle level. This method is well-suited for impact scenarios as it naturally considers discontinuities in materials, making it ideal for simulating impact-induced fractures and wear in Aerospace Industry (Discrete element method to simulate continuous material by using the cohesive beam model—ScienceDirect 2024).

  DEM has been applied to study the high strain rate failure behavior of polycarbonate plates under hypervelocity impact conditions (Callahan et al. 2022). However, the application of DEM for 3-D continuum impact problems is greatly limited due to the lack of a rigorous theoretical foundation and flexibility in modeling arrangement patterns. By combining DEM with FEM, it is easier to solve continuous materials like polycarbonates. Studies have shown that DEM models are capable of three-dimensional impact problems by comparing the numerical results with the corresponding results obtained by the finite element method (FEM) and the experiments (New Discrete Element Models for Three-Dimensional Impact Problems—IOPscience 2024). Without a comprehensive understanding of the continuous behavior of materials, the use of DEM is limited.

In the context of impact simulations involving polycarbonates, the continuous nature of the material's response to dynamic loading makes it challenging for DEM to provide accurate representations. DEM's focus on discrete particle interactions may not adequately capture the complex deformation and fracture behavior exhibited by polycarbonate materials during impact events. Therefore, for simulating impact scenarios involving polycarbonates, alternative modeling approaches that can effectively account for the continuous behavior of the material would be more appropriate.

7.10.4 Particle-Based Modeling

Particle-based modeling is a versatile computational approach that simulates materials and systems by representing them as discrete particles. This method is used across various disciplines, from materials science to aerospace engineering, allowing for the simulation of material interactions at multiple scales—from microscopic to macroscopic (Shilko et al. 2021). While particle-based modeling encompasses various techniques, it differs from the Discrete Element Method (DEM), which focuses specifically on modeling granular materials and brittle fractures at micro- and mesoscopic scales.

Applications of Particle-Based Modeling in Material Sciences

In materials science, particle-based modeling is applied to analyze the behavior of metallic materials at a molecular level. By simulating interactions between atoms, the method captures mechanical behaviors such as elasticity and plasticity (Particle Modeling Based on Interatomic Potential and Crystal Structure: A Multi-Scale Simulation of Elastic-Plastic Deformation of Metallic Material 2024). This approach also finds applications in computer graphics and virtual reality, where it simulates rigid and deformable solids, as well as fluids, for interactive environments and special effects.

Aerospace Applications of Particle-Based Modeling

In the aerospace industry, particle-based modeling is essential for predicting the impact behavior of materials such as polycarbonate. Due to polycarbonate's exceptional impact resistance, it is widely used in aerospace structures. The following are some key areas where particle-based modeling has been applied in aerospace:

- **Monolithic and Laminated Polycarbonate Panels**: Studies using a combination of experimental and numerical methods have investigated the impact resistance and anti-shattering properties of polycarbonate panels. These studies confirm polycarbonate's reliability in aerospace impact protection systems (Ghavanini et al. 2023).
- **Low-Velocity Impact**: Research has shown that the thermal history of polycarbonate, particularly during injection molding, can significantly influence its impact resistance. This understanding is crucial for aerospace applications, where material performance under varying conditions is vital (Xu et al. Dec. 2015) (Ghavanini et al. 2023).
- **Hypervelocity Impact**: **Polycarbonate plates** subjected to hypervelocity impacts have been studied to enhance impact protection systems in aerospace. These studies focus on predicting polycarbonate's behavior under extreme impact conditions (Callahan et al. 2022).
- **Charpy Impact Testing**: Polycarbonate's impact resistance has been evaluated through Charpy impact tests under various conditions, providing valuable data for selecting the appropriate polycarbonate grades for aerospace applications (Charpy impact properties and numerical modeling of polycarbonate composites n.d.).

Advanced Particle-Based Modeling Techniques

- **Bond-Based Peridynamic (PD) Method**: This advanced technique simulates fracture behavior in materials by incorporating crack propagation and dislocation mechanisms. The PD method has been validated against experimental data, demonstrating high accuracy in impact testing (Zhang and Xu 2019).
- **Coarse-Grained Molecular Dynamics (CGMD)**: The CGMD method offers molecular-level insights into deformation and fracture mechanisms in polycarbonate, estimating yield stress under various loading conditions. This technique has shown superior accuracy in predicting polycarbonate failure compared to traditional methods (Azizi et al. 2022; Krop et al. 2016).

7.10.5 Contact and Collision Modeling

Contact and collision modeling techniques play a vital role in improving simulation accuracy and efficiency, especially in materials science and biomechanics. Traditional finite element methods (FEM) often struggle with contact detection, leading to failures in high-energy interactions. Methods like **PolyFEM** offer superior performance in such simulations, delivering accurate results without extensive tuning (Martin et al. 2024). The **Neural Network Finite Element (NNFE)** method accelerates predictions for contact problems, significantly reducing computation time while maintaining accuracy. **Incremental Potential Contact (IPC)** ensures intersection- and inversion-free simulations, offering robust solutions for complex scenarios.

Steps in Contact and Collision Modeling

- **Geometry Creation**:
  - **Modeling the Objects**: At first 3D models of the objects in the collision is made using CAD software. This includes both the impacting object (a projectile) and the target (e.g., a polycarbonate sheet) (Bortoff 2020).
- **Material Properties**:
  - **Defining Material Behavior**: The material properties for each object is specified which includes elastic properties (Young's modulus, Poisson's ratio), plastic properties (yield strength, hardening behavior), and viscoelastic properties (time-dependent behavior) (Lidec et al. 2024).
- **Contact Detection**:
  - **Identifying Contact Points**: Algorithm is used to detect potential contact points between the colliding bodies. This involves checking for intersections between the surfaces of the objects (Contact and Collision ArtiSynth Modeling Guide 2024).

7.10.6 Computational Fluid Dynamics (CFD) for Impact on Aerodynamic Surfaces

Computational Fluid Dynamics (CFD) is extensively utilized to analyze solid–solid interactions during aircraft crash or impact loading tests. This involves creating detailed 3D models of the aircraft and the impacting object, followed by mesh generation to discretize the geometry into finite elements. Material properties, including density, elasticity, and plasticity, are assigned to both the aircraft and the impacting object. Advanced algorithms like Separating Axis Theorem (SAT) and GJK (Gilbert-Johnson-Keerthi) algorithm are employed for collision detection and contact response, ensuring accurate force calculations and constraint enforcement to prevent object penetration. Numerical integration methods, such as Euler and Runge–Kutta, are used to update the positions and velocities of the objects over time, ensuring stability and accuracy in the simulations (Yang et al. 2016).

CFD utilizes numerical algorithms and techniques to solve the governing equations of fluid mechanics. Radioss solver is a leading explicit finite element solver that is commonly used for crash and impact simulations, as it can effectively model the equations of mechanics when particles are free from a meshing grid. Additionally, recent research has explored ways to enhance the computational efficiency of CFD solvers by incorporating machine learning techniques. This approach can aid the pressure solver and improve the accuracy and speed of fluid flow analyses, which is particularly important for simulating the effects of material loads and impacts.

Coupled CFD-FEM approaches, such as the Coupled Eulerian–Lagrangian (CEL) and Arbitrary Lagrangian–Eulerian (ALE) methods, are particularly effective in handling fluid–structure interactions during impact events. These methods combine the advantages of both Eulerian and Lagrangian approaches, allowing for flexible handling of large deformations and complex interactions. Software tools like ANSYS Fluent, LS-DYNA, and ABAQUS are commonly used to execute these simulations, providing robust frameworks for analyzing the behavior of aircraft structures under impact conditions. Studies have demonstrated the efficacy of these methods in improving the resilience and safety of aircraft designs, validating the simulation results against experimental data and enhancing our understanding of impact dynamics (Kloss et al. 2012; Yang et al. 2016; Appa et al. 1996).

The CFD simulation can be used to visualize the various stages of the shock tube experiment by studying the pressure contours of the shock wave. This allows for the visualization and analysis of the pressure waves and their impact on the material under test.

7.11 Assessment of Drag and Lift Forces During Impact

The magnitude of drag and lift forces during impact is influenced by several key factors, including velocity, angle of attack, shape and size of the object, and environmental conditions such as fluid properties and surface roughness. These forces significantly affect the overall dynamics of a collision, impacting the trajectory, stability, and energy dissipation of the colliding bodies. Understanding these factors is crucial for optimizing designs in various applications, from aerodynamics to granular material interactions.

- **Structural Damage and Aerodynamic Performance**: Impact events, such as bird strikes, can lead to severe structural damage, which in turn affects aerodynamic performance. For instance, a study on blast-impacted airfoils revealed that damage leads to a decrease in lift and a reduction in the lift-to-drag ratio due to altered pressure distributions on the wing surface (Xiao and Zhou 2023). Similarly, leading edge damage in wind turbine blades, which can be analogous to aircraft wings, results in performance loss due to surface imperfections like waviness and roughness (Meyer Forsting et al. 2023).
- **Impact Location and Force Distribution**: The location of the impact plays a crucial role in determining the contact force and displacement. Research indicates that the contact force can vary significantly based on where the impact occurs, with maximum forces being up to nine times higher at certain locations. Aerodynamic ingestion, or the "vacuum cleaner effect", is a major concern, where foreign objects can be drawn into aircraft engines.

7.12 Thermal Effect Under Impact

When an aircraft experiences an impact, such as a bird strike or debris collision, the thermal effects can significantly influence the structural integrity and performance of the affected components. The energy from the impact is converted into heat, leading to localized temperature rises that can affect the material properties and cause thermal stresses. This phenomenon is particularly critical in high-speed impacts where the kinetic energy is substantial. When rockets and other aerospace vehicles experience an impact, such as during launch, re-entry, or collision with space debris, the thermal effects can be significant. The high kinetic energy involved in these impacts is often converted into heat, leading to localized temperature rises that can affect the material properties and structural integrity of the vehicle. This is particularly critical for components exposed to extreme thermal environments, such as thermal protection systems (TPS) and engine nozzles (Bouslog n.d.).

Thermal aging can lead to surface cracking and a decrease in mechanical properties such as Young's modulus and resistance at break. From spectroscopic studies it has been found that this happens due to the degradation of chemical bonds within the polymer structure (The effect of the thermal aging on polycarbonate—ScienceDirect 2024; Aghilas et al. 2022). Figure 7.8 it is clearly visible that the stress resistance was reducing with prolonged thermal aging (Aghilas et al. 2022). Despite these changes, PC maintains better thermal aging resistance compared to other materials like epoxy resin, retaining higher volume resistivity and breakdown strength after aging (Fig. 7.9, 7.8, 7.9, 7.10, 7.11, 7.12, 7.13 and 7.14).

7.13 Case Studies

7.13.1 Impact Simulation on Aircraft/Spacecraft Windows (Bird Strikes, Foreign Object and Hailstone Impact)

Aircraft windows are subjected to extreme conditions such as high-altitude pressures, temperature changes, and potential debris impact. **Bird strike and FOD** are one of the major terrors for both civil and military airplanes in low altitude. During 2011 to 2014, the ICAO (International Civil Aviation Organization) recorded 65,139 bird strikes (How Bird Strikes Impact Engines | Aviation Week Network 2024). In recent years FAA reported 14,000 bird strikes only in USA each year. Mainly studies have shown that 88% of such incidents occurred below 2500ft in the USA for the last 30 years mostly happening during take-off and landing in the runway. The most vulnerable parts of the aircraft are windshield/canopy, intake of the engine and leading edge of the wing. International authorities like EASA and FAA have some strict condition over choosing material properties of the windshield for an aircraft depending on the size, passengers etc. The regulation says even with air tightness and better visibility it must withstand a 4 lb (1.8 kg) bird impact at cruise speed without penetration

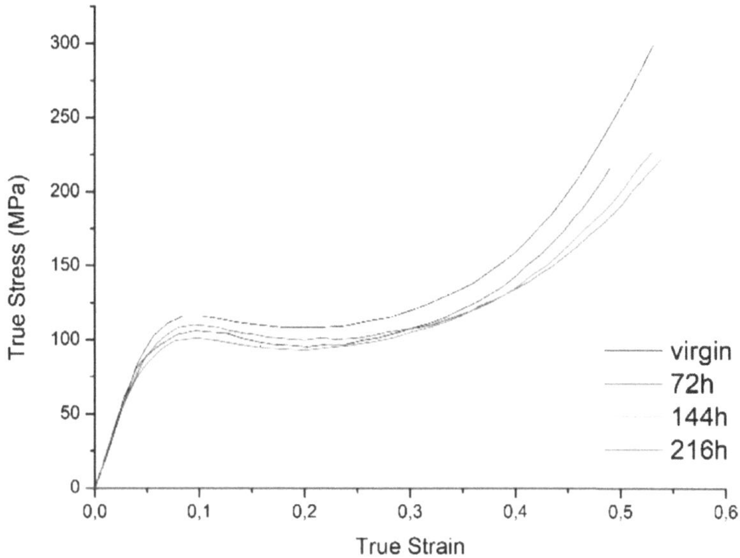

Fig. 7.13 Experimental true stress-true strain curves of a virgin polycarbonate specimen at the following time ageing 72, 144 and 216 h at 100 °C. Adapted from Aghilas et al. (2022)

(FAR/JAR/CS 25.775). Designing the canopy and choosing such material is very costly and time-consuming. Numerical simulation saves the day by meeting design requirement based on the speed, size of the aircrafts and the bird with millions of iterations at a very short time. There are numerous models and among them the designer must choose considering the percentage of error. In Fig. 7.15 an overview is given to incorporate numerical simulation at bird strike phenomena (Figs. 7.16, 7.17 and Table 7.14).

Additionally, numerical methods based on CFD are employed to simulate complex two-phase flow phenomena like rainfall on aircraft/train windshields which showed the effects of flight speed, rainfall intensity, and raindrop temperature on windshield performance (Numerical simulation of rainwater accumulation and flow characteristics over windshield of high-speed trains | Journal of Central South University 2024). These comprehensive analyses will help in designing robust aircraft windows capable of withstanding diverse impact scenarios.

The comparison of the experimental and numerical results reveals close agreement between the two. Particularly relevant is the analysis in terms of energy that shows how, both in the experimental and in the numerical tests, the value of energy lost tends to remain approximately constant and with an error on the calculated energy generally less than 10%. So, it is possible to evaluate the withstanding capability of safety guards by taking into account the amount of energy, depending on the projectile shape, that the plate can absorb. Then, with regard to the purpose of this paper, these results confirm the capability of the numerical model to calculate the solution in a consistent manner.

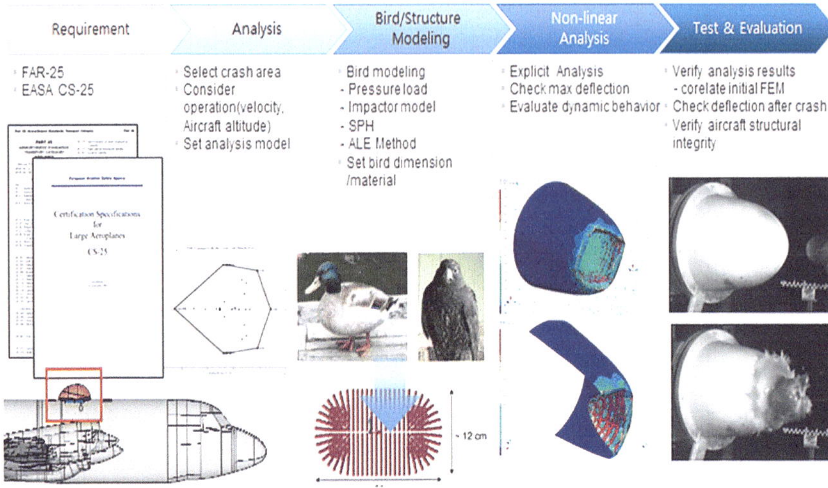

Fig. 7.14 Numerical test procedures of bird strike (Bird-Strike Damage Analysis and Preliminary Design of Composite Radome Structure Using Smoothed Particle Hydrodynamics | Applied Composite Materials 2024)

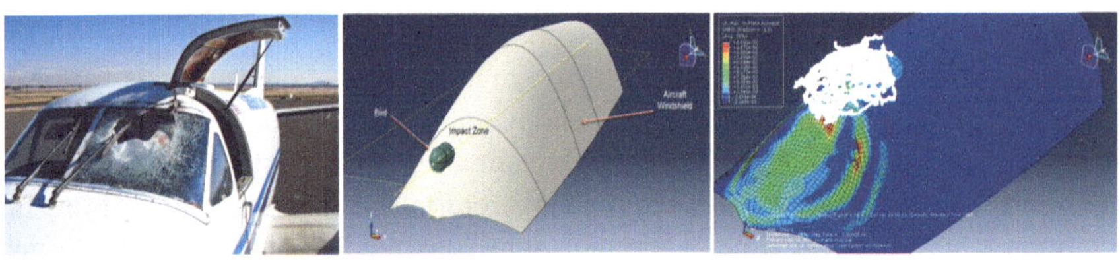

Fig. 7.15 Bird strike on front wind shield with schematic simulation model and Impact Analysis. Picture adapted from +Abaqus (2024)

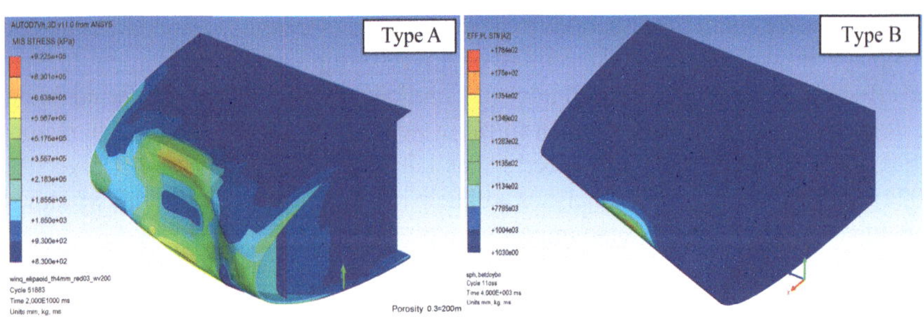

Fig. 7.16 Wing leading edge impact for ellipsoidal bird at 100 m/s impact velocity for 4.0 ms. Adapted from Ugrcic et al. (2015)

Table 7.14 Research involving bird strike/FOD/debris on aircraft windows

Type of impact	Type of polycarbonate	Numerical simulation method	Impact situation	Results/Findings	References
Bird strike	Polycarbonate	SPH + FEM (LS-DYNA)	Bird model in cylindrical shape with hemispherical ends in Polycarbonate target as windshield. For experimental gas gun set-up used	Polycarbonate windshields punctured at speeds of 50 m/s in both experiment and simulation	Ćwiklak et al. (2024), Smetankina et al. (2019)
	PMMA/ Polycarbonate/ PMMA laminate configuration	SPH + FEM (LS-DYNA)	Single and multiple bird strikes at different impact locations (center and edges)	Thicker polycarbonate layers prevent penetration. Laminated PMMA/ Polycarbonate/PMMA configurations exhibit superior resistance. Major stress depends on the initial contact area	Numerical Analysis of Bird Impact Over the Aircraft Windshield|International Journal of Vehicle Structures and Systems (2024), Zhang and Xu (2019), Liu et al. (2012)

(continued)

Table 7.14 (continued)

Type of impact	Type of polycarbonate	Numerical simulation method	Impact situation	Results/Findings	References
FOD impact	Composite plate	Cohesive zone model with 3D FE model in ABAQUS/CAE	Surface-based cohesive zone model is used to find the variation of peanut-shaped delamination with impact energy 1–7.35 J	At low energy impact, this peanut-shaped delamination can be used to design and repair assessment of composite component	Ansari et al. (2024)
	Molecular dynamics simulations with ReaxFF reactive force field	Poly(p-phenylene-terephthalamide)PPTA	Degradation of PPTA is analyzed under atomic oxygen impact by mass loss, temperature spectra, and molecules produced	CO and CO_2 production, temperature distribution varied in different crystal orientations. This model can be used to observe PC or PMMA in spacecraft or aircraft windows in future	Liu et al. (2023)

Based on the excellent results obtained, we can state that the FE analysis can give an important support to the designers of safety guards. The use of accurate numerical models can allow for the approximate prediction of what the ballistic limit of a safety guard might be before experimental testing. Moreover, the numerical simulation can be really helpful for carrying out tests under special conditions not provided by the current standards, such as no perpendicular impact on the plate, the examination of curved design of guards or the impact of projected object with irregular shape.

Using the well-known Recht and Ipson equation as a base for ballistic limit evaluation, a possible new statistical method for evaluating the dispersion of results of impact tests and define a proper confidence interval for ballistic limit is presented.

7.13.2 Impact Analysis on Structural Components

Polycarbonate is commonly used in aircraft canopies and windows, interior panels, cabin components, lighting covers, lenses, instrument panels, display covers, and protective shields. We know that when subjected to impact, polycarbonate panels can absorb significant energy without shattering, often deforming rather than breaking, which helps maintain structural integrity and protect underlying systems. However, even with high resistance to cracking, repeated or extremely high impacts can still lead to crack initiation and propagation. In following topics different structural components will be discussed where polycarbonate composites can be considered because of their material properties.

7.13.3 Impact Analysis on Fuselage

The fuselage of an aircraft is particularly vulnerable to cracks and failures during impact, with specific sections exhibiting higher susceptibility. Research indicates that the fuselage skin, especially around notches such as rivets and holes, is prone to fatigue-induced cracking due to stress concentration under cyclic loading (Aircraft Fuselage Cracking and Simulation—ScienceDirect 2024). Again, the sub-cargo fuselage section has been identified as critical, where bending and twisting of frames can lead to significant structural failure (Mou et al. 024). Through numerical and experimental tests crashworthy designs with composite materials can be developed and validated (Experimental and numerical analysis of composite fuselage panel impacted by tire burst debris under airworthiness clause—ScienceDirect 2024; Rayhan and Pu 2023). Composite materials, such as T800/M21C, are increasingly used in fuselage construction to address these issues. It showed peak strain rate of the fuselage panel during the impact of tire debris is approximately 16 s^{-1} (Lian et al. 2024). Studies showed that PVC foam with aluminum improved energy absorption and reduced stress on the fuselage. It also highlights that challenges of experimental testing due to cost and non-repeatability, as well as the complexity of fuselage structures commercial software codes can be used efficiently for developing crashworthy design solutions (Rayhan and Pu 2023).

Polycarbonate (PC) composites are being studied to be used in aircraft fuselage structures. A constitutive model: ZHU-WANG-TANG was employed to observe mechanical properties under dynamic load and the result was validated with experiment test. Researchers found out the mechanical properties by incorporating composites suitable for withstanding high load optimal composition of Polyethylene Terephthalate (r-PET) and Polycarbonate (PC) ND and organopolysiloxane-polycarbonate block copolymers improves heat aging performance and impact resistance. The yield stress of PC increases with strain rates under dynamic loading conditions (Zhang et al. 2015). Blending PC with other materials, such as Acrylonitrile Butadiene Styrene (ABS) and glass fibers, can further enhance impact performance (Krishna et al. 2011).

However, prolonged exposure to elevated temperatures can degrade PC's mechanical properties, leading to potential vulnerabilities under impact loads. Again, environmental factors, such as immersion in fluids, can further alter its mechanical characteristics, potentially compromising its performance in real-world scenarios.

7.13.4 Impact Analysis on Wing Structures

Polycarbonate (PC) is not typically used in primary structural components of aircraft wings, such as spars or ribs, which require high strength and rigidity—areas where metals like aluminum or advanced metal composites are preferred. However, polycarbonate can be beneficial in non-structural and secondary applications on or around aircraft wings due to its lightweight, impact-resistant, and transparent properties (West et al. 2005). Polycarbonate composites can be used in several portions of an aircraft wing for instance, wingtip navigation light covers, lightweight shielding covers for wing de-icing system, inspection window covers, wing fairings etc. Polycarbonate exhibits high impact strength without adding extra weights makes it ideal for these applications where durability is crucia (Date et al. 2022). By adding ray shielding re-agents or transparent polymeric layers, it can provide near-infrared shielding apabilities and improve thermal efficiency as such it meets General Dynamics Lightweight F-16 aircraft specification (Voss and Raffo 1982). Research indicates that specific types of polycarbonate, like OQ2720, offer enhanced thermal stability, which is vital for wing structures subjected to varying temperatures during fligh0074 (Wang et al. 2014). Even after prolonged exposure to harsh environments, such as ultraviolet radiation in low Earth orbit PC can retain its 70% material properties (Table 7.15).

Table 7.15 The comparison of Mechanical properties after 503 h of UV-C treatment (The Effect of the Ultraviolet Radiation Environment of LEO Upon Polycarbonate Materials 2024)

Property	Polycarbonate sample (Raw)	Sample after 503 Hrs UV-C treatment
Toughness	7.08 MPa	5.380MPA
Young's Modulus	0.875 GPA	1.053 GPa
Ductility	14%	11%
Ultimate Tensile Strength	61.4 MPa	62.6 MPa

7.14 Experimental Correlation

-***Validation of Simulation Results Through Experimental Testing***

Different studies emphasize the importance of experimental testing and numerical simulation to understand how PC reacts under different projectile impacts. Multiple methodologies were explored, including the use of finite element software like LS-DYNA and ABAQUS to model the dynamic behavior of polycarbonate when struck by various projectiles in experiments. For impact testing, most of the papers discuss how factors such as projectile shape, velocity, and PC plate thickness influence penetration depth, energy absorption, and deformation patterns.

Research demonstrated that polycarbonate windshields can withstand bird strikes at speeds up to 50 m/s, with experimental and numerical results showing strong correlation in impact force and deflection (Ćwiklak et al. 2024).

- High-velocity impact tests on PC panels revealed that deformation characteristics are influenced by panel curvature and impact location, emphasizing the need for accurate modeling in design (Mullaoglu et al. 2024).
- The bond-based peridynamic method has been successfully applied to simulate the impact behavior of polycarbonate specimens, showing strong agreement with experimental results by Charpy impact tests in terms of force–time curves and crack shapes proving the reliability of numerical methods in predicting material performance under dynamic loads (Azizi et al. 2022).
- Thermo-mechanical models have been employed to analyze the FSW (Friction Stir welding) analysis of polycarbonate, revealing that increased tool rotational velocity can reduce crack formation during the joining process (Derazkola et al. 2020).

Environmental Factor on Material Properties

Different simulation models, like MIRKHALAF phenomenological model, were validated using Infrared spectroscopy analysis (FTIR) where it indicated the adverse effect of thermal aging for prolonged exposure to high temperatures on polycarbonate components (Aghilas et al. 2022). This highlights the importance of considering environmental factors in simulations.

7.15 Challenges and Limitations

7.15.1 Accuracy of Simulation Models

The accuracy of simulation models for load impact analysis of polycarbonate materials in various industrial applications can vary based on several factors, including the quality of the material data, the chosen constitutive models, and the simulation software used. This involves the use of advanced simulation techniques to predict the material's response to various loads, which is crucial for applications in industries such as aerospace, automotive, and electronics. The following sections delve into the key aspects of simulation models for polycarbonate materials, highlighting the methodologies and findings from recent research.

7.15.2 Constitutive Modeling and Simulation Techniques

- **Tensile Behavior and Constitutive Curves**: Polycarbonate exhibits unique tensile behavior characterized by unconventional deformation mechanisms. Accurate constitutive modeling requires a detailed understanding of these mechanisms, which can be achieved through both experimental and simulation approaches. The true stress–strain curve of PC, derived from experimental tests, serves as a foundation for developing accurate simulation models using both length-based and area-based approaches. The inverse approach used in simulations can effectively replicate the tensile behavior of PC, reducing uncertainties in the constitutive curve estimation (Barbagallo et al. 2024).

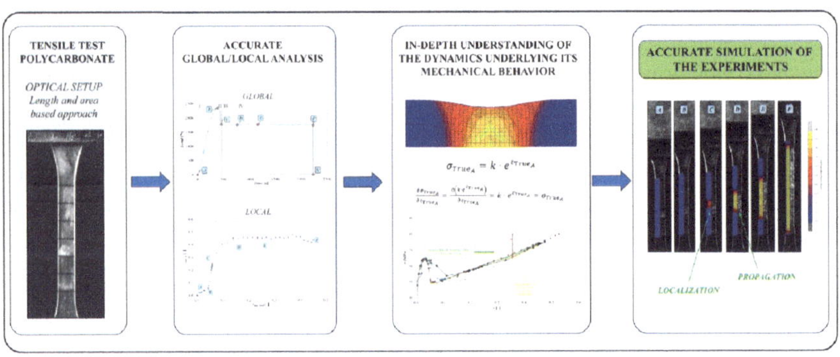

- **Optimization of Constitutive Models**: In the context of impact simulations, the optimization of constitutive models is vital. The use of nonlinear optimization algorithms in constitutive model calibration can significantly reduce experimental requirements while maintaining accuracy. This approach has been shown to reduce errors between experimental and simulated impact strengths to as low as 2.27% for certain polymeric materials

7.15.3 Computational Resources and Time Constraints

The simulation of impact loads on polycarbonate materials involves complex computational resources and time constraints due to the intricate behavior of polycarbonate under dynamic loading conditions. This complexity arises from the material's elasto-plastic nature, temperature and strain rate dependencies, and the need for accurate modeling of deformation and failure mechanisms. The following sections explore the computational aspects and time constraints associated with simulating impact loads on polycarbonate, drawing insights from the provided research papers.

Computational Resources (Model and Techniques):

- **Finite Element Analysis (FEA) Software**: The use of advanced FEA software like LS-DYNA, ABAQUS/Explicit,ANSYS AUTODYN is common for simulating the impact response of polycarbonate plates. These tools allow for detailed modeling of the material's behavior under various impact conditions, including different velocities and projectile types (Mullaoğlu et al. 2016; Fadhel 2011). LS-DYNA, for instance, is used to model the dynamic response of polycarbonate plates, incorporating elastic–plastic material properties to simulate perforation and penetration behaviors.
- **Material Models**: Accurate simulation requires sophisticated material models that capture the complex stress–strain behavior of polycarbonate. Models such as the thermo-elasto-viscoplastic material model and the Piecewise Linear Plasticity (MAT24) model are employed to account for the nonlinear behavior of polycarbonate under impact loading (Low Velocity Impact of Flat and Doubly Curved Polycarbonate Panels 2024). These models are crucial for predicting deformation and failure patterns accurately.

7.16 Future Trends and Innovations

Recent advancements in simulation software have markedly improved the accuracy of impact analyses critical for examining polycarbonate performance in aerospace contexts. Tools such as ANSYS, ABAQUS, and LS-DYNA now feature sophisticated material models for polymers, facilitating precise predictions of impact behavior and optimizing design processes through enhanced computational efficiency and real-time data integration. The incorporation of advanced graphical user interfaces and automation in simulation tools enables researchers to perform intricate simulations with reduced computational times, thus refining design methodologies prior to experimental validation (Dar and Zhang 2015). The incorporation of advanced graphical user interfaces and automation in simulation tools enables researchers to perform intricate simulations with reduced computational times, thus refining design methodologies prior to experimental validation. Again, innovative hybrid techniques, including coupled Eulerian–Lagrangian models are now enhancing accuracy in simulating high-velocity impacts, which is crucial for comprehending strain rate dependence and boundary condition effects on composite panels (Mullaoglu et al. 2024). The forthcoming generation of simulation tools are now aiming to integrate real-time data for immediate design feedback, crucial for adaptive aerospace applications (Application Table: A Bridge Connecting the Designing 'With-The-Material' and 'The-Material' Paradigm's | AIAA SciTech Forum 2024). It aligns with the trend of multiscale modeling and design for example NASA's 2040 Vision, which emphasizes a cyber-physical-social ecosystem to facilitate rapid and cost-effective aerospace material and structure development, thereby enhancing the understanding of material impact behavior and optimizing aerospace engineering design processes (Liu et al. 2040).

7.17 Regulatory Considerations (Compliance and Certification)

The incorporation of polycarbonate materials into aerospace and aviation applications is subject to stringent regulatory oversight. Ensuring compliance with industry standards is crucial for the safety, reliability, and performance of aircraft components subjected to impact events. Compliance involves rigorous testing under controlled conditions to ensure materials can endure operational stresses similar to those experienced in flight. Certification ensures that polycarbonate materials meet all safety and performance requirements.

7.17.1 Compliance with Aerospace Standards for Impact Resistance

The FAA's Title 14 of the Code of Federal Regulations (CFR), particularly the Parts 23, 25, 27, and 29, establishes essential airworthiness criteria for various aircraft categories(14 CFR Part 23—Airworthiness Standards: Normal Category Airplanes 2024). These regulations are essential for materials to withstand specified impact forces. So while working with materials for aviation use like polycarbonate it must meet impact resistance criteria set by authorities like the FAA and EASA. Key regulations and standards include FAA's 14 CFR Part 25.775, which requires aircraft windshields and windows to withstand impacts from birds and other hazards without failure (14 CFR 25.775—Windshields and windows 2024). ASTM D3763 specifies test methods for high-speed puncture properties of plastics, crucial for assessing material's impact resistance (Standard Test Method for High Speed Puncture Properties of Plastics Using Load and Displacement Sensors 2024). SAE AMS-P-83310 details mechanical properties for aerospace-grade polycarbonate materials (AMSP83310: Plastic Sheet, Polycarbonate, Transparent—SAE International 2024), and MIL-STD-810G provides test methods for environmental conditions, including mechanical shock and impact (MIL-STD-810 G ENVIRONMENTAL ENGINEERING LABORATORY 2024).

7.17.2 Certification and Safety Standards in Aerospace Applications

In the context of polycarbonate materials for aerospace use, certification involves demonstrating compliance with a suite of safety standards that address various operational and environmental conditions. Key elements of certification involve environmental testing according to RTCA DO-160G, which outlines methods for evaluating airborne devices under scenarios such as impact and vibration. Flammability regulations in 14 CFR Part 25.853 require that interior materials should self-extinguish and not release toxic gases when burned. The bird strike specifications detailed in FAA 14 CFR Part 25.775 and EASA CS-25.775 comprises that windshields and windows should endure impacts at designated speeds. Following these certifications will guarantee that a material will fulfill all essential safety and performance criteria for impact resistance in aviation uses.

7.18 Conclusion

In conclusion, polycarbonate (PC) materials present a significant opportunity for advancement in aerospace and aviation applications due to their exceptional impact resistance and adaptability under extreme conditions. This review paper has discussed the effectiveness of advanced numerical modeling techniques including finite element analysis (FEA), discrete element methods (DEM), computational fluid dynamics (CFD) etc. to accurately predict PC behavior under diverse impact scenarios such as bird strikes, debris impacts, and high-velocity collisions. The recent incorporation of machine learning techniques emerges as a promising way to further refine simulation accuracy and computational efficiency that can potentially revolutionize design processes with more robust and adaptive solutions. Experimental validations are also crucial to provide deep insights into failure mechanisms, fracture patterns, and energy absorption thereby ensuring the reliability and safety compliance of PC materials in extreme aerospace environments.

Despite these advancements, challenges persist in fully characterizing PC behavior under all possible conditions and in scaling up simulation models for complex real-world applications. Future research should focus on enhancing these models, exploring innovative material modifications such as nano-reinforcements or hybrid composites and fostering interdisciplinary collaborations that bridge material science, computational modeling, and aerospace engineering. The continued evolution of simulation techniques and experimental methodologies will be instrumental in meeting the ever-evolving requirements of aerospace applications, ultimately contributing to safer and more efficient air travel.

References

"14 CFR 25.775—Windshields and windows. Accessed 11 Nov 2024. [Online]. Available: https://www.ecfr.gov/current/title-14/part-25/section-25.775

14 CFR Part 23—Airworthiness Standards: Normal Category Airplanes. Accessed 11 Nov 2024. [Online]. Available: https://www.ecfr.gov/current/title-14/part-23

8 Contact and Collision▶ ArtiSynth Modeling Guide. Accessed 25 Aug 2024. [Online]. Available: https://www.artisynth.org/doc/html/modelguide/Ch8.html

+Abaqus (n.d.) Simulation bird strike impact phenomenon over aircraft cabin in Abaqus | Abaqus tutorials. Accessed 23 Sept 2024. [Online]. Available: https://www.abaqusfem.com/simulation-bird-strike-impact-phenomenon-over-aircraft-cabin-in-abaqus/

Aghilas H, Rabah F, Madjid A (2022) Thermal aging effect on the mechanical behavior of polycarbonate: experimental and modeling. Mater Today Proc 53:202–208. https://doi.org/10.1016/j.matpr.2021.12.551

Ahmed MMZ, Elnaml A, Shazly M, Seleman MME-S (2021) The effect of top surface lubrication on the friction stir welding of polycarbonate sheets. Int Polym Process 36(1):94–102. https://doi.org/10.1515/ipp-2020-3991

Albinmousa J, AlSadah J, Hawwa MA, Al-Qahtani HM (2021) Estimation of mode i fracture of u-notched polycarbonate specimens using the equivalent material concept and strain energy density. Appl Sci 11(8), Art. no. 8. https://doi.org/10.3390/app11083370

Andriß C, Kenf A, Donhauser T, Schmeer S (2022) Characterization and modeling of continuous carbon fiber-reinforced polycarbonate under multiaxial loads. Compos Part B Eng 235:109740. https://doi.org/10.1016/j.compositesb.2022.109740

Ansari MM, Sharad K, Dhayanidhi J, Manju S (2024) An investigation of damage and residual strength of aircraft composite structure caused by foreign object debris impact. In: Velmurugan R, Balaganesan G, Kakur N, Kanny K (eds) Dynamic behavior of soft and hard materials, vol 1. Springer Nature, Singapore, pp 451–458. https://doi.org/10.1007/978-981-99-6030-9_40

Appa K, Argyris J, Guruswamy G (1996) Aircraft dynamics and loads computations using CFD methods. In: 37th structure, structural dynamics and materials conference. American Institute of Aeronautics and Astronautics, Salt Lake City, UT, USA. https://doi.org/10.2514/6.1996-1342

Artero-Guerrero JA, Pernas-Sánchez J, Varas D López-Puente J (2013) Numerical analysis of CFRP fluid-filled tubes subjected to high-velocity impact. Compos Struct 96. https://doi.org/10.1016/j.compstruct.2012.09.020

Azizi MA, Fahad AA, Rahim SA (2022) Peridynamic method for behaviour of polycarbonate specimen in impact test. In: Abdullah S, Karam Singh SS, Md Nor N (eds) Structural integrity cases in mechanical and civil engineering. Springer International Publishing, Cham, pp 29–43. https://doi.org/10.1007/978-3-030-85646-5_3

A two-surface damage-plasticity model based on a Drucker—Prager yield criterion—Zhang—2023—PAMM—Wiley Online Library. Accessed 17 June 2024. [Online]. Available: https://doi.org/10.1002/pamm.202200296

A viscoelastic–plastic constitutive model for uniaxial ratcheting behaviors of polycarbonate—Jiang—2015—Polymer Engineering & Science—Wiley Online Library. Accessed 17 June 2024. [Online]. Available: https://doi.org/10.1002/pen.24148

Accelerating Analysis for Structure Design via Deep Learning Surrogate Models—Shao—2023—Advanced Intelligent Systems—Wiley Online Library. Accessed 21 Sept 2024. [Online]. Available: https://doi.org/10.1002/aisy.202200099

Aircraft Fuselage Cracking and Simulation—ScienceDirect. Accessed 23 Sept 2024. [Online]. Available: https://www.sciencedirect.com/science/article/pii/S2452321620305163?via%3Dihub

Alternative Separation Process of Thermoplastic Polymer for Environmental Assessment | Recent Progress in Science and Technology, vol. 6. Accessed 01 Oct 2024. [Online]. Available: https://stm.bookpi.org/RPST-V6/article/view/9877

AMSP83310: Plastic Sheet, Polycarbonate, Transparent—SAE International. Accessed 11 Nov 2024. [Online]. Available: https://www.sae.org/standards/content/amsp83310/?form=MG0AV3

Application of polycarbonate as ultraviolet photoresist material (2020) | Wu Guangpeng. Accessed 24 Apr 2024. [Online]. Available: https://typeset.io/papers/application-of-polycarbonate-as-ultraviolet-photoresist-47qxkphzac

Application of polycarbonate serving as electron beam photoresist material (2018) | Wu Guangpeng | 1 Citations. Accessed 24 Apr 2024. [Online]. Available: https://typeset.io/papers/application-of-polycarbonate-serving-as-electron-beam-186k9e8cs7

Application Table: A Bridge Connecting the Designing 'With-The-Material' and 'The-Material' Paradigm's | AIAA SciTech Forum. Accessed 11 Nov 2024. [Online]. Available: https://arc.aiaa.org/doi/https://doi.org/10.2514/6.2023-1265

Automotive plastic panel (2015). Accessed 24 Apr2024. [Online]. Available: https://typeset.io/papers/automotive-plastic-panel-32xdi5lgpm

Bachmann R, Michele VD, HEIJL J, COCK F (2022) Method for the preparation of polycarbonate with reduced phosgen excess, EP3719052B1. Accessed 27 July 2024. [Online]. Available: https://patents.google.com/patent/EP3719052B1/en

Ballistic performance of Polycarbonate and Polymethyl methacrylate under normal and inclined dynamic impacts—ScienceDirect. Accessed 28 July 2024. [Online]. Available: https://www.sciencedirect.com/science/article/pii/S2405844021009592

Balaji AB, Pakalapati H, Khalid M, Walvekar R, Siddiqui H (2018) 1—Natural and synthetic biocompatible and biodegradable polymers. In: Shimpi NG (ed) Biodegradable and biocompatible polymer composites, woodhead publishing series in composites science and engineering. Woodhead Publishing, pp 3–32. https://doi.org/10.1016/B978-0-08-100970-3.00001-8

Balamonica K, Goh SH (2019) Characterisation of contact parameters of sand grains to be used for discrete element modelling. E3S Web Conf 92:14002. https://doi.org/10.1051/e3sconf/20199214002

Balasubramanian Kandasubramanian VDJ (2024) Polycarbonate nanocomposites for high impact applications. https://link.springer.com/. Accessed 30 Mar 2024. [Online]. Available: https://doi.org/10.1007/978-981-16-8698-6_22

Barbagallo R, Mirone G, Landi L, Bua G (2024) Tensile behavior of polycarbonate: key aspects for accurate constitutive modelling and simulation. Appl Eng Sci 18:100183. https://doi.org/10.1016/j.apples.2024.100183

Batool SA, Ahmad A, Wadood A, Mateen A, Hussain SW (2018) Development of lightweight aluminum-titanium alloys for aerospace applications. Key Eng Mater 778. https://doi.org/10.4028/www.scientific.net/kem.778.22

Best Material Model for Polycarbonate at High and Low Strain Rates - PolymerFEM.com. Accessed 20 Aug 2024. [Online]. Available: https://polymerfem.com/best-material-model-for-polycarbonate-at-high-and-low-strain-rates/

Bird-Strike Damage Analysis and Preliminary Design of Composite Radome Structure Using Smoothed Particle Hydrodynamics | Applied Composite Materials. Accessed 23 Sept

2024. [Online]. Available: https://link.springer.com/article/https://doi.org/10.1007/s10443-018-9750-9

Bischoff CA, Von Hedenström A (1902) Umsetzungen des Diphenyloxalates. Berichte Dtsch Chem Ges 35(3):3437–3442. https://doi.org/10.1002/cber.190203503159

Bocko J, Nohajová V, Šarloši J (2015) Simulation of material behaviour by bodner-partom model. Am J Mech Eng 3(6), Art. no. 6. https://doi.org/10.12691/ajme-3-6-5

BortoffD SA (2020) Modeling contact and collisions for robotic assembly control. In: Presented at the Proceedings of the american modelica conference, pp 54–63. https://doi.org/10.3384/ecp2016954

Bouslog S (n.d.) High-temperature materials

Brunelle DJ (2005) Advances in polycarbonates: an overview. In: Advances in polycarbonates, vol 898, in ACS symposium series, no 898, vol 898. American Chemical Society, pp 1–5. https://doi.org/10.1021/bk-2005-0898.ch001

Brydson JA (1999) 20—Polycarbonates. In: Brydson JA (ed) Plastics materials, 7th edn. Butterworth-Heinemann, Oxford, pp 556–583. https://doi.org/10.1016/B978-075064132-6/50061-9

Callahan K, Heard WF, Kundu S (2022) High strain rate failure behavior of polycarbonate plates due to hypervelocity impact. Macromolecules 55(21):9640–9649. https://doi.org/10.1021/acs.macromol.2c01151

Cid C, Baldomir A, Rodríguez-Segade M, Hernández S (2022) An open-source framework for aircraft damage simulation in engine failure events. J Comput Sci 62:101682. https://doi.org/10.1016/j.jocs.2022.101682

Charpy impact properties and numerical modeling of polycarbonate composites. https://doi.org/10.1002/mdp2.260

Chemical recycling to monomers: Industrial Bisphenol-A-Polycarbonates to novel aliphatic polycarbonate materials - Liu - 2022 - Journal of Polymer Science - Wiley Online Library. Accessed 26 July 2024. [Online]. Available: https://doi.org/10.1002/pol.20220118

Choi H, Bhowmik P, Rimon T, Bandari Y, Wang Y, Roy S (2024) Exploring the friction and wear mechanisms of nickel titanium alloy fabricated by laser powder and wire-based directed energy deposition. Tribol Trans 67(3):522–537. https://doi.org/10.1080/10402004.2024.2358507

Cleary PW (2004) Large scale industrial DEM modelling. Eng Comput. https://doi.org/10.1108/02644400410519730

Copolymerization of carbon dioxide and epoxide - Inoue - 1969 - Journal of Polymer Science Part B: Polymer Letters - Wiley Online Library. Accessed 27 July 2024. [Online]. Available: https://doi.org/10.1002/pol.1969.110070408

Costa P et al (2021) Polycarbonate based multifunctional self-sensing 2D and 3D printed structures for aeronautic applications. Smart Mater Struct 30:085032. https://doi.org/10.1088/1361-665X/ac0cbe

Couto Carneiro AM, Carvalho Alves AF, Cardoso Coelho RP, Cardoso JS, Andrade Pires FM (2023) A simple machine learning-based framework for faster multi-scale simulations of path-independent materials at large strains. Finite Elem Anal Des 222: 103956. https://doi.org/10.1016/j.finel.2023.103956

Cundall PA, Strack ODL (1979) A discrete numerical model for granular assemblies. Géotechnique 29(1):47–65. https://doi.org/10.1680/geot.1979.29.1.47

Ćwiklak J, Gołda P, Goś A, Kobiałka E, Krasuski K (2024) Experimental and numerical investigations on polycarbonate samples and rotorcraft windshield for bird model impacts. Arch Transp 70(2), Art. no. 2. https://doi.org/10.61089/aot2024.eea5zs04

[citation report] A model of low-velocity impact damage assessment of laminated composite structures. Accessed 05 May 2024. [Online]. Available: https://scite.ai/reports/a-model-of-low-velocity-impact-J1KwzRY

Dar UA, Zhang W (2015) Polymer based aerospace structures under high velocity impact applications; experimental, constitutive and finite element analysis. J Mech Sci Technol 29(10):4259–4265. https://doi.org/10.1007/s12206-015-0922-3

Date S, Abe Y, Okabe T (2022) Effects of fiber properties on aerodynamic performance and structural sizing of composite aircraft wings. Aerosp Sci Technol 124:107565. https://doi.org/10.1016/j.ast.2022.107565

De Macedo TCP et al (2019) Effect of processing conditions on the mechanical and morphological properties of elastomeric poly (Methyl Methacrylate)/polycarbonate blends. Macromol Symp 383(1):1800049. https://doi.org/10.1002/masy.201800049

Decorative Chromium Coatings on Polycarbonate Substrate for the Automotive Industry (2023) Materials 16(6):2315–2315. https://doi.org/10.3390/ma16062315

Degradable polycarbonate sport fishing materials (2017) Accessed 25 Apr 2024. [Online Video]. Available: https://typeset.io/papers/degradable-polycarbonate-sport-fishing-materials-4crupaj9h0

Derazkola HA, Eyvazian A, Simchi A (2020) Modeling and experimental validation of material flow during FSW of polycarbonate. Mater Today Commun 22:100796. https://doi.org/10.1016/j.mtcomm.2019.100796

Dimitrellou S, Iakovidis I, Psarianos D-R (2024) Mechanical characterization of polylactic acid, polycarbonate, and carbon fiber-reinforced polyamide specimens fabricated by fused deposition modeling. J Mater Eng Perform 33(7):3613–3626. https://doi.org/10.1007/s11665-024-09144-9

Discrete element method to simulate continuous material by using the cohesive beam model—ScienceDirect. Accessed 19 June 2024. [Online]. Available: https://www.sciencedirect.com/science/article/abs/pii/S0045782511003811?via%3Dihub

Doubrava R, Oberthor M, Bělský P, Raška J (2018) Bird and hail stone impact resistance analysis on a jet engine composite air inlet. Matec Web Conf. https://doi.org/10.1051/matecconf/201818804006

Drucker-Prager plasticity model in the framework of OSB-PD theory with shear deformation | Engineering with Computers. Accessed 17 June 2024. [Online]. Available: https://doi.org/10.1007/s00366-021-01527-z

Ebert WD et al (2009) Process for the preparation of polycarbonate by the interfacial process, DE102008012613A1. Accessed 27 July 2024. [Online]. Available: https://patents.google.com/patent/DE102008012613A1/en

Emerging Applications of Polymers for Automobile Industries (2022). Accessed 24 Apr 2024. [Online]. Available: https://typeset.io/papers/emerging-applications-of-polymers-for-automobile-industries-m3q83avz

Enhancement of polycarbonate membrane permeability due to plasma polymerization precursors - ScienceDirect. Accessed 01 Oct 2024. [Online]. Available: https://www.sciencedirect.com/science/article/abs/pii/S0169433212021113?via%3Dihub

Experimental and numerical analysis of composite fuselage panel impacted by tire burst debris under airworthiness clause - ScienceDirect. Accessed 23 Sept 2024. [Online]. Available: https://www.sciencedirect.com/science/article/abs/pii/S0734743X23003159?via%3Dihub

Fadhel BM (2011) Numerically study of ballistic impact of polycarbonate. In: 2011 international symposium on humanities, science and engineering research, pp 101–105. https://doi.org/10.1109/SHUSER.2011.6008479

Færgestad RM, Holmen JK, Berstad T, Cardone T, Ford KA, Børvik T (2023) Coupled finite element-discrete element method (FEM/DEM) for modelling hypervelocity impacts. Acta Astronaut 203:296–307. https://doi.org/10.1016/j.actaastro.2022.11.026

Fukuoka S, Fukawa I, Adachi T, Fujita H, Sugiyama N, Sawa T (2019) Industrialization and expansion of green sustainable chemical process: a review of non-phosgene polycarbonate from CO_2. Org Process Res Dev 23(2):145–169. https://doi.org/10.1021/acs.oprd.8b00391

Gao X, Isayev AI, Yi C (2016) Ultrasonic treatment of polycarbonate/carbon nanotubes composites. Polymer 84:209–222. https://doi.org/10.1016/j.polymer.2015.12.051

Ghavanini N, Caporale AM, Astori P, Airoldi A, Panichelli P (2023) Impact response of monolithic and laminated polycarbonate panels: an experimental and numerical investigation. Polymers 15(24), Art. no. 24. https://doi.org/10.3390/polym15244677

Goryl K, Pollák M, Kočiško M, Korol M (2024) Comparison of PLA and PLA carbon fiber materials on tensile test. In: Gapiński B, Ciszak O, Ivanov V, Machado JM (eds) Advances in manufacturing IV. Springer Nature Switzerland, Cham, pp 73–82. https://doi.org/10.1007/978-3-031-56463-5_6
Hamouz OCSA, Sweileh BA, Al-Salah HA (2006) Synthesis and characterization of polycarbonates by melt-phase interchange reactions with alkylene and arylene diphenyl dicarbonates. J Appl Polym Sci 102(4). https://doi.org/10.1002/app.23996
Hassan M, Bhat GA, Darensbourg DJ (2024) Post-polymerization functionalization of aliphatic polycarbonates using click chemistry. Polym Chem 15(18):1803–1820. https://doi.org/10.1039/D4PY00174E
High fidelity computational characterization of the mechanical response of thermally aged polycarbonate—IOPscience. Accessed 18 June 2024. [Online]. Available: https://doi.org/10.1088/1361-651X/aa7341
High-impact-resistance aging-resistant anti-static transparent polycarbonate polymer and preparation method and application thereof (2015) Accessed 25 Apr 2024. [Online Video]. Available: https://typeset.io/papers/high-impact-resistance-aging-resistant-anti-static-1m4jn3ngxu
How Bird Strikes Impact Engines | Aviation Week Network. Accessed 16 July 2024. [Online]. Available: https://aviationweek.com/mro/how-bird-strikes-impact-engines
Huang X, Lewis S, Brittain WJ, Vaia RA (2000) Synthesis of polycarbonate-layered silicate nanocomposites via cyclic oligomers. Macromolecules 33(6):2000–2004. https://doi.org/10.1021/ma991709x
Huang Z, Liu Q (2024) Impact modified polycarbonate composition. WO2017046126A1. Accessed 10 Sept 2024. [Online]. Available: https://patents.google.com/patent/WO2017046126A1/en
Hybrid Physics and Machine Learning Modeling for Material Characterization and Failure Analysis | AIAA SciTech Forum. Accessed 21 Sept 2024. [Online]. Available: https://doi.org/10.2514/6.2024-1526
Ibrahim J, Bhowmik S (2024) Novel recycling of plastic waste into high impact resistant ultralightweight thermally insulated hybrid composite. J Thermoplast Compos Mater 37(4):1462–1479. https://doi.org/10.1177/08927057231196079
Improvement of Finite Element Solutions with Deep Learning (2022) Lecture notes numerical methods engineering and applied sciences, pp 139–166. https://doi.org/10.1007/978-3-031-11847-0_5
Implicit versus Explicit Finite Element Analysis: When to Use Which? Accessed 20 Aug 2024. [Online]. Available: https://simulation-blog.technia.com/simulation/implicit-vs-explicit-finite-element-analysis
Industrialization of automotive glazing by polycarbonate and hard-coating (2019) Polym J 51(12):1249–1263. https://doi.org/10.1038/S41428-019-0240-1
Jaya Christiyan KG, Chandrasekhar U, Venkateswarlu K (2018) Flexural properties of pla components under various test condition manufactured by 3D printer. J Inst Eng India Ser C 99(3):363–367. https://doi.org/10.1007/s40032-016-0344-8
Jia Y, Asahara H, Hsu Y-I, Asoh T-A, Uyama H (2020) Surface modification of polycarbonate using the light-activated chlorine dioxide radical. Appl Surf Sci 530:147202. https://doi.org/10.1016/j.apsusc.2020.147202
Joglekar S, Ranatunga V, Pankow M (2021) Validation of an efficient finite element analysis approach for simulation of low velocity impact and compression strength after impact response. Compos Struct 255:112945. https://doi.org/10.1016/j.compstruct.2020.112945
Juracka D, Kawulok M, Bujdos D, Krejsa M (2023) Mechanical properties of composite samples produced by FFF/FDM 3D printing technology in a three-point flexural test. AIP Conf Proc 2950(1):020007. https://doi.org/10.1063/5.0182644
Kausar A (2017) A review of filled and pristine polycarbonate blends and their applications. J Plast Film Amp Sheeting 34(1). https://doi.org/10.1177/8756087917691088
Kember MR, Buchard A, Williams CK (2011) Catalysts for CO_2/epoxide copolymerization. Chem Commun 47(1). https://doi.org/10.1039/c0cc02207a

Kendall M, Siviour C (2013) Experimentally simulating adiabatic conditions: approximating high rate polymer behavior using low rate experiments with temperature profiles. Polymer 54:5058–5063. https://doi.org/10.1016/j.polymer.2013.06.049
Kendall MJ, Siviour CR (2014) Experimentally simulating high-rate behaviour: rate and temperature effects in polycarbonate and PMMA. Philos Trans R Soc Math Phys Eng Sci 372(2015):20130202. https://doi.org/10.1098/rsta.2013.0202
Khan ZI, Mohamad Z, Rahmat AR, Habib U (2021) Synthesis and characterization of composite materials with enhanced thermo-mechanical properties for unmanned aerial vehicles (Uavs) and aerospace technologies. Pertanika J Sci Technol 29(3). https://doi.org/10.47836/pjst.29.3.15
Kloss C, Goniva C, Hager A, Amberger S, Pirker S (2012) Models, algorithms and validation for opensource DEM and CFD-DEM. Prog Comput Fluid Dyn Int J. https://doi.org/10.1504/pcfd.2012.047457
Knauss WG, Zhu W (2002) Nonlinearly viscoelastic behavior of polycarbonate. II. The role of volumetric strain. Mech Time-Depend Mater 6:301–322. https://doi.org/10.1023/A:1021296110580
Kou M, Han D, Xiao C, Wang Y (2019) Dynamic fracture instability in brittle materials: Insights from DEM simulations. Struct Eng Mech 71(1), Art. no. 1
Krishna TP, Srinivasababu NSN, Babu NR, Adhikari SAS (2011) Assessing the impact performance of PC/ABS blends on addition of PC with glass. Indian J Appl Res 3(2):121–122. https://doi.org/10.15373/2249555X/FEB2013/41
Krop S, Meijer HEH, van Breemen LCA (2016) Global and local large-deformation response of sub-micron, soft- and hard-particle filled polycarbonate. J Mech Phys Solids 87:51–64. https://doi.org/10.1016/j.jmps.2015.11.005
Leelaprachakul T, Kubo A, Umeno Y (2024) Degradation of molecular structure and residual strength of polycarbonate under cyclic loading: insights from coarse-grained molecular dynamics simulation. Comput Mater Sci 240:113028. https://doi.org/10.1016/j.commatsci.2024.113028
Li J, Wang X, Yang C, Yang J, Wang Y, Zhang J (2016) Toughening modification of polycarbonate/poly(butylene terephthalate) blends achieved by simultaneous addition of elastomer particles and carbon nanotubes. Compos Part Appl Sci Manuf 90:200–210. https://doi.org/10.1016/j.compositesa.2016.07.006
Li B, Ma S, Li F, Tang T, Yang S, Zhou J (2024) Finite element simulation analysis of impact signal generation under different conditions. In: Fourth international conference on mechanical, electronics, and electrical and automation control (METMS 2024). SPIE, pp 1398–1402. https://doi.org/10.1117/12.3030327
Lian C et al (2024) Experimental and numerical analysis of composite fuselage panel impacted by tire burst debris under airworthiness clause. Int J Impact Eng 183:104805. https://doi.org/10.1016/j.ijimpeng.2023.104805
Lidec QL, Jallet W, Montaut L, Laptev I, Schmid C, Carpentier J (2024) Contact models in robotics: a comparative analysis. arXiv.2304.06372. Accessed 24 Aug 2024. [Online]. Available: http://arxiv.org/abs/2304.06372
Liu J, Han P (2018) Numerical analyses of a shield building subjected to a large commercial aircraft impact. Shock Vib. https://doi.org/10.1155/2018/7854969
Liu LZ, Ma ZZ, Pang SH, Zang SG, Ou YC, Zuo Y (2012) Investigation of the bird shape effects on the simulation results of bird striking laminated aircraft windshields. Adv Mater Res 418–420:72–76. https://doi.org/10.4028/www.scientific.net/AMR.418-420.72
Liu H et al (2020) A three-dimensional elastic-plastic damage model for predicting the impact behaviour of fibre-reinforced polymer-matrix composites. Compos Part B Eng 201:108389. https://doi.org/10.1016/j.compositesb.2020.108389
Liu L, Miao C, Song D, Shi J, Liu W (2023) A reactive molecular dynamics simulation of the atomic oxygen impact on poly(p-phenylene-terephthalamide). Comput Mater Sci 228:112341. https://doi.org/10.1016/j.commatsci.2023.112341

Liu X, Furrer D, Kosters J, Holmes J (2018) Vision 2040: A Roadmap for Integrated, Multiscale Modeling and Simulation of Materials and Systems, E-19477. Accessed 11 Nov 2024. [Online]. Available: https://ntrs.nasa.gov/citations/20180002010

Liu J et al (2019) The performance of thermoplastic and thermoset composites subjected to low-velocity and high-velocity impact loading

Low Velocity Impact of Flat and Doubly Curved Polycarbonate Panels | J. Appl. Mech. | ASME Digital Collection. Accessed 14 Sep 2024. [Online]. Available: https://asmedigitalcollection.asme.org/appliedmechanics/article-abstract/82/4/041003/443709/Low-Velocity-Impact-of-Flat-and-Doubly-Curved?redirectedFrom=fulltext

Maamoun A, Xue Y, Elbestawi MA, Veldhuis SC (2018) Effect of selective laser melting process parameters on the quality of al alloy parts: powder characterization, density, surface roughness, and dimensional accuracy. Materials 11(12). https://doi.org/10.3390/ma11122343

Mahood JA, Gautam PS (2015) Process to prepare high heat polycarbonates, WO2015048735A1. Accessed 27 July 2024. [Online]. Available: https://patents.google.com/patent/WO2015048735A1/en

Malek-Mohammadi H, Majzoobi GH, Payandehpeyman J (2019) Mechanical characterization of polycarbonate reinforced with nanoclay and graphene oxide. Polym Compos 40(10):3947–3959. https://doi.org/10.1002/pc.25256

Martin L et al (2024) A systematic comparison between FEBio and PolyFEM for biomechanical systems. Comput Methods Programs Biomed 244:107938. https://doi.org/10.1016/j.cmpb.2023.107938

McBride A, Govender I, Powell M, Cloete TJ (2004) Contributions to the experimental validation of the discrete element method applied to tumbling mills. Eng Comput. https://doi.org/10.1108/02644400410519703

Mercier JP, Aklonis JJ, Litt M, Tobolsky AV (1965) Viscoelastic behavior of the polycarbonate of bisphenol A. J Appl Polym Sci 9(2):447–459. https://doi.org/10.1002/app.1965.070090206

Merema BJ, Kraeima J, Glas HH, Spijkervet FKL, Witjes MJH (2020) Patient-specific finite element models of the human mandible: lack of consensus on current set-ups. Oral Dis 27(1). https://doi.org/10.1111/odi.13381

Meyer Forsting A, Olsen AS, Srensen NN, Bak C (2023) The impact of leading edge damage and repair on sectional aerodynamic performance. In: AIAA SCITECH 2023 Forum, in AIAA SciTech Forum. American Institute of Aeronautics and Astronautics. https://doi.org/10.2514/6.2023-0968.

Mills NJ, Wilkes S, Derler S, Flisch A (2009) FEA of oblique impact tests on a motorcycle helmet. Int J Impact Eng. https://doi.org/10.1016/j.ijimpeng.2008.12.011

Mindlin RD (1964) Micro-structure in linear elasticity. Arch Ration Mech Anal 16(1):51–78. https://doi.org/10.1007/BF00248490

Milbradt P, Abed WA (2008) Generalized stabilization techniques in computational fluid dynamics. https://www.researchgate.net/publication/266465949_GENERALIZED_STABILIZATION_TECHNIQUES_IN_COMPUTATIONAL_FLUID_DYNAMICS

Mirone G, Barbagallo R, Bua G (2024) Static and dynamic tensile behaviour of polycarbonate connected to the deforming specimen shape. IOP Conf Ser Mater Sci Eng 1306(1):012006. https://doi.org/10.1088/1757-899X/1306/1/012006

Mou H, Xie J, Feng Z, Shi X (2024) Review on the crashworthiness design and evaluation of fuselage structure for occupant survivability. Prog Aerosp Sci 148:101001. https://doi.org/10.1016/j.paerosci.2024.101001

Mullaoğlu F, Usta F, Türkmen HS, Kazancı Z, Balkan D, Akay E (2016) Deformation behavior of the polycarbonate plates subjected to impact loading. Procedia Eng. 167:143–150. https://doi.org/10.1016/j.proeng.2016.11.681

Mullaoglu F, Usta F, Turkmen HS (2024) The impact-induced deformation characteristics of polycarbonate panels. Polym Eng Sci 64(6):2579–2593. https://doi.org/10.1002/pen.26711

Machine learning based modeling of path-dependent materials for finite element analysis — ScienceDirect. Accessed 21 Sept 2024. [Online]. Available: https://www.sciencedirect.com/science/article/pii/S0266352X23000113?via%3Dihub
Material modeling and solid phase forming of polycarbonate sheet - Lee - 1986 - Polymer Engineering & Science—Wiley Online Library. Accessed 17 June 2024. [Online]. Available: https://doi.org/10.1002/pen.760260906
METHOD FOR MODIFYING POLYCARBONATE | TREA. Accessed: Sep. 10, 2024. [Online]. Available: https://trea.com/information/method-for-modifying-polycarbonate/patent application/31f5e704-a946-48b7-a29e-c495a44a38fc
MIL-STD-810 G ENVIRONMENTAL ENGINEERING LABORATORY. Accessed 11 Nov 2024. [Online]. Available: http://everyspec.com/MIL-STD/MIL-STD-0800-0899/MIL-STD-810G_1 2306/
Narijauskaitė B, Palevičius A, Gaidys R, Janušas G, Šakalys R (2013) Polycarbonate as an elasto-plastic material model for simulation of the microstructure hot imprint process. Sensors 13(9), Art. no. 9. https://doi.org/10.3390/s130911229
Nassauer B, Liedke T, Kuna M (2013) Polyhedral particles for the discrete element method. Granul Matter 15(1):85–93. https://doi.org/10.1007/s10035-012-0381-9
New Discrete Element Models for Three-Dimensional Impact Problems-IOPscience. Accessed 19 June 2024. [Online]. Available: https://doi.org/10.1088/0256-307X/26/12/120202
Nishi Y et al (2018) Increasing Charpy Impact Value of Polycarbonate (PC) Sheets irradiated by electron beam. Mater Trans 59(8). https://doi.org/10.2320/matertrans.m2018095.
Novel aliphatic polycarbonate (2020). Accessed 24 Apr 2024. [Online]. Available: https://typeset.io/papers/novel-aliphatic-polycarbonate-4etks0uz6c
Novel environment-friendly and flame-retardant polycarbonate resin material for electric equipment (2016). Accessed 24 Apr 2024. [Online]. Available: https://typeset.io/papers/novel-environment-friendly-and-flame-retardant-polycarbonate-1bi9h5cagd
Novel polycarbonate block amide copolymer as well as preparation method and application of polycarbonate block amide copolymer (2017) | Dong Xia. Accessed 24 Apr 2024. [Online]. Available: https://typeset.io/papers/novel-polycarbonate-block-amide-copolymer-as-well-as-2j1h8cvjhh
Numerical analysis of ballistic impact through FE and SPH methods | AVESİS. Accessed 21 Sep 2024. [Online]. Available: https://avesis.metu.edu.tr/yayin/ace72cb6-6374-445b-baee-271930cfeec8/numerical-analysis-of-ballistic-impact-through-fe-and-sph-methods
Numerical Analysis of Bird Impact Over the Aircraft Windshield | International Journal of Vehicle Structures and Systems. Accessed 02 July 2024. [Online]. Available: https://yanthrika.com/eja/index.php/ijvss/article/view/1906
Numerical simulation of rainwater accumulation and flow characteristics over windshield of high-speed trains | Journal of Central South University. Accessed 02 July 2024. [Online]. Available: https://doi.org/10.1007/s11771-020-4288-z
Ogończyk D, Jankowski P, Garstecki P (2020) A method for simultaneous polishing and hydrophobization of polycarbonate for microfluidic applications. Polymers 12(11), Art. no. 11. https://doi.org/10.3390/polym12112490
Oñate E et al (2017) Advances in the DEM and coupled DEM and FEM techniques in non linear solid mechanics. https://doi.org/10.1007/978-3-319-60885-3_15
Ono K, Erhard A (2011) Ullmann's encyclopedia of industrial chemistry. https://doi.org/10.1002/14356007.017_002
(Open Access) Study by Thermal Methods of Physico-Mechanical Properties of Polycarbonate used for High Performance Sport Products (2008) | Radu Emil Mărie | 1 Citations. Accessed 25 Apr 2024. [Online]. Available: https://typeset.io/papers/study-by-thermal-methods-of-physico-mechanical-properties-of-26qoc4ag6c
(PDF) Kinked crack paths in polycarbonate samples printed by fused deposition modelling using criss-cross patterns. Accessed 17 June 2024. [Online]. Available: https://www.researchgate.net/publication/350029283_Kinked_crack_paths_in_polycarbonate_samples_printed_by_fused_deposition_modelling_using_criss-cross_patterns?_tp=eyJjb250ZXh0Ijp7ImZpcnN0UGFnZSI6InB1YmxpY2F0aW9uIiwicGFnZSI6InNlYXJjaCIsInBvc2l0aW9uIjoicGFnZUhlYWRlciJ9fQ

Particle Modeling Based on Interatomic Potential and Crystal Structure: A Multi-Scale Simulation of Elastic-Plastic Deformation of Metallic Material. Accessed 20 June 2024. [Online]. Available: https://www.scirp.org/journal/paperinformation?paperid=111302

Pasha M, Hare C, Ghadiri M, Gunadi A, Piccione PM (2016) Effect of particle shape on flow in discrete element method simulation of a rotary batch seed coater. Powder Technol. https://doi.org/10.1016/j.powtec.2015.10.055

Patil N, Bhoopathi S, Chidara V, Hadjichristidis N, Gnanou Y, Feng X (2020) Recycling a borate complex for synthesis of polycarbonate polyols: towards an environmentally friendly and cost-effective process. Chemsuschem 13(18):5080–5087. https://doi.org/10.1002/cssc.202001395

Photo-on-Demand In Situ Phosgenation Reactions That Cross Three Phases of a Heterogeneous Solution of Chloroform and Aqueous NaOH | ACS Omega. Accessed 27 July 2024. [Online]. Available: https://doi.org/10.1021/acsomega.3c04290

Physical characterizations of three phase polycarbonate nanocomposites - SS Ibrahim, Abdullah A Al Jaafari, Ayman S Ayesh, 2011. Accessed 15 July 2024. [Online]. Available: https://doi.org/10.1177/8756087911410592

Pogorelko VV, Mayer AE, Fomin EV, Fedorov EV (2024) Examination of machine learning method for identification of material model parameters. Int J Mech Sci 265:108912. https://doi.org/10.1016/j.ijmecsci.2023.108912

Polycarbonate as well as preparation method and application thereof (2019) | Wang Yamin. Accessed 24 Apr 2024. [Online]. Available: https://typeset.io/papers/polycarbonate-as-well-as-preparation-method-and-application-45beq9qqfk

Polycarbonate block copolymers and methods thereof (2020) | Scott D. Allen. Accessed 24 Apr 2024. [Online]. Available: https://typeset.io/papers/polycarbonate-block-copolymers-and-methods-thereof-55avd3g7zz

Polycarbonate composition with excellent wear resistance and low-temperature impact resistance and preparation method of polycarbonate composition (2019) | Liu Weiwei. Accessed 24 Apr 2024. [Online]. Available: https://typeset.io/papers/polycarbonate-composition-with-excellent-wear-resistance-and-lcdz87qaso

Polycarbonate film applied to speaker network for electronic appliances and manufacturing method thereof (2010) | Wei Luo | 3 Citations. Accessed 24 Apr 2024. [Online]. Available: https://typeset.io/papers/polycarbonate-film-applied-to-speaker-network-for-electronic-14tgmtb5ej

Polycarbonate material selectively shielding near-infrared ray, preparation method, and application thereof (2019). Accessed 24 Apr 2024. [Online]. Available: https://typeset.io/papers/polycarbonate-material-selectively-shielding-near-infrared-3m9jxus18l

Polycarbonate, preparation method therefor and application thereof (2020). Accessed 24 Apr 2024. [Online]. Available: https://typeset.io/papers/polycarbonate-preparation-method-therefor-and-application-13bqv54h9x

Polycarbonate-based composite material and preparation method and application thereof, Jan. 2019, Accessed 24 Apr 2024. [Online]. Available: https://typeset.io/papers/polycarbonate-based-composite-material-and-preparation-5bujjhe3we

Philipp Wellkamp, "Reduction of epistemic uncertainty of a crash box model - experimental and numerical investigations (2019) | Philipp Wellkamp | 1 Citations. Accessed 12 May 2024. [Online]. Available: https://typeset.io/papers/reduction-of-epistemic-uncertainty-of-a-crash-box-model-rhm9zujkek

Polycarbonates Derived from Glucose via an Organocatalytic Approach | Journal of the American Chemical Society. Accessed 27 July 2024. [Online]. Available: https://pubs.acs.org/doi/https://doi.org/10.1021/ja402319m

Pour RH, Hassan A, Soheilmoghaddam M, Bidsorkhi HC (2016) Mechanical, thermal, and morphological properties of graphene reinforced polycarbonate/acrylonitrile butadiene styrene nanocomposites. Polym Compos 37(6):1633–1640. https://doi.org/10.1002/pc.23335

Rafique MMA (2015) Design of an impact testing machine for polymer films by the free falling dart procedure. Mater. Test. 57(9):799–805. https://doi.org/10.3139/120.110779

Rahner C, Al-Qureshi HA, Stainer D, Hotza D, Fredel MC (2014) Numerical evaluation of a light-gas gun facility for impact test. Model Simul Eng 2014(1):501434. https://doi.org/10.1155/2014/501434

Ramakrishnan V et al (2024) Improved impact compositions. EP3765568A1. Accessed 13 Sept 2024. [Online]. Available: https://patents.google.com/patent/EP3765568A1/en

Rayhan SB, Pu X(2023) Crashworthiness study of a newly developed civil aircraft fuselage section with auxiliary fuel tank reinforced with composite foam. Aerospace 10(3), Art. no. 3. https://doi.org/10.3390/aerospace10030314

Recchia G, Magnanimo V, Cheng H, La Ragione L (2020) DEM simulation of anisotropic granular materials: elastic and inelastic behavior. Granul Matter 22(4):85. https://doi.org/10.1007/s10035-020-01052-8

Research on Thin Glass-Polycarbonate Composite Panels (2018) vol 6, pp 271–280. https://doi.org/10.7480/CGC.6.2148

Rodriguez F, Cohen F, Ober CK, Archer L (2003) Principles of polymer systems, 5th edn. CRC Press, Boca Raton. https://doi.org/10.1201/b12837

Rokaya D, Srimaneepong V, Sapkota J, Qin J, Siraleartmukul K, Siriwongrungson V (2018) Polymeric materials and films in dentistry: an overview. J Adv Res 14. https://doi.org/10.1016/j.jare.2018.05.001

Roy R, Sarkar B, Rana A, Bose N (2001) Impact fatigue behaviour of carbon fibre-reinforced vinylester resin composites. Bull Mater Sci 24:79–86. https://doi.org/10.1007/BF02704844

Safety glass with improved impact resistance and safety strength and method for manufacturing the same. Accessed 24 Apr 2024. [Online]. Available: https://typeset.io/papers/safety-glass-with-improved-impact-resistance-and-safety-4cp2p3gsum

Sarva S, Mulliken AD, Boyce MC (2007) Mechanics of Taylor impact testing of polycarbonate. Int J Solids Struct 44(7):2381–2400. https://doi.org/10.1016/j.ijsolstr.2006.07.012

Separation and Solvent Based Material Recycling of Polycarbonate from Electronic Waste | ACS Sustainable Chemistry & Engineering. Accessed 01 Oct 2024. [Online]. Available: https://doi.org/10.1021/acssuschemeng.3c03152

Shilko EV, Smolin AY, Dimaki AV, Eremina GM (2021) Particle-based approach for simulation of nonlinear material behavior in contact zones. In: Ostermeyer G-P, Popov VL, Shilko EV, Vasiljeva OS (eds) Multiscale biomechanics and tribology of inorganic and organic systems: in memory of professor sergey psakhie. Springer International Publishing, Cham, pp 67–89. https://doi.org/10.1007/978-3-030-60124-9_4

Shyamsunder L, Maurya A, Rajan SD, Cordasco D, Revilock D, Blankenhorn G (2022) Impact simulation of composite panels for aerospace applications. Compos Part B Eng 247:110320. https://doi.org/10.1016/j.compositesb.2022.110320

Smetankina N, Malykhina A, Merkulov D (2019) Simulating of bird strike on aircraft laminated glazing. MATEC Web Conf. 304:01010. https://doi.org/10.1051/matecconf/201930401010

Soloukhin VA, Brokken-Zijp JCM, van Asselen OLJ, de With G (2003) Physical aging of polycarbonate: elastic modulus, hardness, creep, endothermic peak, molecular weight distribution, and infrared data. Macromolecules 36(20):7585–7597. https://doi.org/10.1021/ma0342980

Standard Test Method for High Speed Puncture Properties of Plastics Using Load and Displacement Sensors. Accessed 11 Nov 2024. [Online]. Available: https://www.astm.org/d3763-18.html

Stecconi A, Landi L (2020) Finite element analysis for impact tests on polycarbonate safety guards: comparison with experimental data and statistical dispersion of ballistic limit. ASCE-ASME J Risk Uncertain Eng Syst Part B Mech Eng 6(4):041004. https://doi.org/10.1115/1.4047464

Successful polycarbonate synthesis using the photo-on-demand interfacial polymerization method | Kobe University News site. Kobe University. Accessed 27 July 2024. [Online]. Available: https://www.kobe-u.ac.jp/en/news/article/2023_08_08_02/

Thorat S, Phillips P, Semenov V, Gakh A (2003) Physical properties of aliphatic polycarbonates made from CO_2 and epoxides. J Appl Polym Sci 89:1163–1176. https://doi.org/10.1002/app.12355

Torres JP, Frontini PM (2016) Mechanics of polycarbonate in biaxial impact loading. Int J Solids Struct 85–86:125–133. https://doi.org/10.1016/j.ijsolstr.2016.02.010

Tsuda A (2023) In situ photo-on-demand phosgenation reactions with chloroform for syntheses of polycarbonates and polyurethanes. Polym J 55(9):903–912. https://doi.org/10.1038/s41428-023-00800-w

Tailoring polycarbonate surfaces for improved Ge film adhesion: The role of plasma treatments in 50:50 O2/Ar atmospheres - ScienceDirect. Accessed 13 Sept 2024. [Online]. Available: https://www.sciencedirect.com/science/article/pii/S0143749624000733?via%3Dihub

The effect of the thermal aging on polycarbonate—ScienceDirect. Accessed 23 Sept 2024. [Online]. Available: https://www.sciencedirect.com/science/article/pii/S2214785321044266?via%3Dihub

The Effect of the Ultraviolet Radiation Environment of LEO Upon Polycarbonate Materials. Aerospace Sciences Meetings. Accessed 29 Sept 2024. [Online]. Available: https://doi.org/10.2514/6.2005-662

The Growth Behavior of Amorphous Hydrogenated Carbon a-C:H Layers on Industrial Polycarbonates—A Weak Interlayer and a Distinct Dehydrogenation Zone. Accessed 01 Oct 2024. [Online]. Available: https://www.mdpi.com/2311-5629/7/3/59

The impact of graphene oxide on the mechanical and thermal strength properties of polycarbonate - Maziyar Sabet, 2023. Accessed 13 Sept 2024. [Online]. Available: https://doi.org/10.1177/00952443231160236

The relationship between the effect of thermal pre-treatment and the viscoelastic behaviour of polycarbonate in the glassy state | Journal of Materials Science. Accessed 17 June 2024. [Online]. Available: https://doi.org/10.1007/BF00551020

Tracing the History of Polymeric Materials: How polycarbonate came about, virtually simultaneously, through the efforts of two industry giants. Free Online Library. Accessed 24 Feb 2024. [Online]. Available: https://www.thefreelibrary.com/Tracing%20the%20History%20of%20Polymeric%20Materials:%20How%20polycarbonate%20came...-a0678260788

Ugrcic M, Maksimovic S, Stamenkovic D, Maksimovic K, Nabil K (2015) Finite element modeling of wing bird strike. FME Trans. 43(1):76–81. https://doi.org/10.5937/fmet1501076u

Umeno Y, Kubo A, Albina J-M (2019) Coarse-grained molecular dynamics simulation of deformation and fracture in polycarbonate: effect of loading mode, strain rate, temperature and molar mass. Procedia Struct Integr 23:348–353. https://doi.org/10.1016/j.prostr.2020.01.111

Vandenberg LN, Hauser R, Marcus M, Olea N, Welshons WV (2007) Human exposure to bisphenol A (BPA). Reprod Toxicol Elmsford N 24(2):139–177. https://doi.org/10.1016/j.reprotox.2007.07.010

Voss DL, Raffo JA (1982) Recent developments in polycarbonate coatings for advanced aircraft. In: Emerging optical materials, SPIE, pp 52–58. https://doi.org/10.1117/12.932484

Wang XJ, Zheng XT, Zheng W, Wu SZ (2014) Molecular simulation of polycarbonate and thermomechanical analysis. Appl Mech Mater 556–562:441–444. https://doi.org/10.4028/www.scientific.net/AMM.556-562.441

Wang H, Zhou H, Huang Z, Zhang Y, Zhao X (2017) Constitutive modeling of polycarbonate over a wide range of strain rates and temperatures. Mech Time-Depend Mater 21(1):97–117. https://doi.org/10.1007/s11043-016-9320-1

What Is FEA | Finite Element Analysis? Documentation, SimScale. Accessed 13 May 2024. [Online]. Available: https://www.simscale.com/docs/simwiki/fea-finite-element-analysis/what-is-fea-finite-element-analysis/

West M, Ruys A, Bosi S (2005) The effect of the ultraviolet radiation environment of LEO upon polycarbonate materials. In: 43rd AIAA aerospace sciences meeting and exhibit. American Institute of Aeronautics and Astronautics, Reno, Nevada. https://doi.org/10.2514/6.2005-662

Wójcik M, Skrzat A (2022) Numerical modelling of the KOBO extrusion process using the Bodner-Partom material model. Meccanica 57(9):2213–2230. https://doi.org/10.1007/s11012-022-01569-7

Wright SC, Fleck NA, Stronge WJ (1993) Ballistic impact of polycarbonate—an experimental investigation. Int J Impact Eng 13(1):1–20. https://doi.org/10.1016/0734-743X(93)90105-G
Wu Z et al (2021) Influence of regional erosion and sedimentary loading on fault activities in active fold-thrust belts: insights from discrete element simulation and the southern and central longmen shan fold-thrust belt. Front Earth Sci 9. https://doi.org/10.3389/feart.2021.659682
Xiao L, Zhou L (2023) Structural and flow field analysis of an airfoil subjected to blast-impacted. J Phys Conf Ser 2478:072037. https://doi.org/10.1088/1742-6596/2478/7/072037
Xie H, Meng H, Wu L, Li B-G, Dubois P (2022) Aliphatic polycarbonate modified poly(ethylene furandicarboxylate) materials with improved ductility, toughness and high CO2 barrier performance. Polymer 246:124751. https://doi.org/10.1016/j.polymer.2022.124751
Xu Y, Lu H, Gao T, Zhang W (2015) Predicting the low-velocity impact behavior of polycarbonate: influence of thermal history during injection molding. Int J Impact Eng 86:265–273. https://doi.org/10.1016/j.ijimpeng.2015.08.013
Xu J, Feng E, Song J (2014) Renaissance of aliphatic polycarbonates: new techniques and biomedical applications. J Appl Polym Sci 131(5). https://doi.org/10.1002/app.39822
Xu X et al (2024) The effect of notch and molecular weight on the impact fracture behavior of polycarbonate. Polymers 16(8), Art. no. 8. https://doi.org/10.3390/polym16081072
Yang X, Zhang Z, Yang J, Sun Y (2016) Fluid–structure interaction analysis of the drop impact test for helicopter fuel tank. Springerplus 5(1):1573. https://doi.org/10.1186/s40064-016-3040-5
Yang Y, Wang F, Liu J (2022) Application of Honeycomb Structures in Key Components of Space Deployable Structures. Adv Mater Sci Eng. https://doi.org/10.1155/2022/4756272
Yao T-T, Zhang X-F, Zhang W-S, Liu Y-T, Liu Q, Wu G-P (2021) Controlled attachment of polycarbonate nanoparticles on carbon fibers for increased resin impregnation and interfacial adhesion in carbon fiber composites. Compos Part B Eng 224:109218. https://doi.org/10.1016/j.compositesb.2021.109218
Yu C, Kang G, Lu F, Zhu Y, Chen K (2016) Viscoelastic–viscoplastic cyclic deformation of polycarbonate polymer: experiment and constitutive model. J Appl Mech 83(041002). https://doi.org/10.1115/1.4032374
Yu Z, Xue P, Yao P, Zahran MS (2019) Analytical determination of the critical impact location for wing leading edge under birdstrike. Lat Am J Solids Struct 16(1). https://doi.org/10.1590/1679-78255352
Yvonnet J, Chaouch S (2024) Unsupervised machine learning classification for accelerating Fe^2 multiscale fracture simulations. Rochester, NY, p 4822796. https://doi.org/10.2139/ssrn.4822796
Zai L et al (2020) Laser powder bed fusion of precipitation-hardened martensitic stainless steels: a review. Metals. https://doi.org/10.3390/met10020255
Zhang T et al (2020) A discrete element method model of corn stalk and its mechanical characteristic parameters. BioResources. https://doi.org/10.15376/biores.15.4.9337-9350
Zhang S, Wang B, Meng X, Chen Y (2023) Mechanical properties and fracture microstructure of polycarbonate under high strain rate tension. Materials 16(9):3386. https://doi.org/10.3390/ma16093386
Zhang W, Xu Y (2019) 3—Impact Simulation of Polycarbonates in Aeronautical and Aerospace Applications. In: Zhang W, Xu Y (eds) Mechanical properties of polycarbonate. Elsevier, pp 79–112. https://doi.org/10.1016/B978-1-78548-313-4.50003-0
Zhang W, Gao Y, Ye N, Huang W, Li D (2017) Experimental study on dynamic mechanical behaviors of polycarbonate, presented at the Shock Compression of Condensed Matter—2015: Proceedings of the conference of the american physical society topical group on shock compression of condensed matter. Tampa Bay, Florida, USA, p 110014. https://doi.org/10.1063/1.4971677
Zhang L, Gao Z, Xiong Z, Zhang H, Tian Q, Hua W (2016) Method for preparing polycarbonate with continuous two-phase interface phosgene method, EP3040359A1. Accessed 27 July 2024. [Online]. Available: https://patents.google.com/patent/EP3040359A1/en

Zhao Z, Zhang Z, Sun C, Xu M, Li B (2024) A novel macromolecular phosphorus-nitrogen containing flame retardant for polycarbonate. Polym Degrad Stab 220:110648. https://doi.org/10.1016/j.polymdegradstab.2023.110648

Zhou Y, Sun Y, Huang T, Cai W (2019) SPH-FEM simulation of impacted composite laminates with different layups. Aerosp Sci Technol 95:105469. https://doi.org/10.1016/j.ast.2019.105469

Zimmerer C, Ziegler L, Heinrich G, Steiner G (2018) Time resolved characterization of the solid-state reaction between polycarbonate and primary amine. Eur Polym J 98:313–320. https://doi.org/10.1016/j.eurpolymj.2017.11.029

바흐만롤프, 미헬레폴커, 하이일얀, 코크펠릭스, Process for making polycarbonate using reduced phosgene excess, KR20210151796A. Accessed 27 July 2024. [Online]. Available: https://patents.google.com/patent/KR20210151796A/en

孔立明, 陈志伟, 张艳艳, 孔琳仪 (2014) Method for preparing polycarbonate by utilizing ionic liquid as catalyst," CN103936976A. Accessed 27 July 2024. [Online]. Available: https://patents.google.com/patent/CN103936976A/en

Chapter 8
Synthetic Nanofillers in the Aerospace Sector

Rajni Garg, Rishav Garg, Ansar Anjum, and Nnabuk Okon Eddy

Abstract The aviation industry has stringent criteria for the materials used in its components. These materials must be both lightweight and have excellent heat resistance, as well as the ability to withstand various conditions. In addition, they must possess the capability to endure electromagnetic interference, which is another essential requirement. However, only a limited number of materials can meet all of these criteria. In the past, metals were employed for making aeronautical components; but due to the several associated drawbacks, this was not the most advantageous choice. Metals are known for being significantly dense and susceptible to corrosion, both of which can compromise their effectiveness and reduce their durability. To overcome these limitations, researchers have developed synthetic nanofillers-based polymer nanocomposites that meet the industry's requirements as explored in the current chapter. The small size and unique characteristics of the filler material tailors the undesirable properties of the polymer matrix. The fabrication of these materials generally entails altering existing materials or amalgamating two or more materials to generate composites. The review signifies the crucial applications of the functional materials that can be created by combining synthetic nanofillers as reinforcement materials with a polymeric matrix, resulting in improved performance for use in the aerospace industry.

R. Garg (✉)
Department of Applied Sciences and Humanities, Galgotias College of Engineering & Technology, Greater Noida, Uttar Pradesh, India
e-mail: rajnigarg@science.org.in

R. Garg
Department of Civil Engineering, Galgotias College of Engineering and Technology, Greater Noida 201310, Uttar Pradesh, India
e-mail: rishavgarg@science.org.in

A. Anjum
Department of Applied Chemistry, Delhi Technological University, Delhi, India
e-mail: dransaranjum@gmail.com

N. O. Eddy
Department of Pure and Industrial Chemistry, University of Nigeria, Nsukka, Enugu State, Nigeria
e-mail: okon.nnabuk@unn.edu.ng

Visakh P. M. (ed.), *Nanomaterials for Defense, Aeronautical and Aerospace Applications*, Advanced Structured Materials 236,
https://doi.org/10.1007/978-3-031-92723-2_8

Keywords Synthetic nanofillers · Aerospace sector · Aerospace industry · Nano · Inorganic fillers · Polymer matrix · Nanocomposites · Aeronautical components · Electromagnetic interference

8.1 Introduction

In the past, aerospace engines have typically been created using metallic and ceramic materials with severe performance issues. Devices all over the world can create EMI that can impact the electronic devices needed for navigation in aircraft (Ciampa et al. 2018). This interference can come from transmitters used for communication, navigation, or satellite propagation (Fig. 8.1). Aircraft have been protected from electromagnetic radiation using metal materials because of their excellent electrical conductivity and ability to create a shielding effect, similar to the Faraday cage. However, the materials made from metal have some disadvantages owing to their heavy weight and susceptibility to corrosion (Zhang et al. 2019). Thus, the protection from corrosion of such materials plays a crucial role in the aerospace industry since it has a significant impact on different aspects of aircraft, including safety, availability, and structural integrity. Due to this reason, the design of efficient and smart protective coatings is crucial for protection of different parts of aircraft in the harsh environment (Liang et al. 2019). Because of corrosion as a crucial factor, it becomes challenging to account for it due to the convolution of predicting when corrosion will start and the potential damage it can cause to structures (Asif et al. 2022).

Such corrosion-resilient materials can be prepared by nanotechnology using nano-scaled composites. Nanotechnology is an innovative tool that involves modifying bulk materials to nano-scale (Garg et al. 2023). This technique holds the capacity to completely transform numerous industries, including electronics, energy, aerospace, waste management, information technology, medical, textiles, and transportation. By incorporating reinforcing materials such as nanometric-sized fibers, particles, or fillers into polymers, various composites are obtained that exhibit enhanced properties (Anjum et al. n.d.). Researchers worldwide have developed materials that exhibit superior physico-chemical, mechanical, and thermal properties compared to traditional materials. These nano-scaled materials are more durable, lighter, and have a longer lifespan than ever before (Sharma et al. 2023). Damage repair in structural aerospace composites has often necessitated significant quantities of healing agents. Various synthetic nanofillers have been integrated into polymer matrices to successfully obtain self-healing polymer-based nanocomposites for applications in the aerospace industry (Yadav et al. 2020).

The nanoparticles of the metal have been used to strengthen the structure of nanocomposites due to their unique properties and it has been observed that these particles, with dimensions smaller than 100 nm can tailor the properties of the traditional materials and create highly effective and visually appealing materials. Their adaptability and flexibility to be customized as per the desired use these nanoparticles prove to be an attractive choice to create novel materials in several industries

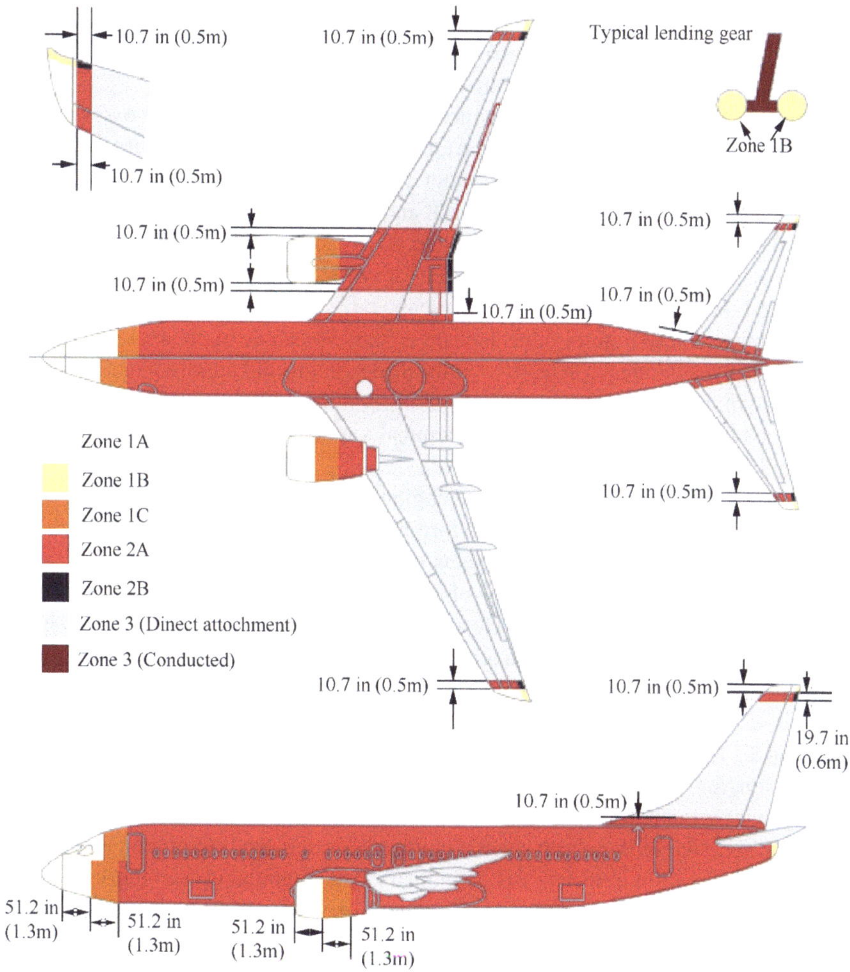

Fig. 8.1 Possible zones in Boeing aircraft under the possible lightning strikes and resulting in electromagnetic induction interference (Yadav et al. 2020)

(Secondary et al. n.d.). The metal nanoparticles have electrical conductivity as well as the capacity to be magnetized by an external field due to which they are highly efficient in their capacity to resist electromagnetic induction. These particles have been prepared using magnetic inorganic substances such as superparamagnetic iron oxide or metallic iron, nickel, cobalt, and various alloys (Garg et al. 2024; Frelon et al. 2018). The physical and chemical properties including the size and distribution of reinforcing nanomaterials of these novel materials are characterized using sophisticated techniques (Guadagno et al. 2019). These nanomaterials possess application in industries as catalysis, sensors, and for bioremediation and in aerospace. This analysis highlights the enhancements provided by the composite material in comparison to its unreinforced counterpart (Meure et al. 2009).

Besides metal nanoparticles, polymers have also been known to possess many advantages as compared to other materials, including their affordability, ability to be synthesized in various ways, and other wide range of characteristics including lightness, electrical conductivity, high heat and corrosion resistance (Li et al. 2021). Self-healing polymers retain the remarkable ability to mend structural damage (Zhu et al. 2019). Several methods have been used for synthesis of nanocomposite materials that depend on the intended application, shape, size, reinforcement type, and available resources. These requirements have been made possible by the recent advancements made by scientists in selective synthesis designing and handling nanoparticles within polymeric matrices (Krishnakumar et al. 2020). The commonly used techniques for synthesis include grafting, chemical vapour deposition, blending, in situ precipitation, and molding.

Lately, high-temperature-resistant jet engine components have incorporated polymeric materials and composites that in turn have the potential to effectively prevent any structural damage caused by impacts or failures. The integration of polymeric matrix composites with various types of synthetic nanofillets has opened up diverse applications in the aerospace industry, particularly in the form of strength enhancement, barrier protection and electromagnetic induction shielding materials. These challenges have been met by using integrated synthetic nanofillers in diverse polymer matrices using the dielectric and magnetic properties of organic and inorganic compounds. This resulted in an effective electromagnetic induction shield, resistant to corrosion, and remarkably lightweight.

This chapter includes investigations of the recent progress in polymer composites using synthetic nanofillers with special emphasis on self-healing polymers for their innovative application in the aerospace industry. The potential uses of these polymer nanocomposites in the aerospace sector have also been studied with their future prospective.

8.2 Synthetic Nanofillers in Polymer Composites

A combination of different materials like polymers, metals, and ceramics make the composite and are responsible to strengthen a matrix material. The matrix fulfils the purpose of securely holding the reinforcement material in place, providing a strong structural foundation. These matrix materials include resins such as polyester, epoxy, and vinyl esters (Zhang et al. 2013). The mechanical characteristics like resistance and rigidity are determined by reinforced materials which transfer the loads to the matrix and also lead to cost reductions (Kumar et al. 2020). The weight of aircraft is significantly reduced by using composite materials like glass fiber, carbon fiber, and aramid fiber which in turn has led to added strength and durability along with lower fuel costs per passenger. The lifespan of these composites depends on how polymers perform and degrade. The degradation of the polymer can take place through various processes, such as biodegradation, oxidation, mechanical stress, catalytic reactions, and pyrolysis. Polymers can be affected by changes in their surroundings because of their molecular makeup. Synthesis processes and mechanical property evaluation of polymeric nanocomposites have not been given the attention they deserve, despite their significance in terms of durability and performance.

Recently, there has been an emergent interest for using synthetic nanofillers for various applications in airplane components. The reinforcement material usually makes up 20–80% of the composite material's volume, typically less than 50%. Iron oxides, such as magnetite and maghemite, have also attracted a lot of attention as nanosized fillers. These materials have been widely used in various fields such as pigments, medical applications, catalysis, photocatalysis, and environmental remediation (Sharma et al. 2023). The tiny magnetic materials such as ferromagnetic and/or ferrimagnetic particles are entrenched within a matrix to form ferrogel nanocomposites. These nanocomposites thus can react to outside influences, like static or alternating magnetic fields displaying fascinating characteristics when subjected to a uniform magnetic field. Ferrites, a compound made from iron oxides and metallic elements, are widely used in magnetic nanocomposites because of their exceptional magnetic and dielectric properties. They possess the properties that are responsible to make them perfect for absorbing electromagnetic waves when used with polymeric matrices. The carbon-based reinforcements like carbon nanotubes, carbon fibers, graphite, and graphene sheets have been used in polymer nanocomposites and it has been found that the electrical conductivity of the matrix can be improved by increasing the reinforcement content in a matrix, an important factor to effectively absorb and shield electromagnetic waves.

The magnetic properties of nanocomposites can be improved by an interaction between organic and inorganic materials, resulting in the emergence of new and unique properties that cannot be achieved by either pristine material alone. Ferromagnetic nanoparticles can be successfully incorporated in polymeric composites because they can serve as templates and self-assemble. The nanoparticles can be coated with a surfactant and then self-assembled or grown in situ within a template (Mendoza et al. 2023). The combination of magnetism with other properties in polymer nanocomposites have resulted in many innovative possibilities like preparation of magnetic conductive materials using magnetic nanoparticles with conductive polymer matrix. This distinctive amalgamation allows the transfer of charge between nanofillers and polymer, resulting in the creation of an electronic system (Romero-Fierro et al. 2022). Similarly transparency can be introduced by using magnetic oxides that possess greater transparency to visible light, at the same time maintaining higher magnetization than conventional transparent magnets (Kobayashi et al. 2016). The dielectric properties and energy storage capabilities of dielectric polymer nanocomposites have been found dependent upon the polarity of polymer matrix (Li et al. 2021). Li et al. (2018) modified the poly(vinylidene fluoride-hexafluoropropylene) matrix by incorporation of $BaTiO_3$ nanofibers. The nanocomposite containing 3 vol% of such nanofibers when subjected to an electric field of 300 MV/m, showed a 43% higher discharged energy density (8.55 J/cm^3) compared to the pristine polymer matrix that in-turn helps in achieving extraordinary energy density at ultrahigh applied electric fields (<400 MV/m) (Li et al. 2018).

8.3 Characteristics of Synthetic Nanofillers Based Polymer Nanocomposites

The polymers can be altered by incorporation with organic or inorganic synthetic nanofillers, resulting in an enhancement of their characteristics and the formation of a strong nanocomposites with enhanced attributes. The combination of polymers and nanotechnology has resulted in innovative applications that set these composites apart. By incorporating synthetic nanofillers into a polymeric matrix, the magnetic, optical, mechanical, and thermal properties of the materials can be enhanced. However, the extent of improvement is still a function of the type of dispersion (homogeneous or diffuse). The two most important characteristics of synthetic nanofillers-based polymer nanocomposites include self-healing and barrier protection.

8.3.1 Self-healing Materials for the Aerospace Sector

There has been significant growth in self-healing materials over the past three decades, with numerous studies conducted on polymeric nanocomposites. Self-healing polymer/fiber composites have been primarily developed through extensive research. As observed in previous studies, nanofillers have been utilized in nanocomposites to improve the self-healing effect (Zhang et al. 2013). The self-healing effect depends a lot on the type of nanofiller and its chemical composition. In this case, self-healing agents along with nanomaterials have been used to enhance the healing process of the matrix. Small capsules or vesicles have been utilized to introduce healing agents, which can rupture while the damaged surface or area is recovering. Self-healing polymeric nanomaterials have been extensively developed for diverse applications in the aerospace sector. These include adhesives and coatings for various engineering structures including engines and fuselage. Pulikkalparambil et al. (2018) integrated a nano-encapsulated healing agent to form epoxy nanocomposites with the self-healing potential. The self-healing epoxy nanocomposite proved to be quite impressive as a coating to safeguard against surface damage. During the healing, the nanocapsules ruptured to release the healing agent that aided in the repair of any related damage. The nanocapsules were incredibly effective in healing the damage, leaving it almost completely healed (Pulikkalparambil et al. 2018).

The continuous impact of environment upon the space vehicles cannot be neglected. The self-healing effect is activated through different stimuli like light, heat, moisture, impact, and more. The prospective structure damage from heat and

UV radiation can be prevented by self-healing of aerospace structures. A study by Zhu et al. (2019) shows successful development of self-healing nanocomposites using UV-responsive nanocapsules. These materials could efficiently protect the aerospace structures from radiation damage and also promoted self-healing of mechanically damaged or scratched aerospace structures (Zhu et al. 2019). An investigation by Paolillo et al. (2021) studied appealing self-healing properties of epoxy/polyethylene-co-methacrylic nanomaterials for aerospace applications and exhibited effective repair damages. The mechanism depends entirely on the ability of matrix to undergo physical or chemical crosslinking in a reversible manner (Paolillo et al. 2021).

Self-healing of many compositions has been observed to involve physico-chemical interactions at the molecular level with physical interactions facilitated by the formation of van der Waals forces and covalent interactions facilitate the chemical interactions. Epoxy composites have been incorporated with microcapsules containing healing agents. Furthermore, the self-healing phenomenon has been observed to occur as a result of nanoparticles, in addition to microcapsules (Turap et al. 2009). These capsules can rupture within the matrix, effectively repairing any damage to the area or surface.

Jin et al. (2011) explored the self-healing properties of epoxy materials by incorporating microcapsules containing $(C_2H_5)_2O{\cdot}BF_3$ as the healing agents. The microcapsules containing 1 wt.% $(C_2H_5)_2O{\cdot}BF_3$ exhibited an impressive self-recovery rate of approximately 80% on the cracked epoxy surface caused by impact. The material was applied as a self-healing adhesive to prevent cohesive loss when applied onto the surface of the steel. Crack tails were observed in the scanning electron microscopic images of the scratched sample with crack-pinning toughening. On the other hand, the healed fracture surface revealed the presence of a crosslinked polymer film possibly due to the in-situ crosslinkage of polymer chains during the self-healing (Jin et al. 2011).

Graphene has also been utilized as a nanofiller in the self-healing of epoxy materials. Vitrimer epoxy/graphene oxide nanocomposites have been signified to exhibit efficient self-healing. These nanomaterials possess remarkable self-healing properties, due to the formation of exchangeable covalent networks within the vitrimer epoxy resin. Hu et al. (2018) conducted tests on self-healing epoxy/graphene oxide nanocomposites obtained by free radical copolymerization of the β-cyclodextrin and azobenzene functional graphene oxide. The self-healing effect was examined by conducting experiments involving ultraviolet curing and thermal treatment. The graphene oxide with β-cyclodextrin/azobenzene could perform as a macro-crosslinking agent and a photothermal agent with appreciable self-healing phenomenon that highlights their remarkable potential to repair themselves. The scratched regions were repaired by the polymer chain flexibility in the affected region, resulting in self-healing with outstanding mechanical properties. The photothermal self-healed samples exhibited higher strength recovery rates compared to the heated samples. The self-healing property was improved with an increase in time (Hu et al. 2018). Krishnakumar et al. (2020) studied the self-healing potential of vitrimer/graphene oxide nanocomposites at a temperature of 60 °C wherein the recovery took

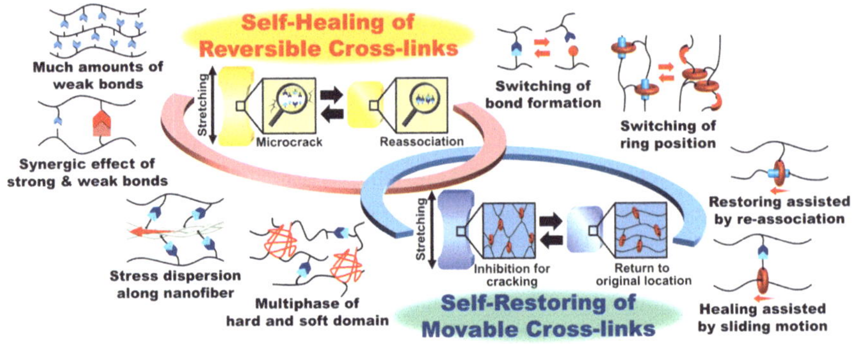

Fig. 8.2 Illustrative mechanism for self-healing (Ikura et al. 2022)

only 5 min. It was observed that by adding 0.5–1 wt.% graphene oxide the self-healing properties of nanocomposites were remarkably improved, resulting in an impressive recovery rate (80–88%) compared to the baseline epoxy matrix (60–73%) (Krishnakumar et al. 2020).

The process of self-healing occurs by the interdiffusion of stimuli-responsive polymer chains resulting in the reorganization of physical links, and the restoration of covalent links (Fig. 8.2). Although external stimuli such as force, temperature, and pressure, can instigate the self-healing but an extensive research has been conducted on this mechanism in the literature. Wang and Urban (2020) studied different physical and chemical methods to prevent structural damage and promote self-healing (Wang and Urban 2020). The healing in polyurethanes has also been observed with reversible chemical bond formation in the polymer chains resulting in interlinking (Xie et al. 2022). Lin et al. (2020) explored the self-healing properties of polyacrylate in which bismaleimide/furfuryl alcohol was used as a self-healing agent. However, the possibility of reversible crosslinking has been suggested due to chemical bond formation via the Diels–Alder reaction (Lin et al. 2020).

The impact of UV curing on self-healing in glass fiber/epoxy composites has also been conducted. Damage repair in hollow glass fibers was achieved after packing with healing agents that were released from the fibers when exposed to UV light (Su et al. 2019). Self-healing nanocomposites have proven to be valuable in aerospace applications, such as engineered structures, adhesives, coatings, and more (Huynh and Haick 2019). Self-healing in epoxy/carbon nanotube nanocomposites has also been explored. Self-healing properties have been observed through the reversible hydrogen bonding between the nanofiller and matrix. The self-healing progressed with the mobility of the polymer chain under the influence of an external stimulus. Both physical and covalent interactions were considered to facilitate the self-healing of thermoplastic or thermosets nanocomposites (Xie et al. 2021).

An interesting observation was made by Es'haghi et al. (2011) in epoxy nanocomposites, where the inclusion of barbiturate-modified nanofiller led to a self-healing effect. The multi-walled carbon nanotube displayed a remarkable self-healing effect due to reversible hydrogen bonding. Additionally, morphological changes in and around interfacial regions were noted in with a fascinating self-healing in the epoxy matrix when integrated with 0.5 wt.% of both functional and non-functional nanofillers. The storage modulus of the nanocomposites as also found to decrease as the temperature increased from –50 to 120 °C due to reversible crosslinking between the nanofiller and the matrix that enhanced the overall performance (Es'haghi et al. 2011). The self-healing phenomenon also depends on repairing damage by forming chemical bonds. It has been investigated that carbon nanotubes possess the potential to create reversible hydrogen bonding interactions within polymers. Functional carbon nanotubes have the ability to form covalent bonds with polymers, expanding their range of physical interactions (Du et al. 2014). Carbon nanotubes can serve as either electron donors or acceptors when interacting with polymers to form covalent bonds, depending on their functionalities. Wang et al. (2016) successfully created self-healing epoxy/carbon nanotube nanocomposite laminates with carbon nanotube sheets and carbon fibers arranged in alternating layers. Additionally, copper foam sheets were used to create composite laminates. The laminates containing carbon-nanotube-based nanosheets exhibited excellent dispersion of nanoparticles (Wang et al. 2016).

Guadagno et al. (2019) synthesized integrated nanocomposites using tetraglycidyl methylene dianiline epoxy resins using a nanofiller, barbiturate functional multi-walled carbon nanotube that demonstrated the healing efficiency of 107.7% and peak load of 96.22% (Guadagno et al. 2019). On the other hand, the composite with copper foam sheet demonstrated better healing efficiency with 63.2% fracture energy, a peak load of 58.8% and maximum healing efficiency of 128%. Thus, the studies show that higher healing efficiency can be obtained by using carbon-nanosheet-based nanocomposite with an improved self-healing effect for epoxy/carbon nanotube nanocomposites due to interactions between the matrix and nanofillers, resulting in the formation of an interface (Murugesan et al. 2020). Another study by Guadagno et al. (2021) developed epoxy nanocomposites using different types of carbon nanotubes, including non-functional, thymine functional, and barbiturate functional carbon nanotubes. The interactions that were found to be appropriate for self-healing included carbon nanotube-thymine ligand interaction with hydrogen bond formation between the C = O and O–H functionalities (epoxy) as well as the N–H functionalities (carbon nanotube). Thus, reversible hydrogen bonding interactions were proposed as a possible explanation for the self-healing in the epoxy nanocomposites containing different functionalities. Out of all the nanocomposite samples, the ones with 0.5 wt.% of thymine functional carbon nanotube showed improved self-healing properties (Guadagno et al. 2021).

8.3.2 Barrier Protection

Various polymer matrix nanocomposites have been reinforced with an inorganic phase to obtain resistant and lightweight materials as well as for the development of coatings and paints with effective barrier protection (Table 8.1). Its relatively easy fabrication and low cost are the additional advantages.

Kang et al. (2021) developed $CoFe_2O_4$/PANI magnetic additives while oleic acid was used as a dispersing agent. These additives have the potential to serve as corrosion protective agents. An epoxy coating was developed with these nanocomposites and exposed to acidic and saline conditions for 100 days. Electrochemical impedance spectroscopic analysis yielded encouraging results with values of 9.47 × 1010 and 4.81 × 1010 Ω, respectively. These results indicate that incorporating the nanocomposite greatly enhances the coating's durability and ability to resist corrosion (Kang et al. 2021). Asif et al. (2022) successfully synthesized magnetic nanocomposites integrating polyurethane (PU) with hydroxyapatite nanoparticles. The magnetic hydroxyapatite showed anticorrosive properties after immersion in a NaCl solution during electrochemical analysis. The contact angle of 105° confirmed its hydrophobic nature that increased with increase of the hydroxyapatite content, reaching its maximum value at a filler concentration of 3% (Asif et al. 2022).

Fazli-Shokouhi et al. (2019) used an in-situ polymerization process to prepare epoxy coatings using polyaniline (PANI)-graphene oxide nanosheets (GONs). Nanocomposites with different amounts of GONs improved the resistance property of epoxy resins to corrosion and fouling, providing valuable protective properties. After 192 h of immersion in saline water the 12 wt.% PANI-GON/epoxy coating demonstrated an impressive corrosion resistance (2.70 × 106 Ω cm^2) and efficiently barred the corrosion even in harsh environments (Fazli-Shokouhi et al. 2019). Zhang et al. (2018) studied nanocomposites made from an amphiphilic copolymer and amino-modified magnetic Fe_3O_4 nanoparticles ($MNP@NH_2$) for their effectivity in preventing ice formation. These nanocomposites were then infused with polyols wherein it showed remarkable anti-icing capabilities, including lowering the freezing point of water to as low as –36.8 °C, thereby postponing the formation of frost for a remarkable 2700 s at a chilly −18 °C, and demonstrating an impressively low ice adhesion strength of 0.1 kPa. The material was signified as an excellent choice for preventing ice buildup and removing ice in the aerospace industry (Zhang et al. 2018). Hou and Choy (2022) successfully created a nanocomposite coating using PVDF-HFP, SiO_2, and CNTs. This coating exhibited remarkable properties such as anti-icing and superhydrophobicity. This product offered outstanding water repellency, with a slide angle lower than 2° and a contact angle of up to 168°. It significantly delayed ice formation on aluminium film, even at temperatures as low as −19 °C, by more than 300 s. This material was considered fit for use in the aerospace industry, especially in harsh polar climates (Hou and Choy 2022).

Coatings for metallic alloys like magnesium, aluminium, and titanium are crucial for both military and civil aircraft. Their parts require effective corrosion protection to ensure their longevity and reliability. Self-healing coatings and adhesives have lately proved to be highly effective at repairing substrate damage (Ramesh et al. 2020). On the other hand, the corrosion resilient can effortlessly repair any damage caused by corrosion. A self-healing coating was prepared by Ahmadi (2019) by incorporating urea–formaldehyde microcapsules that were used as epoxy nanocomposite adhesives to repair aerospace structural damage (Ahmadi 2019). Self-healing

Table 8.1 Properties and applications of synthetic nanofillers in aerospace industry

Material/Coating	Properties	Results/ Applications	References
$CoFe_2O_4$/PANI with epoxy coating	Corrosion protection	Enhanced durability after 100 days of exposure	Kang et al. (2021)
Hydroxyapatite/Polyurethane (MHAP/PU)	Anticorrosive, Hydrophobic	Contact angle: 105°, hydrophobicity increases with 3% filler concentration	Asif et al. (2022)
Epoxy/Polyaniline-Graphene Oxide (PANI-GON)	Corrosion resistance, Fouling protection	Corrosion resistance: $2.70 \times 10^6\ \Omega\ cm^2$ after 192 h in saline water	Li et al. (2021)
Amino-modified Fe_3O_4 nanoparticles with polyols	Anti-icing, Superhydrophobic	Delayed frost formation by 2700 s at −18 °C, freezing point lowered to −36.8 °C, ice adhesion: 0.1 kPa	Li et al. (2018)
PVDF-HFP/SiO_2/CNTs	Anti-icing, Superhydrophobic	Contact angle: 168°, delayed ice formation by 300 s at −19 °C	Liang et al. (2019)
Self-healing Epoxy with urea formaldehyde microcapsules	Damage repair, Self-healing	Effective self-healing, applied on aerospace-grade aluminum	Liu et al. (2023)
Self-healing Epoxy with polyethylene-co-methacrylic acid	Damage repair, Self-healing	Remarkable healing at 150 °C, pressure-driven healing mechanism	de Castilho et al. (2022)
Poly(vinylidene fluoride) (PVDF)/Fe_3O_4/Carbon	EMI shielding	SSE value of 35.6 dB	Tang et al. (2020)
Poly(lactic acid) particles with silver nanoparticles	Corrosion resistance, EMI shielding	SSE value of 22.88 dB	Meure et al. (2009)
Epoxy resin with iron nanoparticles	EMI shielding, High-strength	SSE value of 45 dB	Murugesan et al. (2020)
Polypyrrole/$Ba_{0.6}Sr_{0.4}Fe_{12}O_{19}$ composite	EMI shielding, Microwave absorption	89% microwave absorption, SSE value of 37.49 dB	Paolillo et al. (2021)

(continued)

Table 8.1 (continued)

Material/Coating	Properties	Results/ Applications	References
Polyacrylonitrile (PAN)/ Polyurethane (PU) with Ni–Co nanoparticles	EMI shielding	SSE value of 77.8 dB, SSEt value of 7325.8 dB $cm^2\ g^{-1}$	Parsaee et al. (2020)
Cellulose nanofibril with silver nanoparticles	EMI shielding, High-efficiency	SSE value of 93.8 dB	Pulikkalparambil et al. (2018)
Polylactide-based polyurethane (Fe_3O_4/PLAUs)	Shape-memory, Magnetic response	Rapid shape recovery within 11–15 s in a magnetic field	Qin et al. (2017)
Epoxy-based carbon nanotubes	Intelligent coatings, High mechanical performance	Gage factor of 4.7, suitable for aerospace primary structures	Qing et al. (2019)
Fe_3O_4 decorated carbon nanotubes, reduced graphene oxide foam, and epoxy	EMI shielding	EMISSE value of 36 dB in X-band range	Liang et al. (2019)

graphene nanocomposite-based epoxy adhesives have been successfully used for space structures. The corresponding adhesives and coatings have proven to be effective in curing and preventing aerospace structural damage in a wide range of applications, including primary and secondary structures of helicopters, aero-engines, aircrafts, and spacecraft components (Ciampa et al. 2018).

A study reports an EMI shield composite film prepared by using poly(vinylidene fluoride), Fe_3O_4, and carbon (Tang et al. 2020). The film functions on the principle of conversion of absorbed radiation into thermal energy with specific shielding effectiveness value of 35.6 dB. In the designed composite the ferromagnetic nanoparticles imparted for magnetic loss and the carbon material for dielectric loss. However, the combination of both phases enabled them to achieve a significant specific shielding effectiveness value. Another study used polyethylene-co-methacrylic acid nanocapsules to synthesize self-healing materials that demonstrated healing phenomenon that was pressure-driven at 150 °C in fractured epoxy system. The healing process involved the release of healing agent on bursting (Meure et al. 2009). A study using silver nanoparticles incorporated composite matrix of cellulose nanofibril functionalized with polydopamine was examined in which the matrix showed highly efficient EMI shielding with specific shielding effectiveness value of 93.8 dB (Xu et al. 2020). Gu et al. (2016) developed shape-memory nanocomposites by utilizing polylactide-based polyurethane (Fe_3O_4/PLAUs) and achieved a uniform dispersion of Fe_3O_4 nanoparticles with the assistance of oleic acid. The nanocomposites exhibited a quick magnetic response and regained their original shape rapidly, within 11–15 s, when subjected to a magnetic field, and even faster, within 24–60 s, in an alternating magnetic field. The rapid shape recovery of this material was considered to possess immense potential for applications in the biomedical and aerospace fields, with its high-performance design (Gu et al. 2016).

Multi-walled carbon nanotubes embedded poly(caprolactone) and poly(lactic acid) were coated with silver nanoparticles and studied for its corrosion resistance and EMI shielding properties by Liu et al. (2023). The results showed an increased specific shielding effectiveness value of 22.88 dB along with improved multiple scattering by the metallic interface (Liu et al. 2023). Such type of material was also synthesized for aircraft applications by Jalali et al. (2011) using nickel, cobalt, iron, and iron oxide nanoparticles incorporated carbon mesh and epoxy resin. The material showed specific shielding effectiveness value of 45 dB which proved to be better as compared to carbon nanotubes embedded poly(caprolactone) and poly(lactic acid) nanomaterial (Jalali et al. 2011). Epoxy-based carbon nanotubes were used by Vertuccio et al. (2018) to prepare intelligent coatings for industrial carbon-fibre-reinforced plastics that carry highest value for the gage factor (4.7) among thermosetting resins that possess improved mechanical performance (Vertuccio et al. 2018). Liang et al. (2019) created a nanocomposite by combining reduced graphene oxide foam, Fe_3O_4 decorated carbon nanotubes, and epoxy in a three-dimensional structure. The nanocomposite demonstrated an exceptional EMI with SSE value of 36 dB, indicating its potential as a highly effective solution for EMI shielding applications (Liang et al. 2019).

Another material synthesized using polypyrrole/BaO.6SrO.4$Fe_{12}O_{19}$ nanocomposites by Velhal et al. (2019). This material proved to be cost-effective along with its EMI shielding properties with 89% microwave absorption and specific shielding effectiveness value of 37.49 dB (Velhal et al. 2019). A membrane with effective EMI shielding properties was prepared by electrospinning polyurethane and polyacrylonitrile fibers coating with Ni–Co nanoparticles wherein the intrinsic conductivity and magnetic character of the membrane provided an EMI shielding effect with an SSE value of 77.8 dB. This material was considered a potential material with aerospace applications in protecting electrical devices from radiation pollution (Zhang et al. 2019).

Synthetic nano-fillers have also been incorporated into a polymer matrix, resulting in a barrier that hinders the movement of gases (Bustamante-Torres et al. 2021). In addition, they can influence the size and shape of polymer crystals, as well as limit the movement of polymer chains. Researchers have created models to understand the effects of particle addition on permeability, specifically looking at how it influences membrane tortuosity (Mandal et al. 2024). Research has indicated that incorporating clay into epoxy resin can result in a reduction in its curing reactivity. This, in turn, leads to a decrease in the density of crosslinking and the formation of longer polymer chains. This can result in the composite being more susceptible to degradation (Xu et al. 2024). On the other hand, the silicate layers in the clay create a protective barrier against gases such as oxygen and nitrogen, which helps to slow down the process of degradation. Studies have shown that nanocomposites containing 10 wt.% clay are more susceptible to the barrier effect, whereas materials with 2 wt.% clay that are exfoliated exhibit improved barrier properties (Yadav et al. 2020).

Several factors influence the thermal degradation of composite materials, such as the type of material utilized, the material's composition, and the surrounding environment. Research conducted using the Kissinger method has revealed that modified materials exhibit lower activation energies for thermal degradation during the initial stage, while displaying higher energies in the second stage (Vyazovkin 2020). Scientists have conducted research on the impact of incorporating montmorillonite clay into polymers and how it influences their thermal stability. In order for clay to work well with polymers, it requires some modification using a surfactant. The thermal stability of nanocomposites is influenced by various factors, such as the type of modifier, the inherent thermal resistance of the matrix, and the composition of the nanofiller (Manimaran et al. 2024). Ammonium salts are commonly used to modify clay, enabling it to form strong bonds with polymeric resins. When clay is added to a polymeric matrix, it enhances its performance by providing insulation and acting as a barrier against decomposition products. Improving thermal stability and enhancing carbon formation during decomposition is beneficial. Also, nanocomposites exhibit a higher temperature for volatilization and reduced thermal oxidation when compared to microcomposites. The catalytic effect of the silicate and the barrier effect between the environments contribute to this phenomenon (Bantie et al. 2024). Studies have indicated that epoxy nanocomposites that have been modified exhibit enhanced flame resistance when compared to conventional composites (Zhang et al. 2013). In addition, researchers have discovered that carbon nanotubes exhibit superior flame retardant properties compared to nanoclays when they establish a network within the polymer matrix, resulting in a gel-like behavior (Chen et al. 2024).

8.4 Applications in Aerospace Sector

There is undeniable use of nanocomposites in fields such as military and civil aerospace. Advanced materials play a crucial role in the aerospace industry, as they are essential for the effective design, development, testing, operation, and maintenance of aircraft structures. Synthetic nanofillers based polymer nanocomposites have been used in various sectors of aerospace industry as listed in Table 8.2.

8.4.1 Aerospace Engines

Composite materials have proven to be highly effective in turbine blades for engines. Ceramic fibers can handle extremely high temperatures ranging from 1000 to 1350 °C inside the combustion chamber of space vehicle engines (Castaing et al. 2021). Hence, self-healing nanocomposites have been utilized in the production of both fixed and mobile jet engine components (Qin et al. 2017). Tong et al. (2021) put forward the use of nanomaterials in the compressor casing of gas turbines and aero engines as a means to achieve optimal efficiency and minimize fuel consumption.

Table 8.2 Applications of various nanomaterials in aerospace sector

Category	Materials used	Applications	Performance	References
Engine Components	Ceramic, metallic materials, polymeric materials, composites, self-healing nanocomposites	Turbine blades, combustion chambers, compressor casing	Ceramic fibers handle temperatures from 1000 to 1350 °C	Castaing et al. (2021)
			Self-healing nanocomposites resist high temperature, pressure, and corrosion with improved turbine engine performance and fuel efficiency	Tong et al. (2021)
			Increased porosity after engine use	de Castilho et al. (2022)
Aerospace Fuselage	Epoxy/glass fiber composites, Epoxy/ carbon fiber composites, carbon nanotube, graphene-enhanced nanocomposites	Fuselages of aerospace structures, civil and military aircraft,	Self-healing composites address load, fracture, fatigue, and impact issues	Wang and Li (2021)
			Healing efficiency >90%	Ikumapayi and Akinlabi (2019)
			Enhanced impact properties with NiTi alloy wires	Francesconi et al. (2013)
			Super-elastic materials improve damage tolerance	Balakrishnan et al. (2016)
Space Structures	Self-healing epoxy/ carbon fiber laminates, epoxy/ E-glass fiber composites, nanocarbon nanomaterials	High-performance aerospace structural parts, protection against space debris	Effective repair of microcracks and fatigue damage	Kessler et al. (2003)
			Healing effectiveness enhanced by microcapsules	Guadagno et al. (2010)
			Some mechanical properties compromised, but overall significant progress	Teoh et al. (2010)
			Efficient self-repair observed	Rehman et al. (2022)

(continued)

Table 8.2 (continued)

Category	Materials used	Applications	Performance	References
Advanced Applications	Self-healing nanocomposites, elastomer-based materials, epoxy composites	Propulsive fuselage wake-filling, structural parts with advanced self-healing capabilities	Reduced internal thermal shielding, noise, and vibration along with improved structural safety and buckling resistance	Raichurkar et al. (2017)
			Healing efficiencies up to 70–80% for repeated damage repair	Toohey et al. (2007)

Self-healing nanocomposites have proven to be highly effective in enhancing the scratch force, making them more resistant to vibrations in the blades and engine jams. The epoxy-based materials, containing 31 wt.% hollow microspheres, exhibited exceptional bonding strength. On the other hand, these materials exhibit a porosity of 63% and a hardness of 33. The characteristics of the material in its original state changed after being used in an engine. The material exhibited an increase in porosity with the propagation of the fracture with a significant diameter (Tong et al. 2021). There have been advancements in the development of self-healing nanocomposites for use in aerospace combustion chambers. Self-healing nanomaterials have shown impressive resistance to high temperature, pressure, and damage. Notably, they have also demonstrated corrosion resistance properties that are highly beneficial for engine components. The remarkable qualities of self-healing nanocomposites have greatly improved the performance of turbine engines as well (de Castilho et al. 2022).

8.4.2 *Aerospace Fuselage*

Self-healing composites have been found to overcome fracture, fatigue, load, impact, and sustainability issues (Wang and Li 2021). The fuselage materials with these self-healing compilations can self-heal the various damages in the various body parts of the aircraft. Relative to metal alloy-based structures, self-healing polymer nanocomposites exhibited a healing efficiency of >90% and superior damage-healing properties. This effect has resulted in replacement of the metal-based space fuselage with composite structures (Ikumapayi and Akinlabi 2019). Composites of carbon/glass fiber and epoxy matrix have also been experimented with for application in the aerospace fuselage, for application in the fuselage of aerospace structures, particularly in civil and military crafts (Dong and Kim 2018). Self-healing nanocomposites constituting of epoxy/carbon fiber/carbon nanotubes have also been applied by various researchers to develop aerospace-grade fuselage (Riccio et al. 2019). The insertion of carbon-based nanomaterials such as graphene and carbon nanotube has been reported to enhance the impact resistance of epoxy/fiber nanocomposites (Fig. 8.3).

Various researchers have primarily examined the effects and characteristics of epoxy/carbon fiber self-healing composites. Bond et al. (2008) developed epoxy/carbon fiber laminates with self-healing properties to enhance their impact resistance and conducted flexural strength analysis of a 16-ply aerospace-grade composite with hollow glass fibers. After quasistatic indentation damage, a remarkable 97% strength recovery was observed. The epoxy/glass fiber composites have proven to be highly effective in aerospace fuselage applications, demonstrating an impressive 92% recovery or healing rate following impact damage (Bond et al. 2008). The fiber laminates and resin have been carefully designed for self-healing. To enhance the impact properties of self-healing epoxy composites used in aerospace applications, the addition of super-elastic self-healing NiTi alloy wires, in combination with carbon or glass fiber fillers, has also been employed. The incorporation of self-healing alloy significantly improved the damage tolerance of aerospace parts when subjected to recurrent impacts. Researchers have also explored self-healing nanocomposites as a potential replacement for metal alloys in safeguarding space structures against space debris (Francesconi et al. 2013).

An innovative aircraft design was developed to validate the applications of integrating propulsive fuselage wake-filling. This aircraft was equipped with turbo-electric systems, twin-engine, geared turbofan power plants unitized underneath the wings, and a fuselage fan powered through a generator. These features contributed to

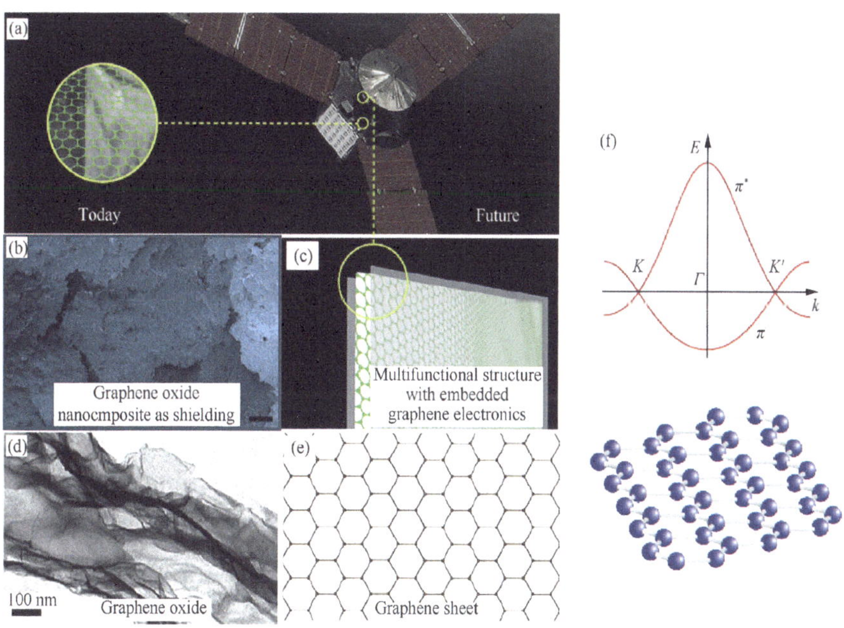

Fig. 8.3 Notional applications of graphene and its derivatives in spacecraft (Yadav et al. 2020)

its impressive performance and efficiency. Advanced self-healing nanocomposites were employed to reduce the internal/external noise, vibrations, and internal thermal shielding requirements of the fuselage. The vertical displacement (57 mm) at the fuselage tip didn't exceed much. Local wrinkling was observed in over 60% of the loads. According to the study, the use of self-healing nanomaterials was found to improve buckling resistance and enhance structural safety. The load-carrying capability was significantly improved through the implementation of an innovative structural design concept. Traditionally, thermosets are not utilized and natural fibers-based polymer composites have been the explored for self-healing in nonstructural parts of the aircraft such as seals and gaskets (Raichurkar et al. 2017). Nanofillers have been used for reinforcement of the elastomers to induce high flexibility and elasticity in addition to high ductility, low density, moderate cost, and excellent corrosion resistance (Balakrishnan et al. 2016).

Aerospace fuselage has greatly benefited from the use of polymer/fiber composites, as demonstrated by extensive research and reports. The epoxy/carbon fiber composites and epoxy/glass fiber composites have been utilized in this case. These materials have found extensive use in the construction of fuselages for both civil and military aircraft, as well as in various aerospace structures (Dong and Kim 2018). Adding nanoparticles like carbon nanotube and graphene can significantly improve the impact resistance of epoxy/fiber nanocomposites. The fuselage materials possess the remarkable ability to autonomously repair different types of damage to the crucial components of an aircraft's main body. Compared to metal alloy-based structures, self-healing polymer nanocomposites have shown an impressive healing efficiency of over 90%. As a result, composite structures with remarkable damage-healing properties have effectively replaced metal-based space fuselage (Ikumapayi and Akinlabi 2019). Self-healing composites in the fuselage have shown promising potential in addressing load, fracture, fatigue, impact, and sustainability concerns. Self-healing epoxy/carbon fiber/carbon nanotube nanocomposites have been utilized to create high-quality fuselage for aerospace applications.

8.4.3 Space Structures

Self-healing nanocomposites have been thoroughly examined for their potential use in aerospace structures. The self-healing epoxy/carbon fiber laminates have also found applications in aerospace structures (Banerjee et al. 2021). Kessler et al. (2003) developed materials made from Bisphenol A diglycidyl ether epoxy with the ability to self-heal. Scanning electron micrographs of fracture surfaces of the delaminated matrix and healed surface were compared. The self-healing agent was observed to effectively bridge the matrix strands to facilitate the self-healing. The loading curves for both the neat and healed samples reflected self-healing at a temperature of 80 °C for a duration of 48 h. In comparison to the neat epoxy sample, the self-healed sample exhibited reduced crack propagation. A significant crack was noticed forming in the middle of the epoxy material (Kessler et al. 2003).

Self-healing polymer/carbon fiber/nanocarbon and epoxy/nanocarbon composites have been extensively utilized in the aerospace industry. Materials with self-healing abilities have made significant progress, but this has often come at the expense of their mechanical properties. Williams et al. (2007) developed composites made of carbon fiber, epoxy, and hollow glass fiber. The addition of glass fibers resulted in an impressive 97% recovery in flexural strength. Unfortunately, the mechanical recovery showed a decrease when tested multiple times. When examining the flexural analysis, it became evident that both the damaged and healed samples experienced a significant decrease in flexural strength compared to the undamaged samples. Significant variation in fiber architecture after healing was found to be the cause of this effect (Williams et al. 2007). Aerospace structures have been formed using epoxy composites that incorporate E-glass fiber. Toohey et al. (2007) have successfully developed a self-healing system using epoxy and fiber materials. This system has the remarkable ability to autonomously repair itself even after multiple instances of damage. The healing agent was carefully injected into the cracks of materials, creating an intricate network of microvessels. The release of the healing agent resulted in the formation of bubbles within the structure. It was discovered that the healing efficiency reached an impressive 70–80%. Upon careful observation, it was noted that there was a healing of 40% after multiple instances of damage (Toohey et al. 2007). The effects of healing micro/nanocapsules on epoxy composite structures have been thoroughly examined for aerospace structural applications. The DGEBA/carbon composite integrated with microcapsules, containing catalyst and a healing agent, exhibited diminished mechanical properties as compared to that of the composites that lack a healing agent and catalyst have experienced a decrease. Nevertheless, the mechanical properties were successfully restored using healing agents (Guadagno et al. 2010). Teoh et al. (2010) introduced a remarkable self-healing epoxy composite that incorporates a healing agent specifically designed for aerospace structures. The specimens suffered from indentation damage, and were subjected to three-point bend flexural tests. Impressive strength recovery was observed after self-healing (Teoh et al. 2010).

Furthermore, the self-healing capability of epoxy-based space materials has been improved by incorporation of the healing agents. The epoxy matrix possesses an ability to undergo reversible crosslinking allowing it to self-repair with the help of micro- or nanocapsules such as urea/formaldehyde as healing agents within the diglycidyl ether of a bisphenol. The integration led to efficient reformation of matrix cracks, effectively restoring microstructural integrity and showcasing a remarkable self-repairing ability (Parsaee et al. 2020). Self-healing fiber-reinforced composites have also found applications in high-performance aerospace structural parts. The composite materials possess a remarkable ability to effectively repair micro-cracks and fatigue damage (Rehman et al. 2022). Not only do these materials possess impressive self-healing capabilities, but they also exhibit exceptional mechanical properties that make them ideal for aerospace structures. Self-healing epoxy/E-glass fiber composites have been utilized to create laminates of aerospace grade.

8.5 Future Prospective

Material defects can arise at any point during the usage or lifespan of a component. Infrared thermography is capable of detecting and quantifying material flaws throughout the entire lifecycle, from manufacturing to maintenance or in-service operations. A wide range of materials and components can be effectively analyzed using infrared thermography. Research studies have shown that there is still a need for further exploration in the field of aerospace adhesives. It is clear that more efforts should be dedicated to developing cutting-edge coatings and adhesives specifically tailored for the aerospace industry (Hu et al. 2009).

Structural Health Monitoring plays a crucial role in the aerospace industry by safeguarding the integrity of structures and averting potential disasters. By combining knowledge in sensors, materials, signal processing, and interpretation, scientists are able to detect even the most subtle anomalies. It goes beyond just detecting structural failures (Qing et al. 2019). It also gives early warnings, enabling quick action to prevent serious consequences. SHM covers a wide range of disciplines, such as sensors, materials, signal processing, and interpretation. This method plays a vital role in assessing the integrity of structures used in aerospace applications. By promptly detecting damage, structural health monitoring allows for the implementation of corrective measures to prevent issues (Dong and Kim 2018). Piezoelectric materials, particularly inorganic/organic composites, are excellent choices for monitoring sensors in the aerospace and aeronautical industries. D'Ambrogio et al. created a piezoelectric composite by combining lead zirconate titanate (PZT) with a poly(dimethyl siloxane) (PDMS) matrix. This new material demonstrates exceptional piezoelectric performance, even when exposed to elevated temperatures of 200 °C. It consists of a 1–3 configuration of PZT/PDMS. The researchers have determined that this composite material shows significant promise for integration into a sensor used in aircraft ball bearings for monitoring their condition (Qing et al. 2019).

Polymer-based nanocomposites have found extensive applications in a range of industries, including aerospace, automotive, and construction, thanks to their remarkable versatility. However, it is important to take into account their performance and durability. Various factors, such as the environment and how long it is used, can impact the composition of the material, potentially causing it to deteriorate (Pulikkalparambil et al. 2018). Over time, the material may start to deteriorate, leading to alterations in its chemical and physical characteristics. This can ultimately lead to a decrease in its mechanical properties and overall lifespan. Developing high-performance polymeric materials is crucial for industrial-level structural applications (Lin et al. 2020). It is also crucial to create polymers that can endure various environmental conditions. Thermal stability and flame retardancy are extremely important properties for aerospace applications (Sahu and Shih 2019). To achieve the desired outcome, enhancing the original polymers' physical properties is necessary. Reforming the damage has been seen as crucial for self-healing, with polymer chain mobilization and crosslinking playing a key role (Rehman et al. 2022).

8.6 Conclusion

This chapter explores the various aspects of incorporating synthetic nanofillers in polymer matrix to obtain polymer nanocomposites with potential use in aerospace sector. The study reveals that when synthetic nanofillers are added to polymers, they create a powerful combination that improves mechanical strength, heat resistance, barrier properties, and electromagnetic performance. The use of polymeric nanocomposites reinforced with these nanofillers offers significant advantages for aeronautical applications. The unique characteristics of magnetic polymeric composites make them well-suited for tackling specific issues in the aerospace industry. Efficient self-healing nanocomposites have found practical applications in aerospace structures. These self-healing materials are designed to restore aircraft structures that have been damaged by various factors such as heat cycles, UV radiation, atomic oxygen, and space debris Barrier properties can be enhanced by incorporating nanoparticles, which is influenced by factors such as particle shape, size, dispersion, and polymer interaction. They can provide solutions for challenges like EMI protection, corrosion resistance, and structural health monitoring. It is important to explore the impact of reinforcement on the mechanical properties of the matrix, including hardness, wear resistance, and traction for further progress of the field.

References

Ahmadi Z (2019) Nanostructured epoxy adhesives: a review. Prog Org Coatings 135:449–453. https://doi.org/10.1016/j.porgcoat.2019.06.028

Anjum A, Anjum A, Kashif M et al (n.d.) Food chemistry advances nano-scale innovations: properties, types, and applications of nanomaterials in packaging

Asif AH, Mahajan MS, Sreeharsha N et al (2022) Enhancement of anticorrosive performance of cardanol based polyurethane coatings by incorporating magnetic hydroxyapatite nanoparticles. Materials (Basel) 15:2308. https://doi.org/10.3390/ma15062308

Balakrishnan P, John MJ, Pothen L et al (2016) 12—Natural fibre and polymer matrix composites and their applications in aerospace engineering. In: Rana S, Fangueiro RBT-ACM for AE (eds) Woodhead Publishing, pp 365–383

Banerjee P, Kumar S, Bose S (2021) Thermoreversible bonds and graphene oxide additives enhance the flexural and interlaminar shear strength of self-healing epoxy/carbon fiber laminates. ACS Appl Nano Mater 4:6821–6831. https://doi.org/10.1021/acsanm.1c00888

Bantie Z, Tezera A, Abera D, Nega T (2024) Nanoclays as fillers for performance enhancement in building and construction industries: State of the art and future trends. In: Developments in clay science and construction techniques. IntechOpen

Bond IP, Trask RS, Williams HR (2008) Self-healing fiber-reinforced polymer composites. MRS Bull 33:770–774. https://doi.org/10.1557/mrs2008.164

Bustamante-Torres M, Romero-Fierro D, Arcentales-Vera B et al (2021) Interaction between filler and polymeric matrix in nanocomposites: magnetic approach and applications. Polymers (Basel) 13:2998. https://doi.org/10.3390/polym13172998

Castaing V, Giordano L, Richard C et al (2021) Photochromism and persistent luminescence in Ni-Doped ZnGa(2)O(4) transparent glass-ceramics : toward optical memory applications to cite this version : HAL Id : hal-03290041

Chen Z, Yu T, Yang Z et al (2024) Superior mechanical behavior and flame retardancy frp via a distribution controllable 1D/2D hybrid nanoclay synergistic toughening strategy. Engineering. https://doi.org/10.1016/j.eng.2024.03.017

Ciampa F, Mahmoodi P, Pinto F, Meo M (2018) Recent advances in active infrared thermography for non-destructive testing of aerospace components. Sensors 18. https://doi.org/10.3390/s18020609

de Castilho BCNM, Rodrigues AM, Avila PRT et al (2022) Hybrid magnetron sputtering of ceramic superlattices for application in a next generation of combustion engines. Sci Rep 12:2342. https://doi.org/10.1038/s41598-022-06131-9

Dey D, De D, Ghaemi F et al (2020) Circuit level modeling of electrically doped adenine-thymine nanotube based field effect transistor. IEEE Access 8:6168–6176. https://doi.org/10.1109/ACCESS.2019.2963225

Dong T, Kim NH (2018) Cost-effectiveness of structural health monitoring in fuselage maintenance of the civil aviation industry. Aerospace 5. https://doi.org/10.3390/aerospace5030087

Du R, Wu J, Chen L et al (2014) Hierarchical hydrogen bonds directed multi-functional carbon nanotube-based supramolecular hydrogels. Small 10:1387–1393. https://doi.org/10.1002/smll.201302649

Es'haghi Z, Rezaeifar Z, Rounaghi GH et al (2011) Synthesis and application of a novel solid-phase microextraction adsorbent: hollow fiber supported carbon nanotube reinforced sol–gel for determination of phenobarbital. Anal Chim Acta 689:122–128. https://doi.org/10.1016/j.aca.2011.01.019

Fazli-Shokouhi S, Nasirpouri F, Khatamian M (2019) Polyaniline-modified graphene oxide nanocomposites in epoxy coatings for enhancing the anticorrosion and antifouling properties. J Coatings Technol Res 16:983–997. https://doi.org/10.1007/s11998-018-00173-3

Francesconi A, Giacomuzzo C, Grande AM et al (2013) Comparison of self-healing ionomer to aluminium-alloy bumpers for protecting spacecraft equipment from space debris impacts. Adv Sp Res 51:930–940. https://doi.org/10.1016/j.asr.2012.10.013

Frelon A, Paruit T, Audichon T et al (2018) Influence of innovative materials incorporation in MEA during its conception on the electrochemical and PEMFC performances. ECS Meet Abstr MA 02:1433–1433. https://doi.org/10.1149/MA2018-02/42/1433

Garg R, Mittal M, Tripathi S, Eddy NO (2024) Core to concept: synthesis, structure, and reactivity of nanoscale zero-valent iron (NZVI) for wastewater remediation. Environ Sci Pollut Res. https://doi.org/10.1007/s11356-024-33197-x

Garg R, Garg R, Sharma R et al (2023) Introduction to nanotechnology and nanomaterials. Nanotechnol Sustain Agric Food Environ 1–15. https://doi.org/10.1201/9781003397861-1

Gu S-Y, Jin S-P, Gao X-F, Mu J (2016) Polylactide-based polyurethane shape memory nanocomposites (Fe_3O_4/PLAUs) with fast magnetic responsiveness. Smart Mater Struct 25:055036. https://doi.org/10.1088/0964-1726/25/5/055036

Guadagno L, Longo P, Raimondo M et al (2010) Cure behavior and mechanical properties of structural self-healing epoxy resins. J Polym Sci Part B Polym Phys 48:2413–2423. https://doi.org/10.1002/polb.22139

Guadagno L, Vertuccio L, Naddeo C et al (2019) Self-healing epoxy nanocomposites via reversible hydrogen bonding. Compos Part B Eng 157:1–13. https://doi.org/10.1016/j.compositesb.2018.08.082

Guadagno L, Vertuccio L, Naddeo C et al (2021) Functional structural nanocomposites with integrated self-healing ability. Mater Today Proc 34:243–249. https://doi.org/10.1016/j.matpr.2020.03.051

Hou Y, Choy KL (2022) Durable and robust PVDF-HFP/SiO_2/CNTs nanocomposites for anti-icing application: water repellency, icing delay, and ice adhesion. Prog Org Coatings 163:106637. https://doi.org/10.1016/j.porgcoat.2021.106637

Hu Z, Zhang D, Lu F et al (2018) Multistimuli-responsive intrinsic self-healing epoxy resin constructed by host-guest interactions. Macromolecules 51:5294–5303. https://doi.org/10.1021/acs.macromol.8b01124

Hu L, Li M, Xu C et al (2009) A polysilazane coating protecting polyimide from atomic oxygen and vacuum ultraviolet radiation erosion. Surf Coatings Technol 203:3338–3343. https://doi.org/10.1016/j.surfcoat.2009.04.019

Huynh T-P, Haick H (2019) Chapter 10—Self-healing materials for analyte sensing. In: Zenkina OV (ed) Nanomaterials design for sensing applications. Elsevier, pp 325–339

Ikumapayi OM, Akinlabi ET (2019) Efficacy of α-β grade titanium alloy powder (Ti–6Al–2Sn–2Zr–2Mo–2Cr–0.25Si) in surface modification and corrosion mitigation in 3.5% NaCl on friction stir processed armour grade 7075-T651 aluminum alloys—insight in defence applications. Mater Res Express 6:76546. https://doi.org/10.1088/2053-1591/ab1566

Ikura R, Park J, Takashima Y (2022) Design of self-healing and self-restoring materials utilizing reversible and movable crosslinks. 1–17. https://doi.org/10.1038/s41427-021-00349-1

Jalali M, Dauterstedt S, Michaud A, Wuthrich R (2011) Electromagnetic shielding of polymer–matrix composites with metallic nanoparticles. Compos Part B Eng 42:1420–1426. https://doi.org/10.1016/j.compositesb.2011.05.018

Jin H, Miller GM, Sottos NR, White SR (2011) Fracture and fatigue response of a self-healing epoxy adhesive. Polymer (Guildf) 52:1628–1634. https://doi.org/10.1016/j.polymer.2011.02.011

Kang Y-T, Wang C-C, Chen C-Y (2021) Corrosion-protective performance of magnetic $CoFe_2O_4$/polyaniline nanocomposite within epoxy coatings. J Taiwan Inst Chem Eng 127:357–366. https://doi.org/10.1016/j.jtice.2021.08.008

Kessler MR, Sottos NR, White SR (2003) Self-healing structural composite materials. Compos Part A Appl Sci Manuf 34:743–753. https://doi.org/10.1016/S1359-835X(03)00138-6

Kobayashi N, Masumoto H, Takahashi S, Maekawa S (2016) Optically transparent ferromagnetic nanogranular films with tunable transmittance. Sci Rep 6:34227. https://doi.org/10.1038/srep34227

Krishnakumar B, Prasanna Sanka RVS, Binder WH et al (2020) Catalyst free self-healable vitrimer/graphene oxide nanocomposites. Compos Part B Eng 184:107647. https://doi.org/10.1016/j.compositesb.2019.107647

Kumar M, Bansal M, Garg R (2020) An overview of beneficiary aspects of zinc oxide nanoparticles on performance of cement composites. Mater Today Proc 43:892–898. https://doi.org/10.1016/j.matpr.2020.07.215

Li Z, Liu F, Yang G et al (2018) Enhanced energy storage performance of ferroelectric polymer nanocomposites at relatively low electric fields induced by surface modified $BaTiO_3$ nanofibers. Compos Sci Technol 164:214–221. https://doi.org/10.1016/j.compscitech.2018.05.052

Li H, Wang L, Zhu Y et al (2021) Tailoring the polarity of polymer shell on $BaTiO_3$ nanoparticle surface for improved energy storage performance of dielectric polymer nanocomposites. Chinese Chem Lett 32:2229–2232. https://doi.org/10.1016/j.cclet.2020.12.032

Liang C, Song P, Ma A et al (2019) Highly oriented three-dimensional structures of Fe_3O_4 decorated CNTs/reduced graphene oxide foam/epoxy nanocomposites against electromagnetic pollution. Compos Sci Technol 181:107683. https://doi.org/10.1016/j.compscitech.2019.107683

Lin C, Ge H, Wang T et al (2020) A self-healing and recyclable polyurethane/halloysite nanocomposite based on thermoreversible Diels-Alder reaction. Polymer (Guildf) 206:122894. https://doi.org/10.1016/j.polymer.2020.122894

Liu T, Feng H, Zeng W et al (2023) Facile fabrication of absorption-dominated biodegradable poly(lactic acid)/polycaprolactone/multi-walled carbon nanotube foams towards electromagnetic interference shielding. J Compos Sci 7:395. https://doi.org/10.3390/jcs7090395

Mandal S, Roy D, Mukhopadhyay K et al (2024) Mechanistic insight into the role of the aspect ratio of nanofillers in the gas barrier properties of polymer nanocomposite thin films. RSC Appl Interfaces. https://doi.org/10.1039/D3LF00214D

Manimaran M, Norizan MN, Kassim MHM et al (2024) Critical assessment of the thermal stability and degradation of chemically functionalized nanocellulose-based polymer nanocomposites. Nanotechnol Rev 13. https://doi.org/10.1515/ntrev-2024-0005

Mendoza C, Nirwan VP, Fahmi A (2023) Nanofabrication of hybrid nanomaterials: macroscopically aligned nanoparticles pattern via directed self-assembly of block copolymers. J Appl Polym Sci 140. https://doi.org/10.1002/app.53409

Meure S, Wu DY, Furman S (2009) Polyethylene-co-methacrylic acid healing agents for mendable epoxy resins. Acta Mater 57:4312–4320. https://doi.org/10.1016/j.actamat.2009.05.032

Murugesan B, Pandiyan N, Kasinathan K et al (2020) Fabrication of heteroatom doped NFP-MWCNT and NFB-MWCNT nanocomposite from imidazolium ionic liquid functionalized MWCNT for antibiofilm and wound healing in Wistar rats: synthesis, characterization, in-vitro and in-vivo studies. Mater Sci Eng C 111:110791. https://doi.org/10.1016/j.msec.2020.110791

Paolillo S, Bose RK, Santana MH, Grande AM (2021) Intrinsic self-healing epoxies in polymer matrix composites (PMCs) for aerospace applications. Polymers (Basel) 13. https://doi.org/10.3390/polym13020201

Parsaee S, Mirabedini SM, Farnood R, Alizadegan F (2020) Development of self-healing coatings based on urea-formaldehyde/polyurethane microcapsules containing epoxy resin. J Appl Polym Sci 137:1–15. https://doi.org/10.1002/app.49663

Pulikkalparambil H, Siengchin S, Parameswaranpillai J (2018) Corrosion protective self-healing epoxy resin coatings based on inhibitor and polymeric healing agents encapsulated in organic and inorganic micro and nanocontainers. Nano-Struct Nano-Objects 16:381–395. https://doi.org/10.1016/j.nanoso.2018.09.010

Qin H, Zhang T, Li H-N et al (2017) Dynamic Au-thiolate interaction induced rapid self-healing nanocomposite hydrogels with remarkable mechanical behaviors. Chem 3:691–705. https://doi.org/10.1016/j.chempr.2017.07.017

Qing X, Li W, Wang Y, Sun H (2019) Piezoelectric transducer-based structural health monitoring for aircraft applications. Sensors 19:545. https://doi.org/10.3390/s19030545

Raichurkar P, Manickam R, Tailor R (2017) Review on non-woven polymeric gaskets their characteristics and applications. Int J Text Eng Prosess 3:1–8. ISSN 23953578

Ramesh M, Arun Ramnath R, Khan A et al (2020) Electrically conductive self-healing materials: preparation, properties, and applications. In: Khan A, Jawaid M, Raveendran SN, Ahmed Asiri AMBT-S-HCM (eds) Woodhead publishing series in composites science and engineering. Woodhead Publishing, pp 1–13

Rehman SU, Javaid S, Shahid M et al (2022) Polystyrene-sepiolite clay nanocomposites with enhanced mechanical and thermal properties. Polymers (Basel) 14:1–15. https://doi.org/10.3390/polym14173576

Riccio A, Saputo S, Sellitto A et al (2019) An insight on the crashworthiness behavior of a full-scale composite fuselage section at different impact angles. Aerospace 6. https://doi.org/10.3390/aerospace6060072

Romero-Fierro D, Bustamante-Torres M, Bravo-Plascencia F et al (2022) Polymer-magnetic semiconductor nanocomposites for industrial electronic applications. Polymers (Basel) 14:2467. https://doi.org/10.3390/polym14122467

Sahu RS, Shih YH (2019) Reductive debromination of tetrabromobisphenol A by tailored carbon nitride Fe/Cu nanocomposites under an oxic condition. Chem Eng J 378:122059. https://doi.org/10.1016/j.cej.2019.122059

Secondary CA, Author C, Sharma R et al (n.d.) Environmental Chemistry letters potential applications of green synthesized iron oxide NPs for environmental

Sharma R, Garg R, Bali M, Eddy NO (2023) Potential applications of green-synthesized iron oxide NPs for environmental remediation. Environ Monit Assess 195. https://doi.org/10.1007/s10661-023-12035-6

Su J-F, Zhang X-L, Guo Y-D et al (2019) Experimental observation of the vascular self-healing hollow fibers containing rejuvenator states in bitumen. Constr Build Mater 201:715–727. https://doi.org/10.1016/j.conbuildmat.2019.01.001

Tang X-H, Tang Y, Wang Y et al (2020) Interfacial metallization in segregated poly (lactic acid)/poly (ε-caprolactone)/multi-walled carbon nanotubes composites for enhancing electromagnetic

interference shielding. Compos Part A Appl Sci Manuf 139:106116. https://doi.org/10.1016/j.compositesa.2020.106116
Teoh SH, Chia HY, Lee MS et al (2010) Self healing composite for aircraft's structural application. Int J Mod Phys B 24:157–163. https://doi.org/10.1142/S0217979210064083
Tong Y-Q, Shi Q-S, Liu M-J et al (2021) Lightweight epoxy-based abradable seal coating with high bonding strength. J Mater Sci Technol 69:129–137. https://doi.org/10.1016/j.jmst.2020.05.077
Toohey KS, Sottos NR, Lewis JA et al (2007) Self-healing materials with microvascular networks. Nat Mater 6:581–585. https://doi.org/10.1038/nmat1934
Turap T, Merupakan TB, Lebih TB, Turap TD (2009) Self-healing materials: fundamentals, design strategies, and applications
Velhal N, Patil ND, Kulkarni G et al (2019) Electromagnetic shielding, magnetic and microwave absorbing properties of Polypyrrole/Ba0.6Sr0.4Fe12O19 composite synthesized via in-situ polymerization technique. J Alloys Compd 777:627–637. https://doi.org/10.1016/j.jallcom.2018.11.041
Vertuccio L, Guadagno L, Spinelli G et al (2018) Smart coatings of epoxy based CNTs designed to meet practical expectations in aeronautics. Compos Part B Eng 147:42–46. https://doi.org/10.1016/j.compositesb.2018.04.027
Vyazovkin S (2020) Kissinger method in kinetics of materials: things to beware and be aware of. Molecules 25:2813. https://doi.org/10.3390/molecules25122813
Wang D, Li S (2021) Collaborative optimization design of lightweight and crashworthiness of the front-end structures of automobile body using HW–GRA for Pareto mining. Proc Inst Mech Eng Part C J Mech Eng Sci 235:095440622199280. https://doi.org/10.1177/0954406221992802
Wang Y, Pham DT, Zhang Z et al (2016) Sustainable self-healing at ultra-low temperatures in structural composites incorporating hollow vessels and heating elements. R Soc Open Sci 3:160488. https://doi.org/10.1098/rsos.160488
Wang S, Urban MW (2020) Self-healing polymers. Nat Rev Mater 5:562–583. https://doi.org/10.1038/s41578-020-0202-4
Williams G, Trask R, Bond I (2007) A self-healing carbon fibre reinforced polymer for aerospace applications. Compos Part A Appl Sci Manuf 38:1525–1532. https://doi.org/10.1016/j.compositesa.2007.01.013
Xie L, Wang Y, Chen G et al (2021) A thermadapt epoxy based on borate ester crosslinking and its carbon fiber composite as rapidly processable prepreg. Compos Commun 28:100979. https://doi.org/10.1016/j.coco.2021.100979
Xie J, Fan L, Yao D et al (2022) Ultra-robust, self-healable and recyclable polyurethane elastomer via a combination of hydrogen bonds, dynamic chemistry, and microphase separation. Mater Today Chem 23:100708. https://doi.org/10.1016/j.mtchem.2021.100708
Xu Y, Qian K, Deng D et al (2020) Electroless deposition of silver nanoparticles on cellulose nanofibrils for electromagnetic interference shielding films. Carbohydr Polym 250:116915. https://doi.org/10.1016/j.carbpol.2020.116915
Xu J, Jia L, Lan Q, Wu D (2024) Enhanced thermal and mechanical properties of cardanol epoxy/clay-based nanocomposite through girard's reagent. Polymers (Basel) 16:1528. https://doi.org/10.3390/polym16111528
Yadav R, Tirumali M, Wang X et al (2020) Polymer composite for antistatic application in aerospace. Def Technol 16:107–118. https://doi.org/10.1016/j.dt.2019.04.008
Zhang X, He Q, Gu H et al (2013) Polyaniline stabilized barium titanate nanoparticles reinforced epoxy nanocomposites with high dielectric permittivity and reduced flammability. J Mater Chem C 1:2886. https://doi.org/10.1039/c3tc30129j
Zhang G, Zhang Q, Cheng T et al (2018) Polyols-infused slippery surfaces based on magnetic Fe_3O_4-functionalized polymer hybrids for enhanced multifunctional anti-icing and deicing properties. Langmuir 34:4052–4058. https://doi.org/10.1021/acs.langmuir.8b00286

Zhang N, Zhao R, He D et al (2019) Lightweight and flexible Ni–Co alloy nanoparticle-coated electrospun polymer nanofiber hybrid membranes for high-performance electromagnetic interference shielding. J Alloys Compd 784:244–255. https://doi.org/10.1016/j.jallcom.2018.12.341

Zhu Y, Cao K, Chen M, Wu L (2019) Synthesis of UV-responsive self-healing microcapsules and their potential application in aerospace coatings. ACS Appl Mater Interfaces 11:33314–33322. https://doi.org/10.1021/acsami.9b10737

Chapter 9
Nanoparticles and Nanocomposites for Aerospace Applications

Adib Bin Rashid, Mahima Haque, S. M. Mohaimenul Islam, K. M. Rafi Uddin Labib, and Pavel Chowdhury

Abstract Nanotechnology in aerospace has brought about a significant transformation in materials science, leading to the creation of advanced materials with improved properties. This section delves into the emerging field of green nanoparticles and nanocomposites for aerospace applications, with a particular emphasis on sustainable and environmentally friendly approaches. It explores the fundamental concepts of nanoparticles and nanocomposites, classifying them into polymer-based, metal matrix, and ceramic matrix nanocomposites commonly employed in the aerospace industry. The discussion also addresses environmental concerns associated with traditional nanomaterials, paving the way for exploring green nanoparticles derived from plant extracts, microorganisms, and other sustainable resources. In terms of practical applications in aerospace, the utilization of green nanoparticles extends to lightweight structural components, thermally conductive coatings, and advancements in electronics and sensing technologies. While highlighting the numerous benefits of green nanomaterials, the section also acknowledges the challenges and potential risks of their implementation in aerospace. Future perspectives focus on innovative trends and the continuous evolution of sustainable nanomaterials to further advance aerospace technology.

Keywords Nanotechnology · Aerospace Engineering · Green nanoparticles · Nanocomposites · Sustainable materials · Green synthesis

---

A. B. Rashid (✉)
Department of Industrial Production Engineering, Military Institute of Science and Technology (MIST), Dhaka, Bangladesh
e-mail: adib@me.mist.ac.bd

M. Haque · S. M. Mohaimenul Islam · K. M. Rafi Uddin Labib · P. Chowdhury
Department of Aeronautical Engineering, Military Institute of Science and Technology (MIST), Dhaka, Bangladesh

Visakh P. M. (ed.), *Nanomaterials for Defense, Aeronautical and Aerospace Applications*, Advanced Structured Materials 236,
https://doi.org/10.1007/978-3-031-92723-2_9

9.1 Introduction

The aerospace industry is characterized by its demanding material requirements, driven by the need for high performance, safety, and efficiency in airplane and spacecraft (Rana and Fangueiro 2016; Parveez et al. 2022). Aerospace materials must endure extreme conditions, including high temperatures, significant stress, and exposure to harsh environments. Traditional materials like aluminum alloys, titanium, and composite polymers have been the mainstays of aerospace engineering due to their durability, lightweight properties, and strength. However, the industry progressively seeks advanced materials that perform better while meeting rigorous environmental and regulatory standards (Bhat et al. 2021; Masrafee et al. 2025).

As global awareness of environmental issues grows, there is an increasing push towards sustainability in all industries, including aerospace (Balakrishnan et al. 2016). Traditional aerospace materials often involve resource-intensive production processes and may contribute to significant environmental impacts. Consequently, there is a shift towards adopting more sustainable materials and practices. Green chemistry and eco-friendly technologies are at the forefront of this shift, aiming to reduce the ecological footprint of material production and usage (Joseph et al. 2021). In this context, green nanoparticles and nanocomposites represent a promising frontier, offering both performance improvements and environmental benefits. Green nanoparticles are engineered at the nanoscale and synthesized through environmentally benign methods. Unlike conventional nanoparticles, which may rely on toxic chemicals and high-energy processes, green nanoparticles are fabricated using sustainable biological or low-impact chemical methods. These methods not only minimize environmental harm but can also enhance the nanoparticles' properties (Barhoum et al. 2022).

Nanocomposites, incorporating nanoparticles into a matrix material, further extend the benefits of nanotechnology (Olaniyan et al. 2024). Compared to their traditional counterparts, these composites can be tailored to exhibit enhanced electrical, thermal, and mechanical properties. In aerospace engineering, nanocomposites can contribute to lighter, stronger, and more efficient materials, potentially transforming the design and performance of aircraft and spacecraft (Rashid and Hoque 2022). Scientists are constantly working to enhance the mechanical characteristics of epoxy composites, with a specific emphasis on fracture toughness. Various toughening agents, including nanoclay, silica nanoparticles, and thermoplastic polymer particles, have been investigated to increase the mechanical characteristics of epoxy. Various scientists have been investigating graphene, carbon nanotubes (CNTs), and, more recently, inorganic two-dimensional (2D) nanomaterials, in order to enhance the mechanical characteristics of epoxy composites. Out of these options, 2D nanomaterials are favored because of their innate mechanical strength and elevated aspect ratio (Wazalwar et al. 2021).

This chapter will portray the innovative role of green nanoparticles and nanocomposites in advancing aerospace materials. It will cover the fundamental aspects of nanoparticle technology, highlight the advantages of green synthesis methods, and discuss how these advanced materials are being integrated into aerospace applications to meet the dual goals of elevated performance and sustainability. By examining current research and real-world uses, this chapter aims to deliver a widespread summary of the potential and challenges associated with these cutting-edge materials.

9.2 Green Nanocomposites in Aerospace

Generally, a composite is a substance that consists of at least two distinct phases: the matrix (the continuous phase) and the reinforcement (the discontinuous phase). Both the reinforcement and matrix serve distinct purposes. The matrix is responsible for both maintaining the composite's structural integrity and transferring mechanical stresses to the reinforcement. Conversely, the reinforcement is designed to withstand these stresses and offer the necessary mechanical strength for the intended function.

A nanocomposite typically has at least one dimension within the range of 1–100 nm (Bogue 2011). Nanocomposites have been formed in many forms based on the kind of matrix employed. They are broadly categorised into metal matrix nanocomposites (MMCs), ceramic matrix nanocomposites (CMNCs), and polymer matrix nanocomposites (PMNCs). PMNC (Polymer Matrix Nanocomposite) and CMNC (Ceramic Matrix Nanocomposite) have been used in aviation and spacecraft applications, but the utilisation of MMCs is limited owing to manufacturing constraints (Jayakrishna et al. 2018).

9.2.1 Polymer Matrix Nanocomposites

In recent times, polymer matrix composites have gained significant importance as essential materials for both functional and structural purposes owing to their characteristics such as low thermal expansion, high strength-to-weight ratio, and lightness (Singh et al. 2013; Tuli et al. 2024). According to research, thermal and mechanical designs that rely only on polymers are considered less trustworthy for aeronautical applications (Haque et al. 2024; Rashid et al. 2024a). As a result, polymeric nanocomposites have gained significant attention in industrial applications because of their exceptional characteristics (Liu et al. 2014). Examples of natural polymers include polyesters, cellulose, starch, chitin, and natural rubber (Rai et al. 2021). In addition, synthetic green polymers such as polycaprolactone, polyurethane, poly(vinyl alcohol), poly(lactic acid) (PLA), and poly(ethylene glycol) (PEG) have been synthesised (Muhammed et al. 2020). Various synthetic polymers have been used using green synthesis methods, including polycarbonates, polyurethane, polystyrene, polyethylene, polyamide, epoxies, and others (Hussain et al. 2013). Eco-friendly polymers are used in several fields, such as engineered constructions, membranes, coatings, textiles, energy/electronic devices, and medicinal techniques (Makhijani et al. 2015).

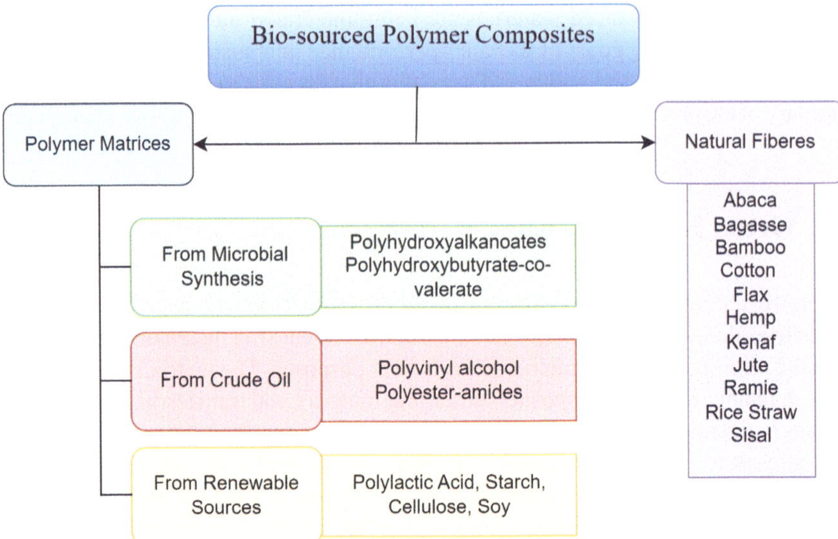

Fig. 9.1 Classification of bio-sourced fiber-reinforced composites (Malucelli and Kausar 2023)

Bio-sourced polymer composites are ingredients that consist of natural fibers and are combined with either nonbiodegradable matrices derived from crude oil (typically epoxy, polyolefins, and unsaturated polyesters) or green alternatives like polyhydroxyalkanoates and polylactic acid. Similarly, biocomposites are formed by combining biodegradable matrices with synthetic fibres such as aramid fibres, carbon, and glass. However, it is significant to note that only one ingredient is derived from biological resources. Composites consisting of natural fibres or bio-sourced polymer matrices reach more eco-friendliness. These composites are often called "green composites," but this term may also be used in other largely bio-sourced systems. Currently, many bio-sourced polymer matrices and fibre reinforcements are accessible, as seen in Fig. 9.1 (Malucelli and Kausar 2023).

A carbon fibre reinforced polymer (CFRP) is the most often used aeroplane structural material, excluding aluminium alloys. It has been used to manufacture key parts of the aircraft's body, including the fuselage, empennage, and wing box. Additionally, it has been used in the construction of control surfaces like the landing-gear doors, flaps, elevator, ailerons, and rudder. The Boeing 777 and 787 airplanes include a significant portion, up to 50%, of their weight in CFRPs (carbon fibre reinforced polymers) material (Zhang et al. 2018; Zhu et al. 2018). While carbon fibres may enhance the strength of polymer matrix composites (PMCs), they are known to induce stress concentration in CFRPs due to their inherent brittleness. Consequently, several investigators are directing their attention towards alternate fibres, including natural fibres, basalt, graphene, and carbon nanotubes (CNT), for use in PMCs (Masrafee et al. 2025).

By including stiff phenyl groups into the matrix, the ablation-resistance implementation of carbon fibre-reinforced poly(silacetylene) composites has been greatly enhanced. These properties would qualify them as suitable candidates for improved thermal protection materials (Kong et al. 2018). Furthermore, reports indicate that polymer composites based on graphene exhibit much superior mechanical characteristics compared to polymer matrix composites reinforced with carbon nanotubes (CNTs). Adding only 0.1 wt% graphene oxide to an epoxy composite significantly improves tensile strength, raising it from 53 to 72 MPa. Basalt is another very efficient reinforcing material used in PMCs. Experimental studies have shown that the addition of basalt fibre content leads to a continual rise in the flexural strength, flexural modulus, tensile strength, and tensile modulus of polybutylene succinate (Wan et al. 2014). In addition, natural fibres are being employed to enhance the mechanical characteristics of PMCs. Bamboo, banana, hemp, flax, and recycled cellulose fibre (RCF) (Saba et al. 2016; Aranno et al. 2019; Haque Protyai et al. 2024) are some examples of natural fibres often used in PMCs. PMCs reinforced with natural fibres and PMCs reinforced with carbon nanotubes have been widely used in military airplanes for structural constituents and as conductive coatings for fighter jets (Masrafee et al. 2025). A significant drawback of PMCs is their limited flame confrontation in the matrix, which may restrict their role in aviation elements in regions where great fire confrontation is essential. Nevertheless, it was noted that including a modest quantity (1–3 wt%) of bentonites that have been altered with quaternary ammonium or phosphonium salts might enhance the mechanical characteristics and fire resistance of PMCs (Oliwa et al. 2016).

9.2.2 Ceramic Matrix Nanocomposites

Ceramic matrix composites (CMCs) have been thoroughly examined owing to their favorable characteristics, including exceptional hardness, remarkable thermal stability, superior corrosion resistance, and commendable adaptability (Rashid et al. 2024b). In addition to alumina, the matrices used for CMCs primarily consist of aluminium titanate, silicon nitride, silicon carbide, zirconia, and aluminium nitride (Zhang et al. 2018). CMCs are utilized in aeroplanes for high-temperature applications, such as exhaust nozzles, due to their competence to endure operating temperatures of up to 1400 °C. Carbon fibre-reinforced silicon carbide is appropriate for aeroplane brakes, particularly in high-temperature environments that may attain up to 1200 °C (Fan et al. 2016). Nanomaterials like graphene and CNTs have been included to enhance the fracture toughness of CMCs. An enhancement in fracture toughness in a silicon nitride-based ceramic composite was reported by incorporating 1.5 vol% of graphene, resulting in a significant increase of 235% (Walker et al. 2011). Additional studies have shown that alumina-based ceramcomposite's fracture toughness improved as the graphene concentration grew from 0 to 0.38 vol%. However, beyond this point, the fracture toughness began to decline with further additions of graphene.

Hence, it is crucial to restrict the amount of graphene to a critical threshold (Liu et al. 2013).

9.2.3 Metal Matrix Nanocomposites

Metal matrix composites (MMC) are also used in the aerospace sector. These materials offer low thermal expansion, fracture toughness, high yield strength, and suitable wear resistance. Metals such as aluminium, nickel copper, magnesium, and titanium are some potential matrices for Metal Matrix Composites (MMCs) (Skoczylas et al. 2019). Aluminum-based composites have improved physical and mechanical qualities compared to the base alloy, making them suitable for aerospace uses (Koli et al. 2015; Adil et al. 2024). The aluminum-based metal matrix composite (MMC) reinforced with 30% silicon carbide (SiC) has a comparable density to 2219 aluminum-based alloys, but exhibits a specific modulus that is 70% higher, an elastic modulus that is 60% higher, and a tensile strength that is 15% higher (Prater 2014). Conventional carbon and ceramic fibres cannot enhance the characteristics of MMCs. The increasing need for fibres that possess enhanced strength reduced thermal expansion coefficient, improved self-lubricating properties, and stronger damping capacity has led to the rising significance of graphene and CNTs (Tabandeh-Khorshid et al. 2020).

GLARE, a glass fibre reinforced aluminium (FML), is a commonly used material in the aerospace industry. It is often employed in the empennage and fuselage skin of the Airbus A380. GLARE is a composite material composed of 2–6 aluminum layers, each with a thickness ranging from 0.2 to 0.5 mm. Layers of fibres and metal are arranged in an alternating pattern and then glued together using an epoxy resin matrix. GLARE has superior properties such as fatigue resistance, high stiffness, low density, high strength, and corrosion resistance compared to monolithic metals. GLARE is a costlier option than aluminium and other composites (Zhu et al. 2018).

Another kind of MMC is known as intermetallic matrix composites (IMC). An instance of using intermetallic compounds (IMCs) is seen in the employment of titanium aluminides in the General Electric GE90. This turbofan engine is manufactured by GE Aviation specifically for the Boeing 777 aircraft. The use of IMCs ensued in a decrease of 150 kg in the overall weight of the engine (Rai et al. 2021).

9.3 Nanoparticles for Aerospace Engineering

The materials used to make aviation industry components must withstand high durability, superior heat resistance, and lightweight, and electromagnetic interference (EMI) resistance. Initially, it was accomplished with composites (CMCs, MMCs, PMCs), which are materials composed of different materials according to anticipated results in application areas, such as an improvement in the material's tribology behavior, an increase in temperature resistance and an intensification in the strength of the base material, alongside other advantages. Despite these composite qualities, they lacked composites' enhanced qualities and multifunctionality. This can be achieved by including various nanoparticles in the matrix material (Bhat et al. 2021). Hence, this section delves into contemporary nanoparticles used in the aerospace sector and green nanoparticles for sustainable air and space travel (Table 9.1).

Modern aviation requires stronger materials that are lower in weight. Nano-size particles have a wider surface area than conventional macroparticles, allowing more contact with surrounding components and causing more reactivity. The variety of nanoparticle classifications is determined by the functional characteristics needed in composite plates. Lamellar nanoparticles are preferred due to their mechanical characteristics, while fibrillar nanoparticles are chosen based on their strength and stiffness (Abbasi et al. 2020). However, carbon and ceramic-based nanoparticles in this industry are ubiquitous (Fig. 9.2).

Table 9.1 Uses of NPs in the aerospace industry (Iqbal et al. 2021)

Application	Used nanoparticles	Nanocomposites
Hard coatings	Alumina (Al_2O_3) Tungsten carbide (WC)	Thin coatings are possible
Catalysts	Alumina (Al_2O_3) Cerium oxide (CeO) Titanium dioxide (TiO_2)	Amplified activity, Better wear resistance
Structural ceramics	Aluminum oxide (Al_2O_3) Aluminum titanate (Al_2O_5Ti) Zirconium oxide (ZrO_2)	Cheap production costs, Enhanced mechanical properties
Conductive inks	Nickel (Ni), Tungsten (W), Silver (Ag)	Electron lithography, Decrease in temperatures, Improved conductivity
Dopants	Wide range of materials based on application	Reduction in operating and capital expenditure, Decrease in processing temperature, Better uniformity
Pigments	Iron oxide (Fe_2O_3) Zirconium silicate ($ZrSiO_4$) Titanium dioxide (TiO_2)	Better physical–optical properties, Lower costs of material costs
Capacitors	Alumina ($Al_2\ O_3$) Barium titanate ($BaTiO_3$) Tantalum (Ta)	Thinner layers, High surface area, Decrease in layer thickness, Small amount of material used to give same level of capacitance
Slurries for polishing	Aluminum oxide (Al_2O_3) Cerium oxide (CeO_2) Tin oxide (SnO_2)	Minimum amount of material improved fnishing Quicker surface removal

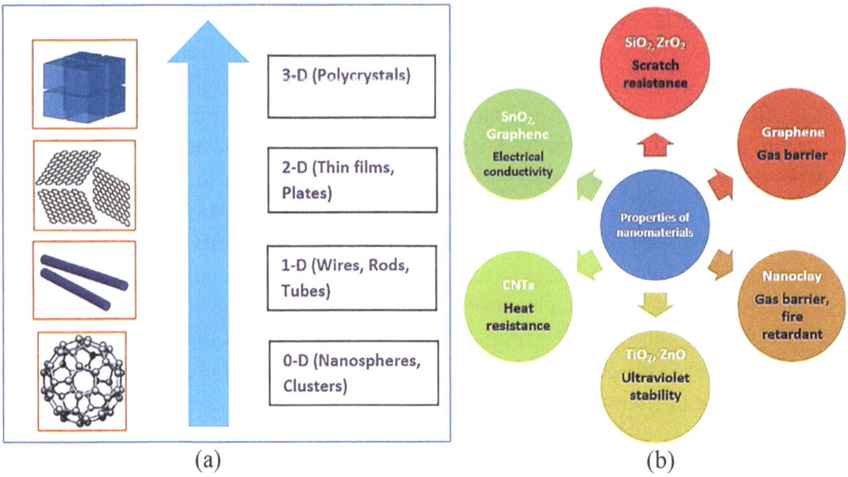

Fig. 9.2 **a** Arrangement of nanomaterials based on number of dimensions, and **b** Properties of nanomaterials (Dhinakaran et al. 2021)

9.3.1 Organic Nanoparticles

Organic nanoparticles include ferritin, micelles liposomes, and dendrimers. They are predominantly used in biomedical applications such as drug and targeted drug delivery. Currently, there are no uses for organic nanoparticles in the aerospace sector.

9.3.2 Inorganic Nanoparticles

Inorganic nanoparticles involve metal oxide and metal-based nanoparticles. Almost all metals can be broken down into their respective nanoparticles. Additionally, metal oxide-based nanoparticles are synthesised by modifying their metal nanoparticle counterparts. They are synthesized to increase reactivity and efficiency. Carbon-based nanoparticles are entirely composed of Carbon. Based on the arrangement of Carbon atoms in the nanoparticle, they can be categorized into Graphene, Carbon Black, Carbon Nanofiber, Carbon Nanotubes (CNTs), and Fullerenes. The properties of different types of nanoparticles have been mentioned in Table 9.2.

9.3.3 Carbon-Based Nanoparticles

Carbon is one of the few elements that may form several physically distinct allotropes because the four valence electrons in its atomic structure create unique properties.

Table 9.2 Characteristics of different types of nanoparticles (Anu Mary Ealia and Saravanakumar 2017)

Carbon-based Nanoparticles	
Nanoparticle	Properties
Graphene	Good light absorption property; Extreme strength, High thermal and electrical conductivity
Carbon Black	High strength and surface area; High electrical conductivity; Resistant to UV degradation
Carbon Nanofiber	High mechanical, electrical, frequency shielding, and thermal properties
Carbon Nano Tubes (CNT)	High tensile strength, flexible and elastic, High thermal and electrical conductivity
Fullerenes	Safe and inert, it transmit light based on the intensity,
Metal-based Nanoparticles	
Aluminium	Large surface area, sensitive to moisture, heat, and sunlight, High reactivity
Iron	Sensitive to air (oxygen) and water, Reactive and unstable
Silver	Stable, anti-bacterial, disinfectant, Absorbs and scatters light
Zinc	Antibacterial, antifungal, anti-corrosive, and UV filtering properties
Copper	Ductile, very high thermal and electrical conductivity, highly flammable solid
Lead	Reactive and highly stable, High toxicity
Cadmium	Insoluble Semiconductor of electricity
Cobalt	Absorbs microwaves, Magnetic, Unstable, and toxic,
Gold	Highly reactive, Interactive with visible light
Metal oxide-based Nanoparticles	
Aluminium oxide	Sensitive to moisture, heat, and sunlight, Large surface area, Increased reactivity
Cerium oxide	Low reduction potential Antioxidant,
Zinc oxide	Antibacterial, antifungal, anti-corrosive and UV filtering
Magnetite	Highly reactive and magnetic,
Silicon dioxide	Able to be functionalize many molecules, Stable and less toxic,
Iron oxide	Reactive and unstable
Titanium oxide	High surface area, Magnetic, Inhibits bacterial growth

Carbon nanotubes (CNT), carbon nanofibers (CNF), diamond, graphene, fullerene, carbon black, and carbon dots are examples of carbon-based nanostructures (Khan and Hossain 2022). GO, graphene, and graphite are members of the graphene family, where graphene is a planar carbon allotrope. Non-covalant forces cause the graphene layers in graphite to adhere to one another. A functionalized sheet of this allotrope is called graphene oxide (GO). The tabular shape of carbom atoms is known as a carbon nanotube (CNT). Researchers introduced three different forms of CNTs. Single-walled nanotubes (SWNTs) are rolled graphene. Double-walled nanotubes (DWNTs) are a form of CNT with two rolled graphene layers. MWNTs are made up of many layers of graphene. The Van der Waals interaction is the primary interlayer force in rolled graphene layers.

9.3.4 Cellulose Nanofibers

Cellulose is a prevalent biopolymer in nature. It has gained material science because they are abundant, eco-friendly, sustainable and renewable (Rashid et al. 2023). Nanocellulose is a category of cellulose that has dimensions in the nanometer range. It can be classified into cellulose nanofibers (CNFs), cellulose nanocrystals (CNCs), and bacterial cellulose (BC). CNFs possess both amorphous regions as well as crystalline regions. CNFs can be converted into crystalline CNCs via controlled hydrolysis. Due to superior tensile properties and low density,, NC-based composites exhibit desirable mechanical properties. They can be used as insulating materials due to their unique dielectric properties. However, the only drawback is that NCs tend to form agglomerates and possess a high degree of supercapacitors, batteries, sensors, and coatings free of water absorption. NC-based composites have been used in solar energy applications (Lasrado et al. 2020).

9.3.5 Halloysite Nanotubes

Halloysite nanotubes (HNTs) are aluminosilicates of the kaolin group possessing a hollow morphology and chemical formula $Al_2Si_2O_5(OH)_4 \cdot nH_2O$. . Classified as dioctahedral 1:1 clay minerals, they are commonly found in soil, particularly in humid tropical and subtropical areas, as well as in weathered igneous and non-igneous rock formations. They are primarily distributed in regions such as the USA, Australia, China, Mexico, Brazil, and New Zealand. The hydration state, characteristic sizes, and purity grade of HNTs may vary depending on the specific deposit. HNTs are considered a valuable alternative to carbon nanotubes (CNTs) for applications as reinforcements in polymers due to their tubular structure and relatively low cost. They have features such as environmental-friendliness, biocompatibility, and high aspect ratio. Most importantly, they have high availability (Massaro et al. 2020).

Due to their high aspect ratio, they are reinforced in polymer matrix composites as they can enhance the fiber-matrix adhesion. The empty lumen in HNTs can be used to load biologically and chemically active substances that exhibit distinct properties in polymers such as corrosion resistance, flame retardancy, and micro-crack self-healing. Hence, HNTs have been used as a filler for different types of polymer matrices (Cavallaro et al. 2013).

9.3.6 Nanoclay

Nanoclays occur naturally, while they can also be obtained via synthetic routes. They are made of SiO_4^{4-} tetrahedra or $[Al_3(OH)_3]^6$ octahedra in the individual nanolayers. Their nano-sized structure has made them excellent nanofillers for polymer matrix composites (PMCs). Polymer nanocomposites incorporated with nanoclay have exhibited superior flame retardancy characteristics, good mechanical properties, and thermal stability owing to their utilizations in tissue engineering, strength applications, drug delivery, and packaging (Kausar et al. 2022). Napier fiber reinforced epoxy composites reinforced with 5 wt% Nanoclay had a 47% increase in tensile strength, 31% enhancement in tensile modulus, 18% intensification of flexural strength, and 19% rise in flexural modulus, exhibiting a significant increase in mechanical characteristics (Wan Ramli et al. 2020). Carbon fiber reinforced composites (CFRCs) strengthened with 1.25 wt% Nanoclay exhibited a 6.3 and 26.06% proliferation in tensile strength and flexural strength, respectively (Agnihotri et al. 2022).

9.4 Environmental Impact of Nanomaterials

9.4.1 Importance of Sustainability in Aerospace Applications

Sustainability considerations must be integrated throughout the entire life cycle of a component, from raw material extraction and fabrication to distribution, usage, maintenance, and end-of-life. These considerations should also be balanced with other factors, such as manufacturability and functionality. Key sustainability issues in product development include evaluating the environmental impact of chemicals and materials used in new products and production processes, assessing the sustainability of manufacturing procedures, improving manufacturing conditions to enhance the working environment, designing products to minimize energy consumption, and ensuring products are designed for recycling and material reuse within the value chain (Hallstedt et al. 2015).

Aircraft materials are confronting a critical sustainability challenge due to their significantly high environmental impact. One of the key reasons behind this is the abundant use of synthetic polymer reinforced composites in major aerospace components. A modern aircraft is composed of roughly 50% by weight of CFRCs in structural parts. The Boeing 787, Bombardier C-Series, and Airbus A350 aircraft are constructed using fiber-reinforced polymer (FRP) composites, comprising 50, 47, and 53% of the aircraft's weight, respectively. In the aviation industry, fiber-metal laminates (FMLs) are commonly utilized in upper fuselage skin panel structures, while CFRCs are employed in aircraft wing panels, wing boxes, and stabilizers. As demonstrated in Fig. 9.3, the significance of composite materials in passenger aviation is rapidly growing (Dyer and Kumru 2023).

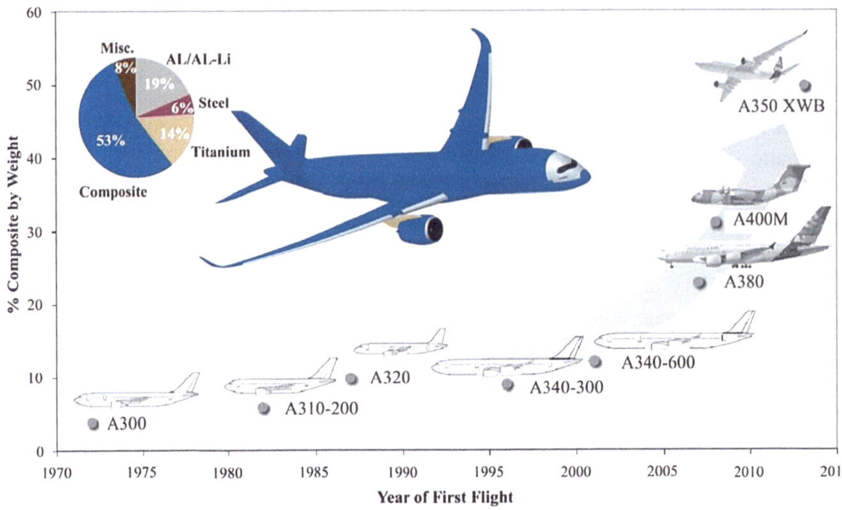

Fig. 9.3 Trends in the utilize of composite materials in Airbus (Dyer and Kumru 2023)

One by one, this subsequent section will delve into the sustainability issues associated with using polymer-reinforced composites in aerospace applications. Epoxy resins are among the most utilised resin systems for composites, but they are primarily made from petroleum-based and toxic chemicals. Consequently, future research should prioritize developing sustainable alternatives to epoxy resins to meet clean aviation objectives (Gonçalves et al. 2022). Other commonly used matrix systems, such as Polyimides and Cyanate Esters suffer from the same problem of being derived from petroleum-based raw materials or chemicals. Hence, using polymer composites in aerospace applications is currently unsustainable in terms of materials and manufacturing processes. The resins used are made from toxic and hazardous chemicals derived from petrochemical refineries, the production of carbon fibers (CF) has a considerable environmental impact, and the manufacturing of composites requires a lot of energy. Additionally, decommissioned aircraft and wind turbines end their life cycle in remote desert graveyards since these materials cannot be recycled. To create a more sustainable aviation industry, an initial step could be to source renewable materials for producing carbon fibers and to develop energy-efficient welding technologies for thermoplastic composites (Dyer and Kumru 2023).

9.4.2 Biobased Resin

Depending on the application, improved starting materials and formulations can generate more viable and higher-performance materials. This is crucial when evaluating biobased materials, as the diverse chemical functionalities in nature can lead to advanced aerospace composites. Contrary to the belief that biobased materials are inferior, their multiple functionalities often enhance performance. Biobased materials are also generally less toxic than petroleum-based ones. Lignin, a phenolic polymer found abundantly in plants and trees, is a significant source of sustainable resins and is valuable for high-performance thermosets due to its chemical properties. A notable research gap links chemistry-focused studies to composite inventing and engineering; resin properties are key to this connection (Schutyser et al. 2018).

9.4.3 Frontal Polymerization

Frontal Polymerization (FP) is a self-sustaining reaction driven by the heat released during polymerization. In FP, resin monomers convert into a solid polymer through a continuously advancing local reaction zone. This can be advantageous, especially from a safety perspective. Production under extreme conditions is possible without the necessity of proximity to the reaction location (Luo et al. 2024). This significantly lowers energy consumption and drastically shortens cure times compared to traditional oven or autoclave methods. Current methods for the fabrication high-performance thermoset and fiber-reinforced polymer composite (FRPC) parts depend on curing in large, costly autoclaves or ovens. For instance, curing a small part of the Boeing 787's carbon fiber/epoxy fuselage is projected to consume 350 gigajoules (GJ) of energy over an eight-hour cure cycle, resulting in the production of over 80 tons of carbon dioxide (Timmis et al. 2015). The polymers and composites produced through this method have comparable mechanical characteristics to those cured using conventional techniques. This approach greatly enhances the manufacturing efficiency of high-performance polymers and composites and is appropriate across various industries. However, most materials used in frontal polymerization are unsuitable for high-performance applications. For instance, while acrylate monomers have the necessary reactivity and energy density for frontal polymerization, the resulting polymers lack the mechanical strength needed for structural FRPCs. On the other hand, epoxy monomers produce mechanically strong polymers but are difficult to polymerize due to their lower reactivity frontally (Robertson et al. 2018). Understanding the chemistry of the thermosetting resin is essential for using FP in aerospace components. Accurate control of polymerization rates at both room and high temperatures enables monomer solutions to turn into fully cured polymers rapidly within seconds.

FP has several advantages, including but not limited to a simple, quick, economical, and environmentally friendly reaction rate without external energy requirements. Hence, they have gained applications in many sectors (Li et al. 2022). The various applications of FP have been highlighted in Fig. 9.4.

Zhang et al. reviewed the effect of integrating CNTs, discontinuous Carbon Fibers (d-CFs), and Graphene Oxide (GO) in epoxy thermosets in order to tune the process of FP. It was observed that the addition of the nanoparticles changed the frontal temperature and frontal velocity, discovering a new pathway for the additive manufacturing

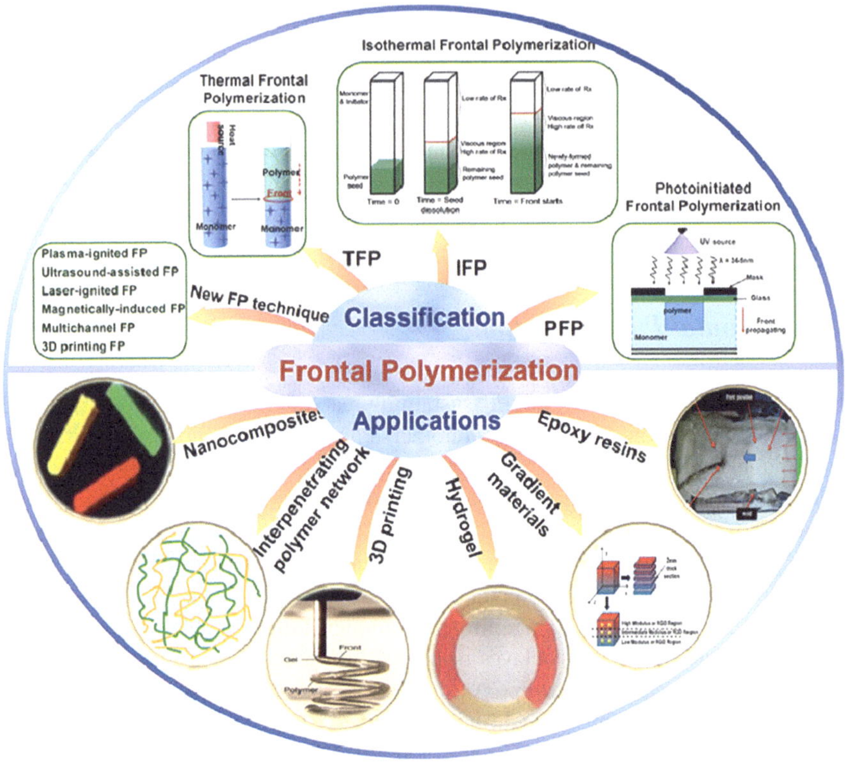

Fig. 9.4 Applications of frontal polymerization (Li et al. 2022)

process (Zhang et al. 2022). Zhang et al. found that frontal polymerization-assisted 3D printing of short carbon fiber/dicyclopenpentadiene composites provided a new method to enhance the interlayer bonding in 3D-printed thermosetting composites (Zhang et al. 2021).

9.4.4 Composite Recycling

The strength of these thermoset composites comes from the cross-linking of polymer chains, but this makes them very challenging to recycle. Recycling requires extreme conditions, and removing fibers under these conditions complicates the process further. This difficulty in recycling the fibers in thermoset polymers leads to high processing costs. The recycling of composites can be divided into three categories: mechanical, thermal, and chemical (Vieira et al. 2017). Table 9.3 illustrates the issues associated with each type of process.

Table 9.3 Different types of composite recycling processes

<table>
<tr><th colspan="2">Recycling process for Composites</th></tr>
<tr><td colspan="2">Mechanical Process: Shredding, chopping or milling the composites to use as fillers or reinforcements in polymers
Disadvantages: Results in deteriorated mechanical properties of the final composite</td></tr>
<tr><td colspan="2">Thermal Process: Involves exposing the composite to high temperatures</td></tr>
<tr><td>• Oxidation in a fluidized bed: The polymer is combusted in the presence of high temperature and high concentration oxygen stream
Disadvantage: Deterioration of mechanical properties of the resulting material</td><td>• Pyrolysis: Combustion of the polymer in the absence of Oxygen
Disadvantage: Risk of damage to the fiber surface and production of toxic gases</td></tr>
<tr><td colspan="2">Chemical Process: Processing the composite in extreme conditions to separate the polymer from the matrix fiber. This can be through supercritical fluids, catalysts or solvents
Disadvantage: Results in reduced adhesion between the fiber and the matrix system</td></tr>
</table>

Perry et al. proposed that the aviation industry should be the starting point for recycling composites due to its need for efficient, durable materials. Their principle suggests that recycled fibers from aviation should then be used in the automotive industry, which has less stringent structural requirements. Eventually, fibers from the sporting goods industry could be recycled for use in civil construction for non-structural applications, such as interior residential uses. This study emphasized managing the recycling process across different industries (Perry et al. 2012).

9.5 Synthesis of Green Nanoparticles for Aerospace

9.5.1 Plant Extracts

Nanoparticles synthesis has become a common term and famous because of its versatility in comparison with macro molecules in many fields, such as more significant surface area ratio, activity of its molecules, and modified structure. One of its most important products is AgNPs, which were synthesized using various chemical products/chemical methods, creating many harmful other bi-products detrimental for humans. The green synthesis of AgNPs involves three main stages: selection of a suitable solvent medium, use of environmentally friendly reducing agents, and incorporation of non-toxic components for the stabilisation of AgNPs. Compared to chemical and physical approaches, sustainable production of AgNPs has several benefits, including being less expensive, environmentally benign, readily scalable for large-scale synthesis, and free of hazardous compounds and high energies, temperatures, and pressures. Plant extracts are a superior synthesis technique for nanoparticles than other biological approaches since they do not require the laborious process to sustain cell cultures. Additionally, it may be appropriately scaled up for AgNPs synthesis on a big scale. When compared to AgNPs made by other techniques, those formed by plant extracts are often more diverse in size and shape and more stable (Ghaffari-Moghaddam et al. 2014). The greatest classes of naturally occurring antioxidants are thought to be found in plants, and they have great potential for use as medications, food additives, and nutraceuticals. The elementary idea behind green production techniques is that the phytochemicals found in plant parts have a dual function as a natural reductant and a stabilizer of nanoparticles. Pd-NPs manufactured using Filicium deciduiens extract were reported to have the desired antibacterial activity, which backs up its application in the field of medicine. Another study revealed that honey might be used to create CeO_2 nanoparticles. K. Rayapa Reddy (Reddy 2017) described the asclepiadaceous plant Calotropis procera as a green method for synthesizing CuO nanoparticles. Thanks to their small band gap and photocatalytic qualities, these NPs are widely employed in numerous applications, including catalysis (Yadi et al. 2018). Gowhar et al. showed in their study that carboxylic acid provides extraordinary reducing reaction conditions that prevent oxidation and modify the shape of Au nanoparticles in Fig. 9.5. In order to prepare plant extracts, acacia gum and gum carbohydrates are frequently used as bio-reducing and bio-stabilizing agents for the manufacture of zero-valent metal nanoparticles. It is also found in another study that Ag nanoparticles attach to the virus particle and stop it from replicating its nucleic acid in the plant cell shown in Fig. 9.6. Additionally, they noticed that the release of reactive oxygen species (ROS) was increased, and that system acquired resistance (SAR) was induced (Naikoo et al. 2021). Faryad et al. saw that the environmentally friendly method of producing titanium NPs involved the use of Albizia saman leaf extract. Anatase crystals of TiO_2 nanoparticles formed when the aqueous TiO_2 solution was introduced dropwise into the leaf extract while stirring at 50 °C. XRD examination determined and verified the synthesised TiO_2 nanoparticles' size of 41 nm (Khan et al. 2022).

9.5.2 Microorganisms

Microorganisms are essential to the environmentally friendly production of nanoparticles. They can function as reducing and stabilizing agents in the synthesis procedure. Microorganisms are ideal contenders for green synthesis methods because they have various benefits in nanoparticle synthesis, such as their capacity to operate under moderate reaction conditions, their potential for large-scale production, and their eco-friendliness. Furthermore, the biomolecules these microbes create can give the manufactured nanoparticles particular characteristics, making them appropriate for various uses in industries, including environmental remediation, medicine, and catalysis.

Microorganisms have special transport mechanisms for intracellular biosynthesis. Because of their negative charge, cell walls play a significant role in this process, allowing positively charged metal ions to be deposited in negatively charged cell walls through electrostatic interactions. Following the entry into the microorganism's cells, ions undergo reduction through metabolic processes facilitated by enzymes

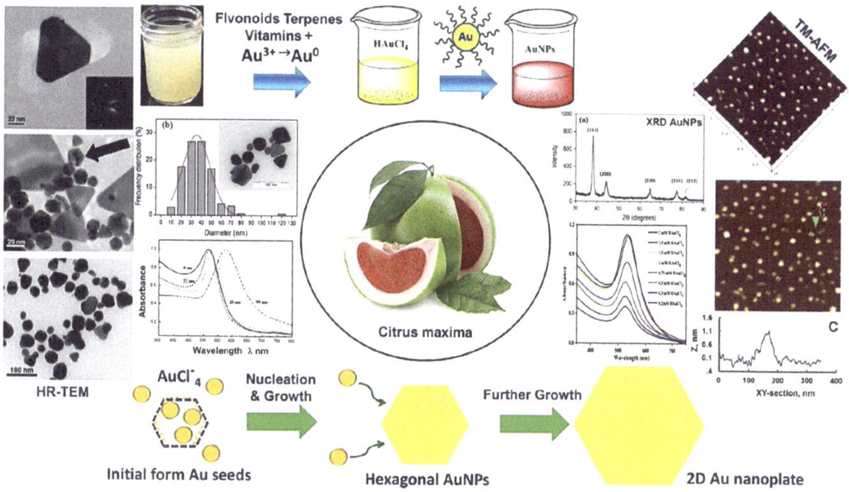

Fig. 9.5 The process of creating gold nanoparticles using citrus fruits such as lemons, illnesses, and reticulate extract (Si et al. 2020)

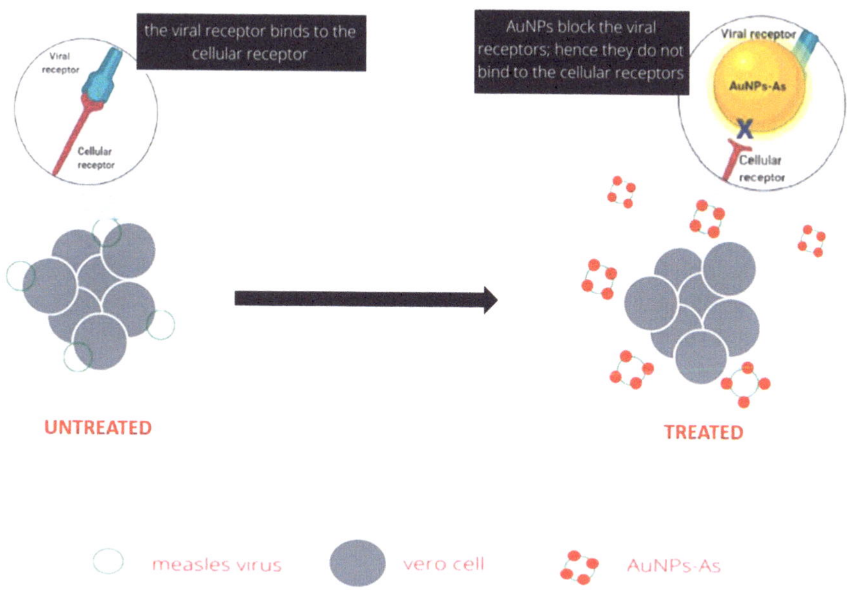

Fig. 9.6 Illustration of the proposed antiviral effect of AuNPs-As on measles (Naikoo et al. 2021)

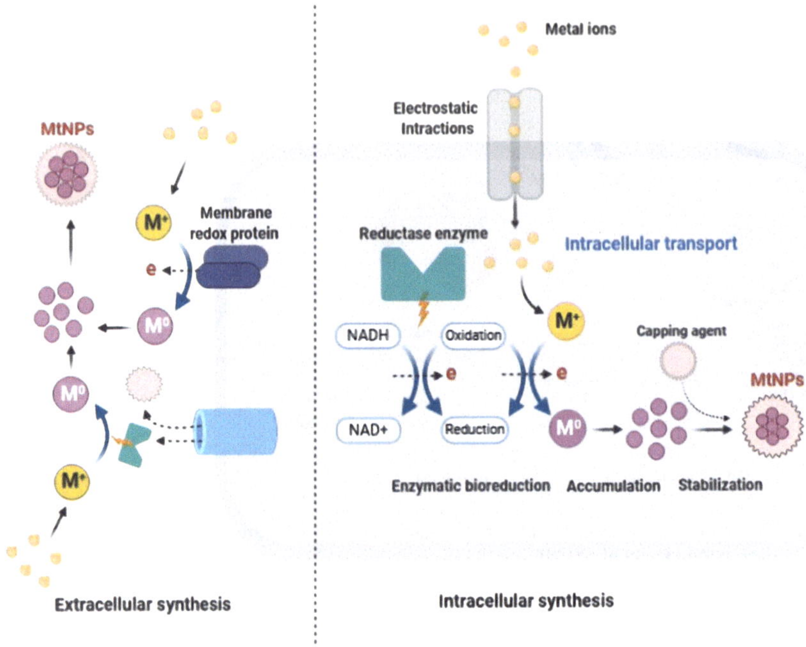

Fig. 9.7 An outline of the biosynthesis processes of MtNPs that take place outside and inside cells (Bahrulolum et al. 2021)

such nitrate reductase, forming MtNPs. The cell wall can then be breached by the MtNPs that have collected in the periplasmic region. Figure 9.7 describes this cellular synthesis (Bahrulolum et al. 2021).

Numerous proteins released by fungi can rapidly and safely hydrolyze metal ions through various mechanisms. Furthermore, fungus may produce high mono-disparity and dimension nanoparticles. Developing natural "nano-factories" by switching from bacteria to fungus makes biomass management and downstream processing considerably easier. Unlike bacterial fermentations, fungal broths may be simply filtered using a filter press or other comparable inexpensive equipment. When Verticillium comes into contact with $AgNO_3$ in water, it reduces the metal ions and forms silver nanoparticles with a diameter of around 25 nm (Talebi et al. 2010).

9.6 Applications of Green Nanoparticles in Aerospace

In some applications within the aerospace industry, achieving high performance is crucial in addition to having a lightweight design (Heidari et al. 2024). These applications include aircraft interiors, cockpit components, equipment enclosures, coatings, solar array substrates, nozzles, shrouds, housings, space-durable mirrors, heat

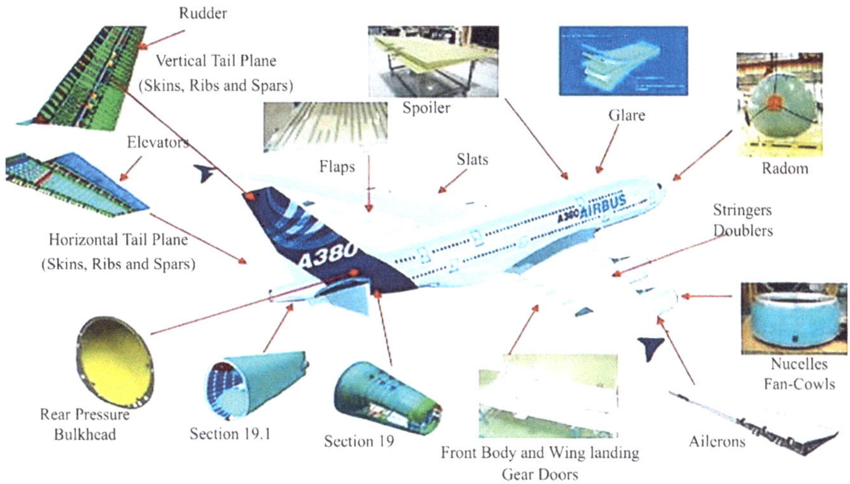

Fig. 9.8 Composites used in aerospace application (Raheem and Subbaya 2021)

shrinkage tubing, and crew gear. Composite materials provide chemical stability and fire resistance in addition to the benefit of cheap running costs resulting from their lightweight nature. Three significant disadvantages restrict the use of composites in most aeronautical constructions. Higher electrical resistance imposes limitations on many applications, such as lightning strike protection, antennas, circuits, and electromagnetic shielding (Olaniyan et al. 2024). Furthermore, the reduced thermal conductivity of composites places a greater burden on de-icing systems that rely on electrical heaters. In addition, composites exhibit lower impact resistance and are susceptible to moisture absorption, environmental deterioration, and ageing over time. Nanocomposites have provided several strategies to circumvent these restrictions, depending on the matrix and filler used. The following section examines several nanocomposites and their uses in the field of aviation.

9.6.1 Aerospace Wing Boxes

Aerospace wing boxes are typically constructed from aluminum alloys or high-performance nonrenewable CFRCs. In this regard, some efforts have been made to use bio-sourced reinforcements, as listed in Table 9.4. Moreover, a Finite Element Analysis indicates that using ramie fibers can reduce the weight of a wing box by approximately 12–14% compared to a standard aluminum wing box (Korba 2024), allowing for the design of a lighter and more biodegradable structure.

Conversely, using composites reinforced with flax or hemp fibers significantly increases wing box weight, as shown in Table 9.5.

Table 9.4 Property of composite made by various bio-sourced reinforcements used for wing boxes of aeroplane (Malucelli and Kausar 2023)

Reinforcement	Yield strength (MPa)	Young's modulus (MPa)	Density (g)
Hemp	550–900	70	1.5
Flax	800–1500	60–80	1.4
Sisal	600–700	38	1.3
Ramie	500	44	1.5

Table 9.5 Assessment of the weight of different polymer composites (Malucelli and Kausar 2023)

Material	Wing box weight in kg
Flax-PLA composite	11,780
Flax-Epoxy composite	11,360
Hemp-PLA composite	10,800
Flax-Epoxy composite	10,500
Ramie-PLA composite	7760
Ramie-PLA composite	7570
Aluminium	8830

9.6.2 Aerospace Radome Applications

Table Y presents the mechanical characteristics of numerous natural reinforcements (pineapple leaf fibers, oil palm empty fruit bunch, kenaf, banana, and bamboo) currently used in aircraft radome applications, compared to E and S-type glass fibers (Table 9.6). The data clearly demonstrate that kenaf exhibits the highest mechanical performance among the bio-sourced reinforcements, with stiffness approaching that of glass fibers. Additionally, due to its low dielectric constant (approximately 3), kenaf is particularly suitable for aircraft radome purposes. Moreover, incorporating kenaf or glass fibers into hybrid composites can improve resistance to rain erosion and overall mechanical performance (Haris et al. 2011).

Table 9.6 Comparison of the mechanical performance between glass and natural reinforcements used in radome applications (Haris et al. 2011)

Fiber	Tensile strength	Young's modulus	Elongation break
Kenaf	930	53	1.6
Banana	500	12	5.9
Pineapple leaf fiber	170	6	3
Oil palm empty Fruit bunch	0.1–0.4	1–9	8–10
Bamboo	503	36	1.4
E-glass	200–3500	70	2.5–3
S-glass	4570	86	2.8

9.6.3 Aerospace Coating Applications

Aerospace structures undergo a wide range of environmental circumstances, including fluctuations in temperature and moisture. They are also exposed to touch with jet fuel, de-icing fluid, and hydraulic fluid. The coatings must be able to endure the impact of lightning strikes, the effects of UV radiation, and the degradation caused by dust streaming at a velocity of 500 mph. An aeronautical coating generally consists of a primer and a topcoat. The primer enhances the substrate's adherence and shields it against corrosion. The topcoat must possess a matte finish, corrosion protection, durability, flexibility, and consistent color in order to have a decent look. The classification of these needs is shown in Fig. 9.9 (Peng 2011).

Lightning strikes may result in direct impacts, indirect effects, and damage to an aircraft. The heat, acoustic shock waves, and electrical charges mentioned may directly inflict significant damage. The heat can reach temperatures as high as 30,000 °C, the acoustic shock waves can exert an impact force of 500 pounds per square inch, and the electrical charge can reach currents of up to 200,000 amperes (Rathod et al. 2017). Indirect impact may result in harm or disruption to technological devices. The generation of heat is mostly due to resistive or Joule heating, which is likely to do the greatest damage by causing the breakdown of the matrix in a composite material. In order to mitigate localized damage, a thin metal mesh is included in the outer layers of the composite material (Gagné and Therriault 2014). Additionally, this functions as a protective barrier for electrical systems. Nonconductive composites are shielded with

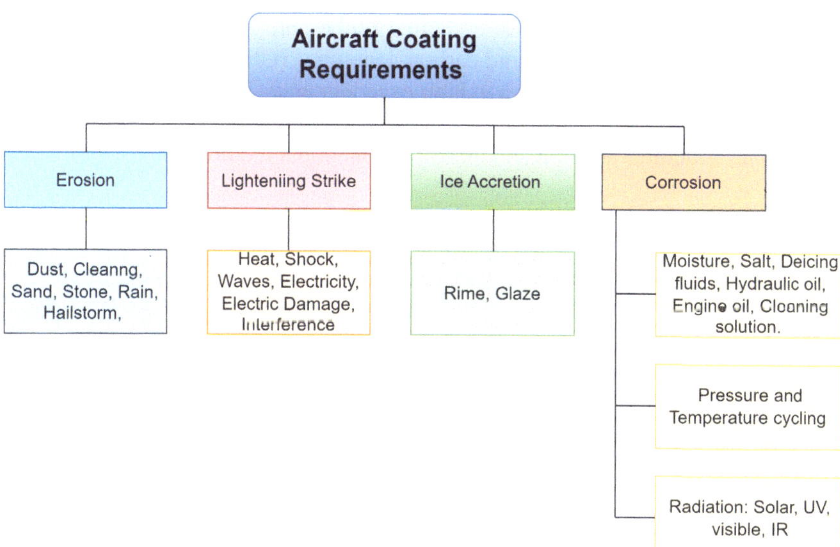

Fig. 9.9 Requirements of aircraft structural coatings (Peng 2011)

external strips to capture lightning, while yet enabling electromagnetic waves radiated from the antennas to pass through the composite panels of the antenna fairings. When there is no need for radio frequency transparency, a conductive substance is put to the outside surface. To enhance resistance to electromagnetic interference (EMI) and lightning strikes, conductive polymers and single-walled carbon nanotubes are used. Bucky-paper materials were included in composite materials (Park et al. 2009). The Bucky-paper materials were composed of densely packed and magnetically oriented carbon nanotubes (CNTs) that exhibited a linear change in conductivity with temperature.

Epoxy coatings are affordable coatings that provide excellent corrosion resistance, chemical resistance, adhesion, and hardness. Turri et al. (2010) created nanostructured coatings using epoxy/clay and epoxy/alumina. These coatings possess enhanced features such as increased scratch resistance, improved thermal properties, and superior barrier properties. These qualities make them suitable for aircraft coatings, meeting the key needs in this field. The indentation hardness and wear resistance values for different kinds of coatings may be found in Table 9.7.

Nanocomposite coatings containing fillers with diverse functions are suitable for use as topcoats on aircraft, effectively meeting their many criteria. Polyurethane elastomers have exceptional durability, resistance to corrosion, elasticity, and adhesive characteristics (Chattopadhyay and Raju 2007). Clay, alumina, CNT, and nanodiamond are essential components used in the production of aircraft coating. Spray painting is a method specifically used for painting airplanes. The use of fillers enhances the viscosity, leading to challenges in spray painting and insufficient wetting of nanoparticles and additives. Alternatively, the electrospinning approach involves propelling a stream of solution via a capillary tube onto a surface (Karchin et al. 2011). Incorporating nanofibers and carbon nanotubes (CNT) into the solution has the potential to create coatings that exhibit excellent area-to-volume ratio, strength, and flexibility. The composition of manufactured coatings for aircraft purposes varies based on the structural location and matching functional requirements.

Table 9.7 Properties of various epoxy nanocomposite coatings (Turri et al. 2010)

Coating type	Scratch hardness (MPa)	Wear index (mg/k cycle)	Indentation hardness	Mean surface roughness (nm)
Epoxy	42.8 ± 3.2	63 ± 15	99 ± 2	0.16 ± 0.02
Hybrid Epoxy/SiO_2	783.6 ± 51.6	35 ± 7	87 ± 11	2.65 ± 0.09
Epoxy/clay	112.3 ± 5.6	33 ± 5	98 ± 3	3.15 ± 0.04
Epoxy/alumina	83.1 ± 3.1	75 ± 13	101 ± 7	0.59 ± 0.04

Table 9.8 Polymer nanocomposites for EMI shielding

Polymer matrix	Reinforcement	Properties	References
Epoxy resin	EDFe_3O_4-CNTs/rGF	High EMISE value	Liang et al. (2019)
	CNTs	High resistance	Vertuccio et al. (2018)
	Iron, cobalt, nickel, and iron oxide	High strength and non-heavy	Jayakrishna et al. (2018)
PLA	Ag	Multiple scattering	Jalali et al. (2011)
PVDF	Fe_3O_4/carbon	Lightweight	Tang et al. (2020)
PLAUs	Fe_3O_4	Shape recovery in a magnetic field	Gu et al. (2016)
PAN and PU	Ni-Co	Intrinsic conductivity and magnetism	Zhang et al. (2019)
PPy	$Ba_{0.6}Sr_{0.4}Fe_{12}O_{19}$	Low-cost and resistant	Velhal et al. (2019)

9.6.4 EMI Shielding

Polymeric matrix composites, which are covered with metallic magnetic nanoparticles, have several applications in the aerospace sector. One of its significant uses is in the development of materials for shielding against electromagnetic interference (EMI). At present, there are many devices worldwide, such as transmitters, used for satellite propagation, communication, or navigation. These devices produce an electromagnetic interference environment that might potentially disrupt the electronic equipment necessary for airplane navigation (Jayakrishna et al. 2018). As a result, metal materials emerged as a viable choice for protecting aircraft from electromagnetic radiation owing to their excellent electrical conductivity (Jalali et al. 2011) and their ability to create a Faraday cage-like effect (Jayakrishna et al. 2018). Nevertheless, the materials used in this context provide a significant drawback due to their high weight values and susceptibility to corrosion (Zhang et al. 2019). To address these issues, a polymer matrix is covered with metallic magnetic nanoparticles. This combination allows for the integration of the electrical insulating properties of the carbon material with the magnetic properties of metals. The result is a highly effective, corrosion-resistant, and lightweight electromagnetic interference (EMI) shield (Tang et al. 2020). Table 9.8 provides a concise summary of the many uses of polymer nanocomposites in the field of electromagnetic interference (EMI) shielding (Romero-Fierro et al. 2022).

9.6.5 Structural Health Monitoring

Structural Health Monitoring (SHM) is a crucial technique used to assess the soundness of structures utilized in aerospace applications. This approach encompasses several fields of study, including sensors, materials, signal processing, and interpretation (Hou and Choy 2022). It is crucial to emphasize that Structural Health Monitoring (SHM) is not just concerned with identifying structural breakdowns in the components, but also offers a timely indication of this damage. These instructions will enable individuals to implement corrective measures to prevent these issues (Rocha et al. 2021). Inorganic/organic composites of piezoelectric materials are used as appropriate constituents for sensors employed in aerospace and aeronautical monitoring. D'Ambrogio et al. fabricated a piezoelectric composite consisting of lead zirconate titanate (PZT) as a filler, incorporated inside a poly (dimethyl siloxane) (PDMS) matrix. The piezoelectric behavior of this novel material, consisting of a 1–3 configuration of PZT/PDMS, is exceptionally superior even at elevated temperatures of 200 °C. Hence, these researchers determined that this composite material has significant promise for incorporation into a sensor used in aviation ball bearings for condition monitoring (Qing et al. 2019).

9.7 Challenges and Limitations

The current adoption of "green" composites in the aerospace engineering has created substantial opportunities for academic and industrial researchers, enabling extensive exploration of innovative materials and processing techniques. The ability to use green composites to design and fabricate a substantial portion of aircraft fuselage components and interiors is a significant achievement in aerospace technology. Aerospace composites' "green" nature primarily comes from using bio-sourced reinforcements. Although these reinforcements have lower thermal stability and mechanical performance than traditional materials, they are biodegradable, cost-effective, and lightweight. However, challenges exist, such as moisture absorption, which reduces mechanical performance and promotes delamination in the manufactured products. Additionally, bio-sourced reinforcements exhibit very limited flame-retardant characteristics (Yahaya et al. 2014).

Moreover, the thermomechanical possessions of bio-sourced reinforcements are inconsistent, as they can vary depending on the harvesting area and time, soil conditions, rain frequency, and sunlight intensity during the day. Furthermore, it is crucial to ponder that bio-sourced reinforcements are typically polar, whereas the polymer matrices are incorporated into are often apolar. This polarity difference can result in poor dispersion homogeneity and insufficient orientation necessary for achieving anisotropy, thereby deteriorating the overall mechanical performance of the fabricated components and parts. This issue can be mitigated through specific chemical treatments on natural reinforcements, such as mercerization, silanization, and acetylation (Sepe et al. 2018; George and Bressler 2017). Additionally, it is important to note that fully replacing synthetic, high-performance assistance with bio-sourced complements is practically impossible, as synthetic materials generally exhibit superior performance. However, in aerospace technology, partial substitution is often feasible, enhancing the resulting composites' biodegradability.

9.8 Future Prospects of Green Nanocomposites

Indeed, the green composites align well with the stringent requirements of the aerospace industry, offering light, cost-effective, renewable, and thermo-mechanically capable solutions. However, some issues remain unresolved, such as ensuring the new green techniques can withstand severe circumstances characterized by high moisture and high temperatures, to which some bio-sourced composites are susceptible. On the other hand, the potential weight reduction accomplished by using these new approaches is highly fascinating to aerospace inventors and producers. Consequently, it is anticipated that new green composite materials with enhanced characteristics will become available on the market and appropriate for deliberate applications in the coming years. The further advancement of green composites for aerospace uses is closely linked to the growing demand for high-performance, recyclable, and durable materials. In this context, composites obtained from natural sources such as animal or plant-based fibers and biodegradable polymer matrices demonstrate significant ability to replace current standard material structures. Moreover, any new improvement to enhance the range and property of green composite materials appropriate for aerospace purposes must address several key considerations. These include the optimal selection of polymer matrix and reinforcement, the enhancement of interfacial strength, achieving effective dispersion and orientation of fibers to minimize porosity, the availability of reliable composite fabrication procedures, and the cost-effectiveness of the new material schemes and associated technologies.

9.9 Conclusion

Aerospace applications of green nanoparticles and nanocomposites have been thoroughly examined in this chapter. A new alternative to conventional nanoparticles, "green" nanoparticles are made in an eco-friendly manner and have the potential to save money while reducing their negative impacts on the ecosystem. Biological, chemical, and physical techniques, such as green production strategies for nanoparticles, were covered in depth in this chapter. Not only do these techniques help create high-performance materials, but they are also in line with environmental ideals. Nanocomposites are ideal for use in the aircraft industry because they are strong, thermally stable, and lightweight thanks to nanoparticles.

The chapter also focused on the environmental advantages of green nanoparticles and nanocomposites over traditional materials, with an eye on how they may lessen the impact of aerospace production and operation on the environment. However, the chapter pointed out a few problems, such as technological issues, economic concerns, and the need for further R&D, notwithstanding these advantages. Consistent progress is expected to address current obstacles and broaden the uses of environmentally friendly nanoparticles and nanocomposites in aircraft, indicating a bright future for these materials. Better and more efficient techniques of synthesis, better material characteristics, and deeper integration of these technologies into commercial aircraft systems are all things that may be expected from ongoing research. Researchers, producers, and regulators must work together to take these materials from the lab to the practical field.

In addition to being in line with larger sustainability objectives, the use of green nanoparticles and nanocomposites signifies a substantial technical change. These cutting-edge materials will be crucial in mitigating the environmental effects of aerospace operations, which are becoming more important as the sector emphasizes ecological responsibility. Finally, a significant step towards a greener future is using nanocomposites and green nanoparticles in aeronautical materials. The aerospace industry has the potential to significantly enhance material performance while reducing environmental effects via the use of green chemistry and sophisticated nanotechnology. The ever-changing nature of these technologies highlights how crucial innovation is for tackling the dual objectives of high performance and environmental responsibility.

Looking forward, aerospace industry players must get behind in creating environmentally friendly materials so that sustainability and technical progress may coexist. Aerospace engineering will surely advance because of the continued study and use of ecologically friendly nanoparticles and nanocomposites, an example for other sectors to emulate.

References

Abbasi S, Peerzada MH, Nizamuddin S, Mubarak NM (2020) Functionalized nanomaterials for the aerospace, vehicle, and sports industries. In: Handbook of functionalized nanomaterials for industrial applications. Elsevier, pp 795–825. https://doi.org/10.1016/B978-0-12-816787-8.00025-9

Adil MM et al (2024) Effect of jute stick charcoal powder on characterization of aluminum metal matrix composite. Results Eng 22:101990. https://doi.org/10.1016/j.rineng.2024.101990

Agnihotri SN, Thakur RK, Singh KK (2022) Influence of nanoclay filler on mechanical properties of CFRP composites. Mater Today Proc 66:1734–1738. https://doi.org/10.1016/j.matpr.2022.05.270

Anu Mary Ealia S, Saravanakumar MP (2017) A review on the classification, characterisation, synthesis of nanoparticles and their application. IOP Conf Ser Mater Sci Eng 263:032019. https://doi.org/10.1088/1757-899X/263/3/032019

Aranno TM et al (2019) Fabrication, experimental investigation of jute fiber reinforced epoxy composites and hybrid composites. IOP Conf Ser Mater Sci Eng 628(1):012011. https://doi.org/10.1088/1757-899X/628/1/012011

Bahrulolum H et al (2021) Green synthesis of metal nanoparticles using microorganisms and their application in the agrifood sector. J Nanobiotechnol 19(1):86. https://doi.org/10.1186/s12951-021-00834-3

Balakrishnan P, John MJ, Pothen L, Sreekala MS, Thomas S (2016) Natural fibre and polymer matrix composites and their applications in aerospace engineering. In: Advanced composite materials for aerospace engineering. Elsevier, pp 365–383. https://doi.org/10.1016/B978-0-08-100037-3.00012-2

Barhoum A et al (2022) Review on natural, incidental, bioinspired, and engineered nanomaterials: history, definitions, classifications, synthesis, properties, market, toxicities, risks, and regulations. Nanomaterials 12(2):177. https://doi.org/10.3390/nano12020177

Bhat A et al (2021) Review on nanocomposites based on aerospace applications. Nanotechnol Rev 10(1):237–253. https://doi.org/10.1515/ntrev-2021-0018

Bogue R (2011) Nanocomposites: a review of technology and applications. Assem Autom 31(2):106–112. https://doi.org/10.1108/01445151111117683

Cavallaro G, Lazzara G, Milioto S (2013) Sustainable nanocomposites based on halloysite nanotubes and pectin/polyethylene glycol blend. Polym Degrad Stab 98(12):2529–2536. https://doi.org/10.1016/j.polymdegradstab.2013.09.012

Chattopadhyay DK, Raju KVSN (2007) Structural engineering of polyurethane coatings for high performance applications. Prog Polym Sci 32(3):352–418. https://doi.org/10.1016/j.progpolymsci.2006.05.003

Dhinakaran V, Sai MS, Shree MV (2021) Some impact of nanomaterials in aerospace engineering. In: Babu BS, Kumar K (eds) Nanomaterials and nanocomposites, 1st edn. CRC Press, Boca Raton, FL, pp 17–29. https://doi.org/10.1201/9781003160946-3

Dyer WE, Kumru B (2023) Polymers as aerospace structural components: how to reach sustainability? Macro Chem Phys 224(24):2300186. https://doi.org/10.1002/macp.202300186

Fan S et al (2016) Progress of ceramic matrix composites brake materials for aircraft application

Gagné M, Therriault D (2014) Lightning strike protection of composites. Prog Aerosp Sci 64:1–16. https://doi.org/10.1016/j.paerosci.2013.07.002

George M, Bressler DC (2017) Comparative evaluation of the environmental impact of chemical methods used to enhance natural fibres for composite applications and glass fibre based composites. J Clean Prod 149:491–501. https://doi.org/10.1016/j.jclepro.2017.02.091

Ghaffari-Moghaddam M, Hadi-Dabanlou R, Khajeh M, Rakhshanipour M, Shameli K (2014) Green synthesis of silver nanoparticles using plant extracts. Korean J Chem Eng 31(4):548–557. https://doi.org/10.1007/s11814-014-0014-6

Gonçalves FAMM, Santos M, Cernadas T, Ferreira P, Alves P (2022) Advances in the development of biobased epoxy resins: insight into more sustainable materials and future applications. Int Mater Rev 67(2):119–149. https://doi.org/10.1080/09506608.2021.1915936

Gu S-Y, Jin S-P, Gao X-F, Mu J (2016) Polylactide-based polyurethane shape memory nanocomposites (Fe_3O_4/PLAUs) with fast magnetic responsiveness. Smart Mater Struct 25(5):055036. https://doi.org/10.1088/0964-1726/25/5/055036

Hallstedt SI, Bertoni M, Isaksson O (2015) Assessing sustainability and value of manufacturing processes: a case in the aerospace industry. J Clean Prod 108:169–182. https://doi.org/10.1016/j.jclepro.2015.06.017

Haque M, Islam SMM, Chowdhury P, Labib KMRU, Bin Rashid A (2024) A comparative study on the influence of MWCNT, GO, and Al(OH)3 gel matrix modification on hierarchical structured composite reinforced with needle-punched jute fiber and glass fiber. Results Eng 21:101671. https://doi.org/10.1016/j.rineng.2023.101671

Haque Protyai MI, Adib FM, Taher TI, Karim MR, Rashid AB (2024) Performance evaluation of kevlar fiber reinforced epoxy composite by depositing graphene/SiC/Al_2O_3 nanoparticles. Hybrid Adv 6:100245. https://doi.org/10.1016/j.hybadv.2024.100245

Haris MY, Laila D, Zainudin ES, Mustapha F, Zahari R, Halim Z (2011) Preliminary review of biocomposites materials for aircraft radome application. KEM 471–472:563–567. https://doi.org/10.4028/www.scientific.net/KEM.471-472.563

Heidari M, Thangavel S, Ghafri KA, Kumar A (2024) Future trends and emerging research in nanofluids for aerospace applications. In: Nanofluids technology for thermal sciences and engineering. CRC Press

Hou Y, Choy KL (2022) Durable and robust PVDF-HFP/SiO_2/CNTs nanocomposites for anti-icing application: water repellency, icing delay, and ice adhesion. Prog Org Coat 163:106637. https://doi.org/10.1016/j.porgcoat.2021.106637

Hussain ST, Abbas F, Kausar A, Khan MR (2013) New polyaniline/polypyrrole/polythiophene and functionalized multiwalled carbon nanotube-based nanocomposites: Layer-by-layer in situ polymerization. High Perform Polym 25(1):70–78. https://doi.org/10.1177/0954008312456048
Iqbal A, Saeed A, Ul-Hamid A (2021) A review featuring the fundamentals and advancements of polymer/CNT nanocomposite application in aerospace industry. Polym Bull 78(1):539–557. https://doi.org/10.1007/s00289-019-03096-0
Jalali M, Dauterstedt S, Michaud A, Wuthrich R (2011) Electromagnetic shielding of polymer–matrix composites with metallic nanoparticles. Compos B Eng 42(6):1420–1426. https://doi.org/10.1016/j.compositesb.2011.05.018
Jayakrishna K, Kar VR, Sultan MTH, Rajesh M (2018). Materials selection for aerospace components. In: Sustainable composites for aerospace applications. Elsevier, pp 1–18. https://doi.org/10.1016/B978-0-08-102131-6.00001-3
Joseph B et al (2021) Green chemistry approach for fabrication of polymer composites. Sustain Chem 2(2):254–270. https://doi.org/10.3390/suschem2020015
Karchin A, Simonovsky FI, Ratner BD, Sanders JE (2011) Melt electrospinning of biodegradable polyurethane scaffolds. Acta Biomater 7(9):3277–3284. https://doi.org/10.1016/j.actbio.2011.05.017
Kausar A, Ahmad I, Maaza M, Eisa MH (2022) State-of-the-art nanoclay reinforcement in green polymeric nanocomposite: from design to new opportunities. Minerals 12(12):1495. https://doi.org/10.3390/min12121495
Khan F, Shariq M, Asif M, Siddiqui MA, Malan P, Ahmad F (2022) Green nanotechnology: plant-mediated nanoparticle synthesis and application. Nanomaterials 12(4):673. https://doi.org/10.3390/nano12040673
Khan S, Hossain MK (2022) Classification and properties of nanoparticles. In: Nanoparticle-based polymer composites. Elsevier, pp 15–54. https://doi.org/10.1016/B978-0-12-824272-8.00009-9
Koli DK, Agnihotri G, Purohit R (2015) Advanced aluminium matrix composites: the critical need of automotive and aerospace engineering fields. Mater Today Proc 2(4–5):3032–3041. https://doi.org/10.1016/j.matpr.2015.07.290
Kong L et al (2018) Novel carbon-poly(silacetylene) composites as advanced thermal protection material in aerospace applications. Compos Sci Technol 162:163–169. https://doi.org/10.1016/j.compscitech.2018.04.038
Korba P (2024) Structural design and material comparison for aircraft wing box beam panel
Lasrado D, Ahankari S, Kar K (2020) Nanocellulose-based polymer composites for energy applications—a review. J ApplPolym Sci 137(27):48959. https://doi.org/10.1002/app.48959
Li Q, Shen H-X, Liu C, Wang C-F, Zhu L, Chen S (2022) Advances in frontal polymerization strategy: From fundamentals to applications. Prog Polym Sci 127:101514. https://doi.org/10.1016/j.progpolymsci.2022.101514
Liang C et al (2019) Highly oriented three-dimensional structures of Fe_3O_4 decorated CNTs/reduced graphene oxide foam/epoxy nanocomposites against electromagnetic pollution. Compos Sci Technol 181:107683. https://doi.org/10.1016/j.compscitech.2019.107683
Liu J, Yan H, Jiang K (2013) Mechanical properties of graphene platelet-reinforced alumina ceramic composites. Ceram Int 39(6):6215–6221. https://doi.org/10.1016/j.ceramint.2013.01.041
Liu Y, Du H, Liu L, Leng J (2014) Shape memory polymers and their composites in aerospace applications: a review. Smart Mater Struct 23(2):023001. https://doi.org/10.1088/0964-1726/23/2/023001
Luo T, Ma Y, Cui X (2024) Review on frontal polymerization behavior for thermosetting resins: materials, modeling and application. Polymers 16(2):185. https://doi.org/10.3390/polym16020185
Makhijani K, Kumar R, Sharma SK (2015) Biodegradability of blended polymers: a comparison of various properties. Crit Rev Environ Sci Technol 45(16):1801–1825. https://doi.org/10.1080/10643389.2014.970682

Malucelli G, Kausar A (2023) Green composites and nanocomposites for aerospace applications. In: Polymeric nanocomposites with carbonaceous nanofillers for aerospace applications. Elsevier, pp 297–316. https://doi.org/10.1016/B978-0-323-99657-0.00014-4

Masrafee MR, Rashid AB, Hoque E (2025) Innovations in 3D printing-assisted biopolymer composites for aerospace applications

Massaro M, Lazzara G, Noto R, Riela S (2020) Halloysite nanotubes: a green resource for materials and life sciences. Rend Fis Acc Lincei 31(2):213–221. https://doi.org/10.1007/s12210-020-00886-x

Muhammed NS, Haq MdB, Al-Shehri D, Rahaman MM, Keshavarz A, Hossain SMZ (2020) Comparative study of green and synthetic polymers for enhanced oil recovery. Polymers 12(10):2429. https://doi.org/10.3390/polym12102429

Naikoo GA et al (2021) Bioinspired and green synthesis of nanoparticles from plant extracts with antiviral and antimicrobial properties: a critical review. J Saudi Chem Soc 25(9):101304. https://doi.org/10.1016/j.jscs.2021.101304

Olaniyan OT, Adeyomoye A, Adetunji CO, Oloruntoba A (2024) Biocomposites for aerospace engineering applications. In: Advances in biocomposites and their applications. Elsevier, pp 285–298. https://doi.org/10.1016/B978-0-443-19074-2.00010-1

Oliwa R, Heneczkowski M, Oleksy M, Galina H (2016) Epoxy composites of reduced flammability. Compos B Eng 95:1–8. https://doi.org/10.1016/j.compositesb.2016.03.074

Park JG et al (2009) Electromagnetic interference shielding properties of carbon nanotube buckypaper composites. Nanotechnology 20(41):415702. https://doi.org/10.1088/0957-4484/20/41/415702

Parveez B, Kittur MI, Badruddin IA, Kamangar S, Hussien M, Umarfarooq MA (2022) Scientific advancements in composite materials for aircraft applications: a review. Polymers 14(22):5007. https://doi.org/10.3390/polym14225007

Peng H-X (2011) Polyurethane Nanocomposite coatings for Aeronautical Applications. In: Leng J, Lau AK-T (eds) Multifunctional polymer nanocomposites. Taylor & Francis Group, pp 337–388

Perry N, Bernard A, Laroche F, Pompidou S (2012) Improving design for recycling—application to composites. CIRP Ann 61(1):151–154. https://doi.org/10.1016/j.cirp.2012.03.081

Prater T (2014) Friction stir welding of metal matrix composites for use in aerospace structures. Acta Astronaut 93:366–373. https://doi.org/10.1016/j.actaastro.2013.07.023

Qing X, Li W, Wang Y, Sun H (2019) Piezoelectric transducer-based structural health monitoring for aircraft applications. Sensors 19(3):545. https://doi.org/10.3390/s19030545

Raheem A, Subbaya KM (2021) A review on hybrid composites used for marine propellers. Mat Sci Res India 18(1):01–06. https://doi.org/10.13005/msri/180101

Rai P, Mehrotra S, Priya S, Gnansounou E, Sharma SK (2021) Recent advances in the sustainable design and applications of biodegradable polymers. Biores Technol 325:124739. https://doi.org/10.1016/j.biortech.2021.124739

Rana S, Fangueiro R (2016) Advanced composites in aerospace engineering. In: Advanced composite materials for aerospace engineering. Elsevier, pp 1–15. https://doi.org/10.1016/B978-0-08-100037-3.00001-8

Rashid AB et al (2023) Synthesis, properties, applications, and future prospective of cellulose nanocrystals. Polymers 15(20):4070. https://doi.org/10.3390/polym15204070

Rashid AB, Hoque ME (2022) Polymer nanocomposites for defense applications. In: Advanced polymer nanocomposites. Elsevier, pp 373–414. https://doi.org/10.1016/B978-0-12-824492-0.00015-5

Rashid AB, Haque M, Islam SMM, Uddin Labib KMR (2024) Nanotechnology-enhanced fiber-reinforced polymer composites: Recent advancements on processing techniques and applications. Heliyon 10(2):e24692. https://doi.org/10.1016/j.heliyon.2024.e24692

Rashid AB, Haque M, Islam SMM, Uddin Labib KMR, Chowdhury P (2024b) Breaking boundaries with ceramic matrix composites: a comprehensive overview of materials, manufacturing techniques, transformative applications, recent advancements, and future prospects. Adv Mater Sci Eng 1–33. https://doi.org/10.1155/2024/2112358

Rathod VT, Kumar JS, Jain A (2017) Polymer and ceramic nanocomposites for aerospace applications. Appl Nanosci 7(8):519–548. https://doi.org/10.1007/s13204-017-0592-9
Reddy KR (2017) Green synthesis, morphological and optical studies of CuO nanoparticles. J Mol Struct 1150:553–557. https://doi.org/10.1016/j.molstruc.2017.09.005
Robertson ID et al (2018) Rapid energy-efficient manufacturing of polymers and composites via frontal polymerization. Nature 557(7704):223–227. https://doi.org/10.1038/s41586-018-0054-x
Rocha H, Semprimoschnig C, Nunes JP (2021) Sensors for process and structural health monitoring of aerospace composites: a review. Eng Struct 237:112231. https://doi.org/10.1016/j.engstruct.2021.112231
Romero-Fierro D, Bustamante-Torres M, Bravo-Plascencia F, Esquivel-Lozano A, Ruiz J-C, Bucio E (2022) Recent trends in magnetic polymer nanocomposites for aerospace applications: a review. Polymers 14(19):4084. https://doi.org/10.3390/polym14194084
Saba N, Jawaid M, Alothman OY, Paridah MT (2016) A review on dynamic mechanical properties of natural fibre reinforced polymer composites. Constr Build Mater 106:149–159. https://doi.org/10.1016/j.conbuildmat.2015.12.075
Schutyser W, Renders T, Van Den Bosch S, Koelewijn S-F, Beckham GT, Sels BF (2018) Chemicals from lignin: an interplay of lignocellulose fractionation, depolymerisation, and upgrading. Chem Soc Rev 47(3):852–908. https://doi.org/10.1039/C7CS00566K
Sepe R, Bollino F, Boccarusso L, Caputo F (2018) Influence of chemical treatments on mechanical properties of hemp fiber reinforced composites. Compos B Eng 133:210–217. https://doi.org/10.1016/j.compositesb.2017.09.030
Si A, Pal K, Kralj S, El-Sayyad GS, De Souza FG, Narayanan T (2020) Sustainable preparation of gold nanoparticles via green chemistry approach for biogenic applications. Mater Today Chem 17:100327. https://doi.org/10.1016/j.mtchem.2020.100327
Singh AP, Sharma M, Singh I (2013) A review of modeling and control during drilling of fiber reinforced plastic composites. Compos B Eng 47:118–125. https://doi.org/10.1016/j.compositesb.2012.10.038
Skoczylas J, Samborski S, Kłonica M (2019) The application of composite materials in the aerospace industry. JTEME 5(1). https://doi.org/10.35784/jteme.73
Tabandeh-Khorshid M, Kumar A, Omrani E, Kim C, Rohatgi P (2020) Synthesis, characterization, and properties of graphene reinforced metal-matrix nanocomposites. Compos Part B Eng 183:107664. https://doi.org/10.1016/j.compositesb.2019.107664
Talebi S, Ramezani F, Ramezani M (2010) Biosynthesis of metal nanoparticles by microorganisms
Tang X-H, Tang Y, Wang Y, Weng Y-X, Wang M (2020) Interfacial metallization in segregated poly (lactic acid)/poly (ε-caprolactone)/multi-walled carbon nanotubes composites for enhancing electromagnetic interference shielding. Compos A Appl Sci Manuf 139:106116. https://doi.org/10.1016/j.compositesa.2020.106116
Timmis AJ et al (2015) Environmental impact assessment of aviation emission reduction through the implementation of composite materials. Int J Life Cycle Assess 20(2):233–243. https://doi.org/10.1007/s11367-014-0824-0
Tuli NT, Khatun S, Rashid AB (2024) Unlocking the future of precision manufacturing: a comprehensive exploration of 3D printing with fiber-reinforced composites in aerospace, automotive, medical, and consumer industries. Heliyon 10(5):e27328. https://doi.org/10.1016/j.heliyon.2024.e27328
Turri S, Torlaj L, Piccinini F, Levi M (2010) Abrasion and nanoscratch in nanostructured epoxy coatings. J Appl Polym Sci 118(3):1720–1727. https://doi.org/10.1002/app.32309
Velhal N et al (2019) Electromagnetic shielding, magnetic and microwave absorbing properties of Polypyrrole/Ba0.6Sr0.4Fe12O19 composite synthesized via in-situ polymerization technique. J Alloy Compd 777:627–637. https://doi.org/10.1016/j.jallcom.2018.11.041
Vertuccio L et al (2018) Smart coatings of epoxy based CNTs designed to meet practical expectations in aeronautics. Compos B Eng 147:42–46. https://doi.org/10.1016/j.compositesb.2018.04.027

Vieira DR, Vieira RK, Chang Chain M (2017) Strategy and management for the recycling of carbon fiber-reinforced polymers (CFRPs) in the aircraft industry: a critical review. Int J Sustain Dev World Ecol 24(3):214–223. https://doi.org/10.1080/13504509.2016.1204371

Walker LS, Marotto VR, Rafiee MA, Koratkar N, Corral EL (2011) Toughening in graphene ceramic composites. ACS Nano 5(4):3182–3190. https://doi.org/10.1021/nn200319d

Wan Y-J, Gong L-X, Tang L-C, Wu L-B, Jiang J-X (2014) Mechanical properties of epoxy composites filled with silane-functionalized graphene oxide. Compos A Appl Sci Manuf 64:79–89. https://doi.org/10.1016/j.compositesa.2014.04.023

Wan Ramli WMA, Abdul Majid MS, Ridzuan MJM, Sultan MTH, Amin NAM, Gibson AG (2020) The effect of nanomodified epoxy on the tensile and flexural properties of Napier fiber reinforced composites. Polym Compos 41(3):824–837. https://doi.org/10.1002/pc.25413

Wazalwar R, Sahu M, Raichur AM (2021) Mechanical properties of aerospace epoxy composites reinforced with 2D nano-fillers: current status and road to industrialization. Nanoscale Adv 3(10):2741–2776. https://doi.org/10.1039/D1NA00050K

Yadi M et al (2018) Current developments in green synthesis of metallic nanoparticles using plant extracts: a review. Artif Cells Nanomed Biotechnol 46(sup3):336–343. https://doi.org/10.1080/21691401.2018.1492931

Yahaya R, Sapuan S, Jawaid M, Leman Z, Zainudin E (2014) Mechanical performance of woven kenaf-Kevlar hybrid composites. J Reinf Plast Compos 33(24):2242–2254. https://doi.org/10.1177/0731684414559864

Zhang X, Chen Y, Hu J (2018) Recent advances in the development of aerospace materials. Prog Aerosp Sci 97:22–34. https://doi.org/10.1016/j.paerosci.2018.01.001

Zhang N et al (2019) Lightweight and flexible Ni–Co alloy nanoparticle-coated electrospun polymer nanofiber hybrid membranes for high-performance electromagnetic interference shielding. J Alloy Compd 784:244–255. https://doi.org/10.1016/j.jallcom.2018.12.341

Zhang Z et al (2021) Frontal polymerization-assisted 3D printing of short carbon fibers/dicyclopentadiene composites. J Manuf Process 71:753–762. https://doi.org/10.1016/j.jmapro.2021.10.014

Zhang Z, Gao C, Liu R, Li W, Qiu J, Wang S (2022) Catalyzed frontal polymerization-aided 3D printing of epoxy thermosets. Addit Manuf Lett 2:100030. https://doi.org/10.1016/j.addlet.2022.100030

Zhu L, Li N, Childs PRN (2018) Light-weighting in aerospace component and system design. Propul Power Res 7(2):103–119. https://doi.org/10.1016/j.jppr.2018.04.001

Chapter 10
Green Nanoparticles and Nanocomposites for Aerospace Applications

Ansar Anjum, Rajni Garg, Rishav Garg, and Nnabuk Okon Eddy

Abstract The aerospace industry is on the cusp of a sustainable revolution with the integration of green nanoparticles and nanocomposites for potential applications in sustainable spacecraft and aircraft design. This innovative approach combines the benefits of nanotechnology with environmentally friendly materials, enhancing electrical conductivity, mechanical properties, thermal stability, durability, and while minimizing ecological footprint. Green nanoparticles have especially shown a transformative potential in shaping a more sustainable and technologically advanced aerospace sector. This chapter highlights the requirements and key dimensions of the aerospace industry for excellent performance under deteriorating and adverse environmental conditions. The multidimensional characteristics of green synthesized nanomaterials for integration with traditional aerospace materials have been explored. The synergistic advancements and the associated challenges to attain sustainable and environment-friendly goals have been signified.

Keywords Green nanoparticles · Nanocomposites · Aerospace · Nanotechnology · Environment-friendly polymers · Aerospace industry · Nanoparticles

---

A. Anjum
Department of Applied Chemistry, Delhi Technological University, Delhi, India

R. Garg (✉)
Department of Applied Science and Humanities, Galgotias College of Engineering and Technology, Greater Noida, India
e-mail: rajnigarg@science.org.in

R. Garg
Department of Civil Engineering, Galgotias College of Engineering and Technology, Greater Noida, Uttar Pradesh, India

N. O. Eddy
Department of Pure and Industrial Chemistry, University of Nigeria, Enugu, Nigeria

Visakh P. M. (ed.), *Nanomaterials for Defense, Aeronautical and Aerospace Applications*, Advanced Structured Materials 236,
https://doi.org/10.1007/978-3-031-92723-2_10

10.1 Introduction

Aerospace involves a broad range of activities including, the design and manufacturing of aircraft, spacecraft, and missiles with diverse applications in transportation (air travel), navigation (GPS and other navigation systems), communication (satellites and satellite communications), national security (military aircraft, missile defence), weather forecasting (weather satellites), scientific research (space exploration, earth observation, manned and unmanned, interplanetary and orbital spacecraft), drones (unmanned aerial vehicles, UAVs), and rockets (launch vehicles, missiles) (Liang et al. 2009). The success of aerospace engineering is highly dependent upon the development of innovative and advanced materials. Aircraft and spacecraft components must be exceptionally lightweight to maximize their payload capacity and enhance fuel efficiency. On a parallel side, they also should possess immense strength and flexibility to withstand the harsh environmental conditions encountered during their launch, operation, and re-entry (Verma et al. 2018). These conditions include mechanical stress, extreme temperatures, micrometeoroid impacts, action of atomic oxygen, and radiation. Aerospace vehicles are also subjected to drastic temperature variations during different phases of their missions (Ince et al. 2023). The structures must endure intense heat during atmospheric re-entry, extreme cold in outer space, and significant mechanical stress during take-off, landing, and flight maneuvers without compromising their structural integrity. In space, vehicles are at risk of collisions with tiny micrometeoroids traveling at high velocities due to which they need to be resilient enough to resist penetration from these impacts and protect the occupants of the vehicle and its critical systems (Laurenzi et al. 2019). Traditional materials such as metals and conventional composites generally fail to meet the demanding combination of these desired properties. Thus, the development and utilization of advanced materials play a crucial role in the success of designing and constructing aircraft and spacecraft (Hołyńska et al. 2018).

To achieve the necessary balance between lightweight construction, strength, and durability required for aerospace applications advanced materials are being investigated by researchers. However, advancements in nanotechnology offer a transformative approach to developing next-generation aerospace materials because by manipulating matter at the nanoscale (1–100 nm), scientists can create novel materials with exceptional properties (Parveez et al. 2022). One such advancement is the emergence of green nanoparticles and green nanocomposites. Green nanoparticles, specifically tailored for aerospace applications, prioritize environmental sustainability throughout their lifecycle and hold immense potential to revolutionize the future of aerospace materials. This encompasses (Balakrishnan et al. 2016):

- Sustainable Source Materials: Utilizing renewable or responsibly sourced materials for nanoparticle production (Hussain et al. 2009).
- Non-toxic and Eco-friendly Synthesis Methods: Employing techniques that minimize environmental impact and ensure worker safety.
- Biodegradable or Recyclable Nanoparticles: Designing nanoparticles that readily degrade or can be efficiently reintroduced into the manufacturing process.

Nanoscale technology unveils a realm where materials exhibit extraordinary properties unseen in their bulk counterpart (Romero-Fierro et al. 2022) covering the areas of intricate design, characterization, precise production, and ingenious application of structures, devices, and systems at the nanoscale level (Bhat et al. 2021). In the field of aerospace research, nanotechnology emerges as a transformative force, poised to revolutionize materials engineering, central to this is the advent of green nanoparticles and nanocomposites, which is leading into the new era of aerospace innovation.

The role of nanotechnology in aerospace technology includes the following:

- *Improved Mechanical Strength*: The integration of nanoparticles into traditional aerospace materials has led to impressively improved mechanical properties, including strength, stiffness, and toughness. The lightweight yet robust materials produced using nanoparticles can be vital for optimizing fuel efficiency and performance (Gautier et al. 2019).
- *Improved Thermal Stability*: One of the major limitations of conventional counterparts in aerospace is thermal resilience. This is crucial for components facing the rigors of extreme temperature fluctuations during flight. The incorporation of nanomaterials like graphene, carbon nanotubes, and nanoceramics in aerospace not only improves overall thermal performance but also reduces defects and grain boundaries that usually serve as points of thermal weakness. Also, the nano-thermal barrier coatings offer good insulation properties preventing the transfer of heat to underlying areas (Uddin et al. 2019).
- *Durability*: The nanoparticles in aerospace materials help in increasing durability by enhancing the physical, mechanical, thermal, and chemical properties. Graphene and carbon nanotubes help in distributing the stress evenly and inhibit the cause of cracks. The nanofillers also bridge the cracks and prevent their propagation which translates to prolonged service life and diminished maintenance demand, enhancing operational efficiency and cost-effectiveness (Yates et al. 2021).
- *Better Electrical Conductivity*: Nanoparticles such as carbon nanotubes and graphene provide exceptional conductivity to aerospace materials, for applications such as lightning strike protection, electromagnetic interference shielding, and efficient energy transfer in aerospace systems by providing a smoother and longer conductive path (Faraji et al. 2023).
- *Environment-Friendly Approach*: Green nanoparticles, synthesized through environmentally benign methodologies, offer a sustainable approach in aerospace manufacturing (Fig. 10.1), without compromising on performance metrics (Li et al. 2023).

Thus, there are bountiful avenues that can be achieved by using nanotechnology in aerospace materials (Patel 2023). The green nanoparticles are environmentally friendly and sustainable in their synthesis, application, and impact. Hence, these are designed to meet the preferred requirements of the aerospace industry. These nanoparticles are synthesized using eco-friendly methodologies, using solvent-free or water-based synthesis routes, and bio-based materials as precursors for nanoparticle synthesis. Incorporating bio-based materials into nanoparticle synthesis can

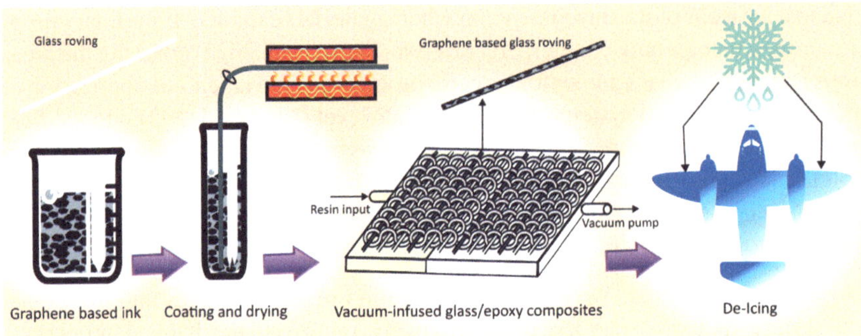

Fig. 10.1 Application of nanocomposites in aviation components

significantly reduce the environmental impact of aerospace applications (Zipay et al. 2016). This minimizes the generation of waste and consumption of energy. The green nanoparticles are designed to have minimal environmental impact throughout their lifecycle, from synthesis to disposal, thus contributing to sustainable aerospace practices and helping mitigate the ecological footprint of aerospace technologies.

10.2 Challenges and Requirements in Aerospace Materials

The major challenges that researchers and manufacturers usually come across in the manufacturing of aerospace materials are weight reduction and mechanical strength, upon which the efficacy and safety of aerospace research are dependent.

10.2.1 Challenges

i *Weight Reduction*

In aerospace engineering, the weight and efficiency of aerospace materials are paramount, and thus achieving the required weight becomes a pivotal challenge to improve fuel efficiency and expanding payload capacity. Since aircraft travels through vast distances, so every additional kilogram of weight imposes a burden on propulsion systems, demanding greater fuel consumption to maintain optimal performance. Thus, reducing the weight of aircraft is not merely an engineering aspect but a basis of sustainable aviation practices (Norkhairunnisa Mazlan 2022). Ranging from commercial aircraft to space exploration endeavors, minimizing vehicle weight maximizes the ability to carry cargo and essential supplies. This constant push for efficiency drives innovative design solutions that optimize performance while keeping weight to a minimum. The payload capacities can be increased by lightening the load, whether it is ferrying passengers across continents or launching satellites into orbit, the ability to carry more payload transforms aspirations into realities, empowering aerospace ventures to reach new heights of ambition and achievement (Zhu et al. 2018).

However, it is important to ensure that reducing weight does not compromise the structural integrity, durability, and operational safety of the materials. For over 80 years, aluminum alloys have been used as the primary material in structural components of aircraft due to their exceptional properties, including good strength-to-weight ratio, and excellent corrosion resistance (Fu et al. 2023). Due to their significant advantages over steel and titanium alloys, including cost-effectiveness, they are extensively used in the fabrication of structural elements across various industries, including defense, automotive, and aerospace. However, in the aerospace sector, there is a particular demand for advanced aluminum alloys to meet stringent requirements for high fracture toughness, superior fatigue performance, enhanced formability, and superplasticity. These attributes are essential for achieving reduced structural weight, improved damage tolerance, and enhanced durability in aircraft structures (Nakai and Eto 2000).

- *Robust Mechanical Strength for Launch and Operation*

The process of launching a spacecraft involves a combination of engineering innovation and material durability. Throughout the launch phase, materials endure a variety of intense mechanical stresses. Initially, the powerful thrust from rocket engines creates significant axial and lateral loads on the spacecraft's structure. These loads challenge the material's strength and flexibility, ensuring the spacecraft remains structurally sound. As the spacecraft progresses through different atmospheric conditions post-launch, it experiences rapid deceleration and heating. During this phase, materials must demonstrate exceptional fracture toughness to withstand the formation and spread of cracks under the combined influence of mechanical stresses and thermal loads (Boyer et al. 2015).

The success of a launch hinges on the ability of materials to withstand this multifaceted mechanical challenge. Their performance in this crucible lays the foundation for a successful mission and underscores the critical role of material science in space exploration. Also, in the harsh expanse of space or the turbulent skies above, aerospace materials are subjected to a relentless onslaught of forces including gravitational, thermal, and mechanical. From the relentless pull of Earth's gravity to the ferocious buffeting of atmospheric turbulence, materials must stand firm, their mechanical strength serving as a bulwark against the vicissitudes of celestial exploration (Pisacane 2003).

Yet, achieving exceptional mechanical strength is no trivial task. Aerospace materials must navigate a labyrinth of competing priorities, balancing the imperatives of strength, weight, and cost against the backdrop of ever-evolving technological landscapes (Ganpatye and Kinra 2006). The quest for materials capable of withstanding the extreme rigors of aerospace environments remains an ongoing odyssey—a testament to the unyielding spirit of human ingenuity and exploration. Thus, the challenges of weight reduction and mechanical strength demand unwavering resolve, unparalleled ingenuity, and boundless innovation (Vaz et al. 2023).

- *High Thermal Stability for Extreme Environments*

Aerospace materials face a thermal gauntlet, enduring a colossal swing in temperatures from the blistering inferno of re-entry to the frigid depths of space. During atmospheric re-entry, spacecraft experience a hypersonic descent, generating extreme frictional heating. Temperatures can surpass 3,000 °C, pushing materials to their thermal limits (Mankodi et al. 2019). The materials used in spacecraft structures must maintain their integrity and strength when exposed to extreme temperatures. To achieve this, heat-resistant alloys containing elements such as tantalum and niobium are utilized. These alloys exhibit remarkable strength even at elevated temperatures, making them suitable for critical spacecraft components (Shulkin et al. 2022). Thermal protection systems utilize ablative materials that strategically char and decompose on the surface, absorbing heat and creating a protective barrier against the re-entry plasma. These materials, often composed of carbon composites or phenolic resins, sacrifice themselves in a controlled manner, safeguarding the spacecraft interior from the intense heat (Davis 2017). Conversely, the vast expanse of space presents an opposing challenge. The near-vacuum environment offers minimal thermal conductivity, leading to a rapid loss of heat. Temperatures can plummet to near absolute zero (–273 °C), requiring exceptional thermal insulation (Torres-Herrador et al. 2021). Also, materials like advanced aerogels with their extremely low density and remarkable insulating properties are crucial in maintaining critical spacecraft systems within operational temperature ranges. They are required to demonstrate thermal versatility, capable of withstanding extreme heat of re-entry as well as the cold temperatures of space. These materials trap air molecules within their porous structure, significantly hindering heat transfer and preserving warmth (Piacquadio et al. 2023).

Thus, using the nanoparticles, the thermal performance of aerospace materials can be pushed further for improved success of missions.

- *Robustness Against Harsh Conditions*

The cosmic hazards, including UV radiation, micrometeoroids, and corrosive space-borne chemicals, pose a threat to the structure of aircraft. The UV radiation from the sun causes degradation, discoloration, and embrittlement of aircraft. The materials used must be capable of shielding against UV radiation and hence UV-resistant coatings, polymers, and composites are used to preserve functionality and appearance (Yates et al. 2021).

Additionally, they must be capable of resisting corrosion from exposure to moisture, salt, and different atmospheric pollutants. The mechanical stress from aerodynamic forces during take-off, flight, and landing can cause strain upon the materials. The collision of aircraft with micrometeoroids can easily puncture the spacecraft hulls and compromise mission integrity. The aerospace materials must be strong enough to bear against such assaults due to which impact-resistant structures and shielding are employed to mitigate the risks posed by cosmic debris. Advanced materials like composites, titanium alloys, and high-performance polymers are engineered for enhanced durability, but balancing lightweight properties with the need for strength and longevity remains complex (Davis 2017).

- *Fire Resistance for Enhanced Safety*

A critical aspect of aerospace materials engineering, to ensure safety during adverse conditions, is the resistance to ignition to protect passengers and crew during emergencies. Aerospace materials are subjected to rigorous testing to withstand fire hazards, including ignition, propagation, combustion, and extreme temperatures of re-entry (Parveez et al. 2022). Thus, materials such as specialized composites, metals, and fire-retardant polymers are designed to restrict the spread of flames and reduce the production of toxic smoke and fumes, and also help to extend evacuation time, allowing passengers and crew to escape safely. From the intense heat of re-entry to the harsh conditions of space, aerospace materials rise to the challenge, paving the way for a future characterized by safety, reliability, and resilience against adversity.

10.3 Future of Green Nanocomposites for Improved Performance

10.3.1 Limitations of Conventional Methods

Metals and polymers have been a significant part of aerospace engineering for decades, accompanied by their limitations. Thus, with the increasing demand for aerospace materials, conventional materials are increasingly strained to meet the strict requirements for space exploration and aviation. Some limitations of metals and polymers in meeting the multifaceted demands of aerospace applications are as follows:

- **Metals**: A major disadvantage of using metals in aerospace applications is their relatively high density which leads to increased weight. Since the reduction in weight of aerospace materials is important for fuel efficiency and payload capacity, the use of metals becomes less ideal to achieve these objectives (Fu et al. 2023).

Though metals possess good thermal conductivity their ability to withstand high temperatures, as during re-entry, is inadequate and can lead to thermal expansion, deformation, and even structural failure. They are also vulnerable to corrosion in the presence of atomic oxygen and other corrosive agents that can degrade the surface over time. Corrosion compromises structural integrity and necessitates costly maintenance and replacement efforts. Aerospace metals are subjected to cyclic loading during flight, leading to fatigue failure over time (Pisacane 2003). The repetitive stress cycles experienced during take-off, flight, and landing can induce microstructural damage, ultimately culminating in catastrophic failure if not mitigated. Titanium and its alloys are known to boast numerous desirable characteristics for structural applications; however, their primary drawback lies in their high initial cost. Despite this drawback, titanium and its alloys find widespread use in airframes, gas turbine engines, and rocket engines. The elevated cost poses a significant deterrent, especially in airframe applications, where competing alloys generally offer substantially lower costs (Williams and Boyer 2020).

- **Polymers**: While polymers offer advantages in terms of weight and flexibility, their temperature resistance is often limited. Exposure to high temperatures, such as those encountered during re-entry, can result in thermal degradation, melting, or combustion, compromising structural integrity and safety (Davis 2017). Polymers generally exhibit lower mechanical strength compared to metals, making them less suitable for applications requiring exceptional load-bearing capabilities. This limitation restricts their use in critical structural components subjected to high mechanical stresses. They are also susceptible to degradation upon exposure to ultraviolet (UV) radiation, which is prevalent in the space environment. UV-induced degradation can lead to embrittlement, discoloration, and loss of mechanical properties, posing durability and reliability concerns. However, certain polymers exhibit outgassing phenomena, wherein volatile compounds are released from the material's surface under vacuum conditions. Outgassing can contaminate sensitive equipment, interfere with scientific instruments, and compromise mission objectives in aerospace applications (Dyer and Kumru 2023).

Thus, while metals and polymers have played pivotal roles in aerospace engineering, their inherent limitations pose significant challenges in meeting the multifaceted demands of modern aerospace applications. As the quest for lightweight, durable, and high-performance materials intensifies, the aerospace industry continues to explore alternative materials and advanced composites that offer superior properties and overcome the shortcomings of conventional materials (Rathod et al. 2017). Epoxy composites are commonly produced from prepreg systems, where an initial curing step is sometimes employed to optimize the processing properties of the prepreg. Final curing typically occurs at cost-effective temperatures (100–180 °C) and durations, while still achieving excellent performance. These dual-curing systems often utilize different curing agents with distinct curing kinetics. The curing kinetics are determined by the chemical structure of the hardener, resulting in formulations for both rapid curing (e.g., linear amines) and slow curing (e.g., aromatic amines). The liquid nature of epoxies also renders them suitable for Resin Transfer Molding, as the two components can be blended and injected into the mold under high pressure (Bekyarova et al. 2007).

Nanotechnology stands as a pivotal force in revolutionizing the automotive industry, offering substantial strides in the optimization of vehicular performance (Dhinakaran et al. 2021). Amidst this escalating landscape, a multitude of forthcoming developments are foreseen, each leveraging the capabilities of nanotechnology to drive innovation forward. This entails a progressive enhancement in the functionalities of modern automobiles, including electronically regulated components such as fuel injection, exhaust emission control, antilock brakes, automatic air-conditioning, headlight management, mechanical seat adjustments, lateral control, and electronic suspension systems (Fig. 10.2) (Levchenko et al. 2018).

- **Green Nanoparticles**

Synergistic Advancements: Green Nanoparticles in Aerospace Materials Engineering

Nanocomposites represent a class of advanced materials where nanoparticles are integrated into polymer matrices to enhance their properties and performance (Njuguna

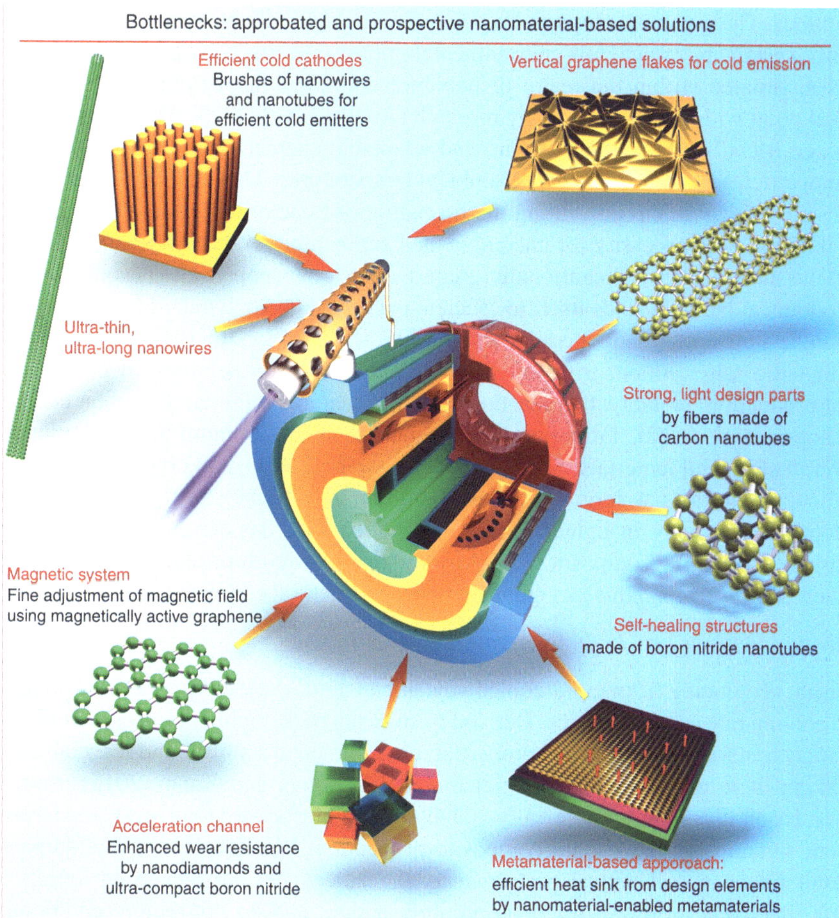

Fig. 10.2 Advanced applications of nanomaterials in the aerospace industry (Levchenko et al. 2018)

and Pielichowski 2003). These nanocomposites offer unique advantages over traditional materials due to the synergistic effects arising from the combination of nanoparticles and polymers (Table 10.1). The primary aim of composite material synthesis is to capitalize on the combined strengths of its constituent components—the reinforcement and matrix. For instance, the carbon-epoxy composite in aircraft structures merges carbon fibers (reinforcement) within epoxy resin (matrix). While carbon fibers contribute stiffness and strength, epoxy provides ductility. Alone, carbon fibers and epoxy are unsuitable for aircraft structures due to brittleness and weakness respectively, but as a composite, they yield a high-performance material with superior properties (Malucelli and Kausar 2023).

The incorporation of green nanoparticles into polymer matrices involves the utilization of environmentally friendly nanomaterials, such as biodegradable nanoparticles derived from natural sources or synthesized using eco-friendly methods. These green nanoparticles possess desirable attributes such as biocompatibility, sustainability, and low environmental impact. The green synthesis of nanoparticles, situated within the realm of nanobiotechnology, has emerged as a pivotal focal point within nanotechnology research (Abuzeid et al. 2023). Distinguished by its eco-friendly, pure, cost-efficient, and adaptable characteristics, biosynthesis of nanoparticles is often conducted at ambient temperatures. The integration of biosynthesis is critical in circumventing the formation of toxic or hazardous by-products, highlighting the necessity for uncomplicated and benign production methodologies (Matussin et al. 2020). Operationally, green synthesis of nanoparticles not only proves economical but also lends itself to efficient execution under standard conditions. In a single-step synthesis procedure, a natural bio-reductant extract is combined with a metallic salt solution, triggering a swift redox reaction resulting in nanomaterial production. Notably, this reaction typically requires minimal activation energy (Chopra et al. 2022). Fiber-polymer composites and aluminum alloys dominate aircraft structural materials. Since the 1990s, composite usage in civil, military, and helicopter aircraft has surged, rivaling aluminum in airframe construction. Military adoption, especially in fighter aircraft and helicopters, has driven composite use. Carbon fiber-epoxy composites, employed in fighter aircraft primary structures like wings and fuselage, reduce weight and enhance structural efficiency (Mallick 2007).

Advancements in Eco-Friendly Aerospace Materials

Basalt, originating from rapid-cooled volcanic magma, varies in performance based on cooling rate and lava source. Derived from basalt rock, highly fibrous basalt fibers display remarkable mechanical properties, environmental stability, and recyclability, making them eco-friendly industrial materials (Mohan and Kanny 2015). Moreover, basalt fibers are economically viable and straightforward to process, rendering them suitable for large-scale composite production. While carbon fibers surpass basalt fibers in mechanical performance, the latter offer advantages in terms of cost-effectiveness. Basalt fibers are entirely natural and do not require additional materials for reinforcement. They demonstrate high strength and modulus, exceptional resistance to temperature and corrosion, and remarkable chemical stability.

Table 10.1 Applications of green nanomaterials in the aerospace industry

Type of nanomaterial	Properties	Application	References
Cellulose Nanoparticles	Cellulose nanoparticles can be used as reinforcements in polymer-based nanocomposites to enhance mechanical properties, leading to lightweight and durable structural components for aerospace applications	Structural Components (e.g., aircraft fuselages, wings)	Leao et al. (2023)
Boron Nitride Nanoparticles	Boron nitride nanoparticles have high thermal conductivity, making them ideal for improving heat dissipation and thermal stability in aerospace electronics and communication systems	Thermal Management (e.g., electronic components, avionics)	Li et al. (2020)
Carbon-Based Nanoparticles (e.g., Graphene, Carbon Nanotubes)	Carbon-based nanoparticles can be used to enhance EMI shielding capabilities, ensuring reliable and secure data transmission in aerospace applications	Electromagnetic Interference (EMI) Shielding (e.g., aerospace electronics, communication systems)	Zecchi et al. (2024)
Metal Nanoparticles	Metal nanoparticles can be used to improve durability, corrosion resistance, and anti-icing properties in aerospace coatings and paints, extending the service life of aerospace products	Coatings and Paints (e.g., aircraft surfaces, spacecraft components)	Hołyńska et al. (2018)
Plant Extract-Based Nanoparticles	Plant extract-based nanoparticles can be used as sustainable reinforcements in polymer-based nanocomposites, enhancing mechanical properties while reducing environmental impact	Structural Components (e.g., composite materials for aircraft, spacecraft)	Balakrishnan et al. (2016)
Hybrid Nanoparticles	Hybrid nanoparticles have high thermal conductivity, making them suitable for improving heat dissipation and thermal stability in aerospace electronics and communication systems	Thermal Management (e.g., electronic components, avionics)	Faraji et al. (2023)

Despite sharing a similar composition with glass fibers, basalt fibers outperform E-type glass fibers across various properties, including breaking load, Young's modulus, and environmental resilience (Dhand et al. 2015).

Green nanoparticles thus offer numerous benefits to nanocomposites, including enhanced mechanical strength, improved thermal stability, superior electrical conductivity, and increased resistance to UV radiation and chemical corrosion.

The Hubble space telescope's main truss utilizes a lightweight carbon fiber-epoxy composite for stiffness and low thermal expansion. Similarly, sandwich composites form the space shuttle orbiter's cargo doors, while the remote manipulator system's arm employs carbon fiber composite (Mathur et al. 2022). Incorporating green nanoparticles into polymer matrices involves dispersion techniques to ensure even distribution. Methods like solution blending, melt mixing, and in situ polymerization optimize dispersion and interaction with polymer chains (Table 10.2). In the advanced manufacturing sector, aerospace component manufacturers are proactively reducing energy consumption by transitioning to lighter materials such as lightweight metals, polymers, and composites. Utilizing these materials allows for compliance with rigorous weight reduction standards in aeronautical applications. Sustainability is a key consideration, focusing on material recyclability, emission reduction in aircraft, and life cycle analysis. In aircraft polymerization, homopolymers consist of one monomer, while copolymers involve two. Copolymers, used as elastomers in seals, exhibit a random or regular distribution of monomers along the chain, with block copolymers featuring long sequences of one type followed by another, termed graft copolymers when chains attach across monomer types. Beyond plant oils, lignin and proteins hold promise as renewable raw materials for bio-based thermosetting materials due to their abundance and functionality. Lignin, the second most abundant natural macromolecule, offers phenolic properties suitable for phenol–formaldehyde resin substitutes (Bismarck et al. 2006). Thermo-chemical conversion enhances lignin reactivity with formaldehyde. Plant proteins like soy and wheat gluten offer the potential for thermosetting biomaterials, serving as templates for temporary replacement implants due to processability, substrate adhesion, and surface-active traits (Wang et al. 2018). Protein processing involves heat treatment leading to cross-linked systems, sometimes aided by cross-linkers for density control. Wood plastic composites find extensive use in aerospace sectors, primarily serving as eco-friendly substitutes. Plant fibers are increasingly favored for thermo-acoustic insulation, with flax fibers replacing asbestos in car disk brakes, though exterior parts made from plant fiber composites remain limited (Raquez et al. 2010).

A significant concept arising from nanocomposites and nanoarchitectures is the substantial increase in the interfacial area between nano-reinforcements and polymeric matrices when at least one filler size is reduced to the nanoscale. The incorporation of small amounts of nanoparticles into polymeric melts has been shown to significantly enhance the mechanical, thermal, electrical, and optical properties of these polymeric nanoarchitectures (PNC) compared to unreinforced materials. This improvement in material characteristics stems from the interface formed between the nano-reinforcement and the bulk polymeric matrices, leading to changes in chain dynamics compared to the pristine material. Consequently, the spatial arrangement or organization of nano-reinforcements influences the three-dimensional structure of the interface, resulting in improved control over macroscopic properties (Idumah 2023). A study reveals two significant findings when employing the intercalation method. Firstly, they observed that viscosity and non-Newtonian behavior increase compared to micro-range composites. Secondly, they noted that viscosity changes depending on the method used, even when the same nanofiller is employed.

Table 10.2 Synthetic techniques for developing advanced materials with aerospace applications

Name of Technique	Brief description	Example	Reference
Solution-induced intercalation	Mixing nanoparticles with a solvent to create a uniform dispersion	Nylon polymer solution with clay nanoparticles, vaporizing the solvent to create a nylon-6 polymer/clay nanocomposite	Wang and Xiao (2018)
Co-deposition	Depositing nanoparticles and a polymer matrix simultaneously	Suspended Al_2O_3 nanoparticles in a nickel plating solution, using high-speed plating to embed them in a metal matrix for cost-effective turbine blade production	Hussain et al. (2009)
Vacuum-assisted resin transfer molding	Injecting a polymer resin into a mold under vacuum	Sheet-based nanocomposites by applying resin to pre-laid composite fabrics (glass fiber and carbon nanopaper) in a sealed vacuum bag, resulting in a material with high damping characteristics for structural applications	Liang et al. (2009)
Single screw extruder	Melting and mixing nanoparticles with a polymer using a single screw extruder	PET/HDPE nanocomposites, with barrel temperatures ranging from 200 to 250 °C	Rosnan and Arsad (2013)
Nanocomposite coatings	Depositing nanoparticles onto a surface using techniques like sputtering or evaporation	Adding graphene and ITO to an epoxy system and spray coating it onto carbon fiber reinforced plastics (CFRP) to improve resistance to lightning strikes	Zhang et al. (2013)

(continued)

Table 10.2 (continued)

Name of Technique	Brief description	Example	Reference
Cross-accumulative roll bonding	Alternately rolling and folding a nanocomposite material to enhance properties	A356 aluminum alloy matrix and SiC nanocomposite, involving furnace treatment, addition of SiC particles, casting, and roll bonding	Kamali et al. (2014)
Melt compounding	Mixing nanoparticles with a molten polymer	LDPE/nano TiO_2-based bio nanocomposite system, involving the addition of TiO_2 nanoparticles to an LDPE matrix using a single screw extruder	Katbab et al. (2014)
Vibrational Casting	Using vibrations to mix and cast a nanocomposite material	Nanocomposite system consisting of PEI/Cloisite (30B), providing vibration and heat simultaneously to melt and stir the mixture, with the ability to recover and recycle the solvent	Dwivedi et al. (2014)
Ball mill fragmentation	Breaking down nanoparticles using a ball mill	Aluminum and NbB_2, and ZrB_2 nanoparticles, involving sintering at high temperatures and cold forming to draw aluminum wires	Florián-Algarín et al. (2015)
Melt extrusion	Extruding a molten polymer-nanoparticle mixture	PP matrix with reinforcement fibers filled with nanoclay, involving a single screw extruder and winding fibers around a rotating mandrel	Mohan and Kanny (2015)

(continued)

Table 10.2 (continued)

Name of Technique	Brief description	Example	Reference
Drop casting	Depositing a nanocomposite material onto a surface using a droplet	Polymer/MWCNTs-based nanocomposite system, involving mixing PES with epoxy resin and MWCNTs, and casting the mixture drop-by-drop onto a glass slide	Farahani et al. (2015)
Powder metallurgy	Consolidating nanoparticles using heat and pressure	Graphene nanosheet/SiC-based nanocomposite system, involving ball milling graphene nanosheets and SiC, and injecting the particulate system into molten A357 aluminum alloy	Fadavi Boostani et al. (2015)
Fused Deposition Modeling	3D printing a nanocomposite material using melted polymer	ABS/OMMT nanocomposites, involving mixing ABS pellets with montmorillonite, extruding filaments, and melting and extruding through an FDM printer nozzle	Weng et al. (2016)
Compression molding	Shaping a nanocomposite material under heat and pressure	Nanocomposite system for proton shielding, involving melt-mixing PMMA powders with MWCNTs and compression molding with PMMA pellets	Li et al. (2016)
Hot filament chemical vapor deposition	Depositing nanoparticles onto a surface using a hot filament	Tungsten oxide (WO2)-glass fiber mat nanocomposite, involving a tungsten filament as the heating source in a stainless steel chamber	Shahidi et al. (2016)

(continued)

Table 10.2 (continued)

Name of Technique	Brief description	Example	Reference
Ultrasonic cavitation	Using ultrasound to mix and disperse nanoparticles	A356 alloy matrix and Al_2O_3 nanoparticle composites, involving ultrasonic stirring technology (UST) to distribute nanoparticles uniformly in the alloy	Xuan et al. (2017)
Flux-assisted liquid state processing	Using a flux to enhance nanoparticle dispersion	Aluminum matrix, TiB_2 reinforced nanocomposite system, involving melting aluminum ingots, adding TiB_2 nanoparticles and a flux (KAlF4), and stirring	Javadi et al. (2017)
Selective laser melting	Melting and fusing nanoparticles using a laser	Nanocomposite system consisting of IN718 (a nickel superalloy) as the matrix with TiO_2 as the reinforcement, involving blending the matrix and reinforcement, and using the mixture as powdered particles in the SLM process	Yao et al. (2017)
Magnetic field induced alignment	Aligning nanoparticles using a magnetic field	Nanocomposite system by tethering Fe_3O_4 nanoparticles onto MWCNTs and orienting them in epoxy resin using a low magnetic field	Huang et al. (2018)
Stir casting	Mixing nanoparticles with a molten polymer using a stirrer	Aluminum metal matrix nanocomposites (AMMNCs) by melting $AlSi_9Cu_3$ alloy, adding Al_2O_3 nanoparticles, and stirring till well blended	Ünal and Diler (2018)

(continued)

Table 10.2 (continued)

Name of Technique	Brief description	Example	Reference
High frequency induction heat sintering	Sintering nanoparticles using high-frequency induction heat	Alumina/graphene nanoplatelet system by thermally exfoliating graphene platelets, dipping them into alumina nanoparticles, and consolidating under high temperature and pressure	Ahmad et al. (2018)
Thin-film bonding	Bonding nanoparticles onto a surface using a thin film	GO/laminated composite panel system by fabricating a thin film of GO and bonding it to a laminated panel using a suitable adhesive	Uddin et al. (2019)
Dip coating	Depositing nanoparticles onto a surface using a dip coating process	Thermal coating for a spacecraft by dispersing MWCNTs into a polymer solution, dipping an aluminum alloy substrate into the solution, and drying	Verma et al. (2018)
Pultrusion	Extruding a nanocomposite material using a pultrusion process	Composite rods by mixing epoxy resin with a hardener, adding CNTs, coating dry glass fibers with the matrix, and pulling them through a glass tube mold	Venkatesan et al. (2018)
Ultrasonication	Using ultrasound to mix and disperse nanoparticles	Well-dispersed nanocomposite system consisting of GO and MWCNTs	Gao et al. (2019)
Vacuum sintering	Sintering nanoparticles under vacuum	Nanocomposite system consisting of copper powder and TiB_2 nanoparticles, followed by hot extrusion for densification	Li et al. (2019)

(continued)

Table 10.2 (continued)

Name of Technique	Brief description	Example	Reference
Physical vapor deposition	Depositing nanoparticles onto a surface using vapor deposition	Ceramic coatings, AlSiTiN and AlSiCrN, on high-speed steel (HSS) samples to assess wear resistance	Gautier et al. (2019)
Spray coating	Depositing nanoparticles onto a surface using a spray coating process	Mylar substrates coated with MWCNTs/epoxy nanocomposites	Laurenzi et al. (2019)
Solvent exchange and sol–gel techniques	Using solvent exchange and sol–gel processes to create nanoparticles	SiO_2 aerogel/TiO_2 nanocomposite system, retaining structural characteristics	Yi et al. (2020)
Melt blending	Mixing nanoparticles with a molten polymer	Nanocomposite system consisting of polybutylene succinate (PBS) and graphene nanoplatelets	Platnieks et al. (2020)
Spark plasma sintering	Sintering nanoparticles using spark plasma	Aluminum matrix composite reinforced with graphene nanoparticles, possessing high densities, suitable for aerospace applications	Khoshghadam-Pireyousefan et al. (2020)
Electrospray deposition	Depositing nanoparticles onto a surface using electrospray	Woven carbon fiber (CF) with MWCNTs deposited on its surface, potentially improving flexural strengths for avionic components	Zakaria et al. (2020)

Challenges and Future Prospective

The scalability of green synthesis methods is the foremost challenge in using green nanomaterials for aerospace applications, typically performed on a laboratory scale. Scaling up these processes involves synthesizing materials on a large scale for consistent and reliable production of green nanoparticles in sufficient quantities to meet the demands of the aerospace industry (Yi et al. 2020). The green nanomaterials are morphologically heterogeneous and their composition depends upon the synthesis method and precursors used leading to inconsistencies in the performance and properties. Thus, it is difficult to ensure the reproducibility and reliability of the materials required for real-life aerospace applications. However, the integration of these materials with the conventionally existing materials and processes is a significant challenge that requires vigorous research and development efforts. It is found that the assimilation of new materials in aerospace applications faces obstacles due to the stringent safety and performance regulations within the aerospace sector (Davis 2017). The process of approval and defining clear benchmarks, along with characterizing methods and standardizing testing protocols, can prove to be intricate and time-consuming. Though the long-term effects of nanomaterials on the environment and human health are not completely understood due to which continuous research and comprehensive risk assessments are required to ensure safe and enduring utilization (Malucelli and Kausar 2023).

The use of improved, eco-friendly nanocomposites engineered to exhibit characteristics like high thermal stability, lightweight, and mechanical strength, all of which are indispensable for the challenging conditions of aerospace environments. The by-products from the manufacturing and disposal of these materials present notable difficulties (Javadi et al. 2017). Existing waste disposal methods may not be entirely capable of efficiently managing the recycling of nanoparticles which raises worries about the accumulation of these nanoparticles in landfills and the potential uncontrolled release of non-recyclable nanomaterials into the environment (Yao et al. 2017). Biodegradable polymer nanocomposites offer a solution, as they are designed to degrade under certain conditions, reducing long-term environmental impact. Yet, the integration of these materials into existing waste management systems and the technical challenges of recycling remain critical issues to be addressed. Thus, green nanoparticles in aerospace applications are multifaceted and align well with sustainability principles, offering high performance while reducing environmental footprints (Huang et al. 2018). These nanoparticles and nanocomposites developed for aerospace applications prioritize environmental considerations. However, their life cycle, from production to disposal, poses environmental challenges. Improvements in waste management strategies and the development of fully biodegradable materials are crucial to reduce potential negative impacts on the environment (Gao et al. 2019).

10.4 Conclusion

The aerospace research sector greatly depends on the development of innovative and advanced materials to achieve specific physical, mechanical, and thermal properties that cannot be met by conventional materials used. Metals, while offering good thermal conductivity, pose challenges due to their high density, which increases weight and impacts fuel efficiency and payload capacity. Additionally, they struggle to withstand extreme temperatures experienced during re-entry, leading to potential thermal expansion, deformation, and structural failure. These limitations highlight the necessity for innovative materials that combine lightweight properties with thermal and mechanical resilience. Green nanotechnology introduces a new approach to developing next-generation aerospace materials, including green nanoparticles and nanocomposites, with the potential to transform the future of aerospace materials. Green nanoparticles are cost-effective, environmentally friendly and sustainable in their synthesis, application, and impact. The smooth integration of green nanoparticles into aerospace applications can result in improved mechanical properties, electrical conductivity, enhanced thermal stability, and increased durability. Thus, the fusion of nanotechnology with aerospace materials with the use of green nanoparticles in aerospace applications can significantly reduce environmental harm and promote sustainability embodying the pinnacle of scientific ingenuity.

References

Abuzeid HM, Julien CM, Zhu L, Hashem AM (2023) Green synthesis of nanoparticles and their energy storage, environmental, and biomedical applications. Crystals 13(11):1576. https://doi.org/10.3390/cryst13111576

Ahmad I, Subhani T, Wang N, Zhu Y (2018) Thermophysical properties of high-frequency induction heat sintered graphene nanoplatelets/alumina ceramic functional nanocomposites. J Mater Eng Perform 27(6):2949–2959. https://doi.org/10.1007/s11665-018-3395-6

Balakrishnan P, John MJ, Pothen L, Sreekala MS, Thomas S (2016) Natural fibre and polymer matrix composites and their applications in aerospace engineering. In: Advanced omposite materials for aerospace engineering. Elsevier, pp 365–383. https://doi.org/10.1016/B978-0-08-100037-3.00012-2

Bekyarova E, Thostenson ET, Yu A, Kim H, Gao J, Tang J, Hahn HT, Chou T-W, Itkis ME, Haddon RC (2007) Multiscale carbon nanotube−carbon fiber reinforcement for advanced epoxy composites. Langmuir 23(7):3970–3974. https://doi.org/10.1021/la062743p

Bhat A, Budholiya S, Aravind Raj S, Sultan MTH, Hui D, Md Shah AU, Safri SNA (2021) Review on nanocomposites based on aerospace applications. Nanotechnol Rev 10(1):237–253. https://doi.org/10.1515/ntrev-2021-0018

Bismarck A, Baltazar-Y-Jimenez A, Sarikakis K (2006) Green composites as panacea? socio-economic aspects of green materials. Environ Dev Sustain 8(3):445–463. https://doi.org/10.1007/s10668-005-8506-5

Boyer RR, Cotton JD, Mohaghegh M, Schafrik RE (2015) Materials considerations for aerospace applications. MRS Bull 40(12):1055–1065. https://doi.org/10.1557/mrs.2015.278

Chopra H, Bibi S, Singh I, Hasan MM, Khan MS, Yousafi Q, Baig AA, Rahman MM, Islam F, Emran TB, Cavalu S (2022) Green metallic nanoparticles: biosynthesis to applications. Front Bioeng Biotechnol 10. https://doi.org/10.3389/fbioe.2022.874742

Davis D (2017) Fundamentals of launch vehicle ablative thermal protection system (TPS) materials

Dhand V, Mittal G, Rhee KY, Park S-J, Hui D (2015) A short review on basalt fiber reinforced polymer composites. Compos B Eng 73:166–180. https://doi.org/10.1016/j.compositesb.2014.12.011

Dhinakaran V, Sai MS, Shree MV (2021) Some impact of nanomaterials in aerospace engineering. In: Nanomaterials and nanocomposites. CRC Press, pp 17–29. https://doi.org/10.1201/9781003160946-3

Dwivedi M, Gaur KK, Alam S, Bhatnagar N, Ghosh AK (2014) Processing of polyetherimide/cloisite 30B nanocomposite by vibration casting method. Int Polym Proc 29(5):635–640. https://doi.org/10.3139/217.2952

Dyer WE, Kumru B (2023) Polymers as aerospace structural components: how to reach sustainability? Macromole Chem Phys 224(24). https://doi.org/10.1002/macp.202300186

Fadavi Boostani A, Yazdani S, Taherzadeh Mousavian R, Tahamtan S, Azari Khosroshahi R, Wei D, Brabazon D, Xu JZ, Zhang XM, Jiang ZY (2015) Strengthening mechanisms of graphene sheets in aluminium matrix nanocomposites. Mater des 88:983–989. https://doi.org/10.1016/j.matdes.2015.09.063

Farahani RD, Klemberg-Sapieha JE, Therriault D (2015) Enhanced conductivity of nanocomposite films through heterogeneous distribution of nanofillers during processing. Mater des 88:1175–1182. https://doi.org/10.1016/j.matdes.2015.09.088

Faraji H, Yıldız Ç, Arshad A, Arıcı M, Choukairy K, El Alami M (2023) Passive thermal management strategy for cooling multiple portable electronic components: hybrid nanoparticles enhanced phase change materials as an innovative solution. J Energy Storage 70:108087. https://doi.org/10.1016/j.est.2023.108087

Florián-Algarín D, Padilla A, López NN, Suárez OM (2015) Fabrication of aluminum wires treated with nanocomposite pellets. Sci Eng Compos Mater 22(5):485–490. https://doi.org/10.1515/secm-2013-0193

Fu R, Guo Y, Cui Y, Wang J, Lei H, Liu C (2023) Large-size ultra-high strength-plasticity aluminum alloys fabricated by wire arc additive manufacturing via added nanoparticles. Mater Sci Eng, A 864:144582. https://doi.org/10.1016/j.msea.2023.144582

Ganpatye AS, Kinra VK (2006) Fracture toughness of space shuttle external tank insulation foam

Gao Y, Jing H, Zhou Z (2019) Fractal analysis of pore structures in graphene oxide-carbon nanotube based cementitious pastes under different ultrasonication. Nanotechnol Rev 8(1):107–115. https://doi.org/10.1515/ntrev-2019-0010

Gautier G, Faga MG, Tebaldo V (2019) Impact wear resistance of nanocomposite coatings for aircraft components. Key Eng Mater 813:387–392. https://doi.org/10.4028/www.scientific.net/KEM.813.387

Hołyńska M, Tighe A, Semprimoschnig C (2018) Coatings and thin films for spacecraft thermo-optical and related functional applications. Adv Mater Interfaces 5(11). https://doi.org/10.1002/admi.201701644

Huang Y, Jiao W, Niu Y, Ding G, Wang R (2018) Improving the mechanical properties of Fe_3O_4/carbon nanotube reinforced nanocomposites by a low-magnetic-field induced alignment. J Polym Eng 38(8):731–738. https://doi.org/10.1515/polyeng-2017-0257

Hussain M, Alswailem S, Hala A (2009) Advanced nanocomposites for high temperature aero-engine/turbine components. Int J Nanomanuf 4. https://doi.org/10.1504/IJNM.2009.028132

Idumah CI (2023) Influence of interfacial engineering on properties of polymeric nanoarchitectures. Polym Plast Technol Mater 62(14):1844–1877. https://doi.org/10.1080/25740881.2023.2240393

Ince JC, Peerzada M, Mathews LD, Pai AR, Al-qatatsheh A, Abbasi S, Yin Y, Hameed N, Duffy AR, Lau AK, Salim NV (2023) Overview of emerging hybrid and composite materials for space applications. Adv Compos Hybrid Mater 6(4):130. https://doi.org/10.1007/s42114-023-00678-5

Javadi A, Cao C, Li X (2017) Manufacturing of Al–TiB_2 nanocomposites by flux-assisted liquid state processing. Procedia Manuf 10:531–535. https://doi.org/10.1016/j.promfg.2017.07.038

Kamali MR, Khorsand S, Amirkhanlou S, Nayyeri M (2014) Application of compocasting and cross accumulative roll bonding processes for manufacturing high-strength, highly uniform and ultra-fine structured Al/SiCp nanocomposite. Mater Sci Eng A 592:121–127. https://doi.org/10.1016/j.msea.2013.11.006

Katbab P, Alizadeh M, Kaffashi B, Katbab AA (2014) Bionanocomposites with enhanced antimicrobial activity and photodegradability based on low density polyethylene and nano TiO_2/Organoclay. e-Polymers 14(1):43–55. https://doi.org/10.1515/epoly-2013-0009

Khoshghadam-Pireyousefan M, Rahmanifard R, Orovcik L, Švec P, Klemm V (2020) Application of a novel method for fabrication of graphene reinforced aluminum matrix nanocomposites: synthesis, microstructure, and mechanical properties. Mater Sci Eng A 772:138820. https://doi.org/10.1016/j.msea.2019.138820

Laurenzi S, Clausi M, Zaccardi F, Curt U, Santonicola MG (2019) Spray coating process of MWCNT/Epoxy nanocomposite films for aerospace applications: effects of process parameters on surface electrical properties. Acta Astronaut 159:429–439. https://doi.org/10.1016/j.actaastro.2019.01.043

Leao AL, Cesarino I, Chanes M, Botelho EC, Dias OAT, Jawaid M (2023) Ecologically enhanced natural/synthetic polymer hybrid composites for aviation-interior and secondary structures. In: Khan T, Jawaid M (eds) Green hybrid composite in engineering and non-engineering applications. Springer Nature, Singapore, pp 43–59. https://doi.org/10.1007/978-981-99-1583-5_4

Levchenko I, Xu S, Teel G, Mariotti D, Walker ML R, Keidar M (2018) Recent progress and perspectives of space electric propulsion systems based on smart nanomaterials. Nat Commun 9(1). https://doi.org/10.1038/s41467-017-02269-7

Li Z, Chen S, Nambiar S, Sun Y, Zhang M, Zheng W, Yeow JTW (2016) PMMA/MWCNT nanocomposite for proton radiation shielding applications. Nanotechnology 27(23):234001. https://doi.org/10.1088/0957-4484/27/23/234001

Li S, Guo X, Zhang S, Feng J, Song K, Liang S (2019) Arc erosion behavior of TiB_2/Cu composites with single-scale and dual-scale TiB_2 particles. Nanotechnol Rev 8(1):619–627. https://doi.org/10.1515/ntrev-2019-0054

Li Q, Guo Y, Yi S, Xu Z, Duan W, Liu S (2023) Fluorinated polyimide/sepia eumelanin nanocomposites for aerospace applications. ACS Appl Polym Mater 5(2):1520–1529. https://doi.org/10.1021/acsapm.2c01999

Li H, Ren L, Zhou Y, Yao B, Wang Q (2020) Recent progress in polymer dielectrics containing boron nitride nanosheets for high energy density capacitors. High Voltage 5(4);365–376. https://doi.org/10.1049/hve.2020.0076

Liang F, Tang Y, Gou J, Gu H, Song G (2009) Multifunctional nanocomposites with high damping performance for aerospace structures. In: ASME international mechanical engineering congress and exposition, proceedings, vol 11. https://doi.org/10.1115/IMECE2009-12542

Mallick PK (2007) Fiber-reinforced composites. CRC Press. https://doi.org/10.1201/9781420005981

Malucelli G, Kausar A (2023) Green composites and nanocomposites for aerospace applications. In: Polymeric nanocomposites with carbonaceous nanofillers for aerospace applications. Elsevier, pp 297–316. https://doi.org/10.1016/B978-0-323-99657-0.00014-4

Mankodi TK, Bhandarkar UV, Puranik BP (2019) Hypersonic flow over stardust re-entry capsule using Ab-initio based chemical reaction model. Acta Astronaut 162:243–255. https://doi.org/10.1016/j.actaastro.2019.06.021

Mathur J, Goswami P, Gupta A, Srivastava S, Minkina T, Shan S, Rajput VD (2022) Nanomaterials for water remediation: an efficient strategy for prevention of metal (Loid) hazard. https://doi.org/10.3390/w14243998

Matussin S, Harunsani MH, Tan AL, Khan MM (2020) Plant-extract-mediated SnO_2 nanoparticles: synthesis and applications. ACS Sustain Chem Eng 8(8):3040–3054. https://doi.org/10.1021/acssuschemeng.9b06398

Mazlan N, Sapuan SM, Ilyas RA (2022) Advanced composites in aerospace engineering applications. Springer International Publishing, Cham. https://doi.org/10.1007/978-3-030-88192-4

Mohan TP, Kanny K (2015) Preparation and characteristics of polypropylene-clay nanocomposite fibers. J Polym Eng 35(8):773–784. https://doi.org/10.1515/polyeng-2014-0273

Nakai M, Eto T (2000) New aspect of development of high strength aluminum alloys for aerospace applications. Mater Sci Eng, A 285(1–2):62–68. https://doi.org/10.1016/S0921-5093(00)00667-5

Njuguna J, Pielichowski K (2003) Polymer nanocomposites for aerospace applications: properties. Adv Eng Mater 5(11):769–778. https://doi.org/10.1002/adem.200310101

Parveez B, Kittur MI, Badruddin IA, Kamangar S, Hussien M, Umarfarooq MA (2022) Scientific advancements in composite materials for aircraft applications: a review. Polymers. MDPI. https://doi.org/10.3390/polym14225007

Patel P (2023) Materials for space exploration take a giant leap. ACS Central Sci 9(4):582–585. https://doi.org/10.1021/acscentsci.3c00376

Piacquadio S, Pridöhl D, Henkel N, Bergström R, Zamprotta A, Dafnis A, Schröder KU (2023) Comprehensive comparison of different integrated thermal protection systems with ablative materials for load-bearing components of reusable launch vehicles. Aerospace 10(3). https://doi.org/10.3390/aerospace10030319

Pisacane VL (2003) Spacecraft systems design and engineering. In: Encyclopedia of physical science and technology. Elsevier, pp 463–483. https://doi.org/10.1016/B0-12-227410-5/00888-7

Platnieks O, Gaidukovs S, Neibolts N, Barkane A, Gaidukova G, Thakur VK (2020) Poly(Butylene Succinate) and graphene nanoplatelet-based sustainable functional nanocomposite materials: structure-properties relationship. Mater Today Chem 18:100351. https://doi.org/10.1016/j.mtchem.2020.100351

Raquez J-M, Deléglise M, Lacrampe M-F, Krawczak P (2010) Thermosetting (Bio)materials derived from renewable resources: a critical review. Prog Polym Sci 35(4):487–509. https://doi.org/10.1016/j.progpolymsci.2010.01.001

Rathod VT, Kumar JS, Jain A (2017) Polymer and ceramic nanocomposites for aerospace applications. Appl Nanosci 7(8):519–548. https://doi.org/10.1007/s13204-017-0592-9

Romero-Fierro D, Bustamante-Torres M, Bravo-Plascencia F, Esquivel-Lozano A, Ruiz JC, Bucio E (2022) Recent trends in magnetic polymer nanocomposites for aerospace applications: a review. Polymers 14(19):4084. https://doi.org/10.3390/polym14194084

Rosnan RM, Arsad A (2013) Effect of MMT concentrations as reinforcement on the properties of recycled PET/HDPE nanocomposites. J Polym Eng 33:1–9. https://doi.org/10.1515/polyeng-2013-0107

Shahidi S, Jafari A, Sharifi SD, Ghoranneviss M (2016) Deposition of nano tungsten oxide on glass mat using hot filament chemical vapor deposition for high catalytic activity. High Temp Mater Processes (London) 35(5):515–521. https://doi.org/10.1515/htmp-2015-0051

Shulkin V, Bogdanova N, Elovskiy E (2022) Influence of clogging at the filtration on analysis of dissolved and particulate forms of chemical elements in boreal rivers of the Russian Far East. Minerals 12(6). https://doi.org/10.3390/min12060773

Torres-Herrador F, Eschenbacher A, Coheur J, Blondeau J, Magin TE, Van Geem KM (2021) Decomposition of carbon/phenolic composites for aerospace heatshields: detailed speciation of phenolic resin pyrolysis products. Aerosp Sci Technol 119:107079. https://doi.org/10.1016/j.ast.2021.107079

Uddin MN, Le L, Nair R, Asmatulu R (2019) Effects of graphene oxide thin films and nanocomposite coatings on flame retardancy and thermal stability of aircraft composites: a comparative study. J Eng Mater Technol 141(3). https://doi.org/10.1115/1.4042663

Ünal TG, Diler EA (2018) Properties of $AlSi_9Cu_3$ metal matrix micro and nano composites produced via stir casting. Open Chem 16(1):726–731. https://doi.org/10.1515/chem-2018-0079

Vaz RF, Garfias A, Albaladejo V, Sanchez J, Cano IG (2023) A review of advances in cold spray additive manufacturing. Coatings. MDPI. https://doi.org/10.3390/coatings13020267

Venkatesan M, Palanikumar K, Boopathy SR (2018) Experimental investigation and analysis on the wear properties of glass fiber and CNT reinforced hybrid polymer composites. Sci Eng Compos Mater 25(5):963–974. https://doi.org/10.1515/secm-2017-0068

Verma P, Anoop S, Sasidhara Rao V, Sharma AK, Uma Rani R (2018) Multiwalled carbon nanotube-poly vinyl alcohol nanocomposite multifunctional coatings on aerospace alloys. Mater Today Proc 5(10):21205–21216. https://doi.org/10.1016/j.matpr.2018.06.520

Wang J, Tian L, Luo B, Ramakrishna S, Kai D, Loh XJ, Yang IH, Deen GR, Mo X (2018) Engineering PCL/Lignin nanofibers as an antioxidant scaffold for the growth of neuron and schwann cell. Colloids Surf B 169:356–365. https://doi.org/10.1016/j.colsurfb.2018.05.021

Wang Z, Xiao H (2018) Nanocomposites: recent development and potential automotive applications. 1(1);631–640

Weng Z, Wang J, Senthil T, Wu L (2016) Mechanical and thermal properties of ABS/montmorillonite nanocomposites for fused deposition modeling 3D printing. Mater des 102:276–283. https://doi.org/10.1016/j.matdes.2016.04.045

Williams JC, Boyer RR (2020) Opportunities and issues in the application of titanium alloys for aerospace components. Metals 10(6):705. https://doi.org/10.3390/met10060705

Xuan Y, Jia S, Nastac L (2017) Processing and microstructure characteristics of As-Cast A356 alloys manufactured via ultrasonic cavitation during solidification. High Temp Mater Processes (London) 36(4):381–387. https://doi.org/10.1515/htmp-2016-0147

Yao X, Moon SK, Lee BY, Bi G (2017) Effects of heat treatment on microstructures and tensile properties of IN718/TiC nanocomposite fabricated by selective laser melting. Int J Precis Eng Manuf 18(12):1693–1701. https://doi.org/10.1007/s12541-017-0197-y

Yates SF, Isella G, Rahislic E, Barbour S, Tiznado L (2021) Effects of ultraviolet-C radiation exposure on aircraft cabin materials. J Res Nat Inst Stand Technol 126:126019. https://doi.org/10.6028/jres.126.019

Yi Z, Jiang T, Cheng Y, Tang Q (2020) Effect of SiO_2 aerogels loading on photocatalytic degradation of nitrobenzene using composites with tetrapod-like ZnO. Nanotechnol Rev 9(1):1009–1016. https://doi.org/10.1515/ntrev-2020-0081

Zakaria MR, Akil HM, Omar MF, Abdullah MMAB, Rahman AAA, Othman MBH (2020) Improving flexural and dielectric properties of carbon fiber epoxy composite laminates reinforced with carbon nanotubes interlayer using electrospray deposition. Nanotechnol Rev 9(1):1170–1182. https://doi.org/10.1515/ntrev-2020-0090

Zecchi S, Cristoforo G, Bartoli M, Tagliaferro A, Torsello D, Rosso C, Boccaccio M, Acerra F (2024) A comprehensive review of electromagnetic interference shielding composite materials. Micromachines 15(2). https://doi.org/10.3390/mi15020187

Zhang B, Patlolla VR, Chiao D, Kalla DK, Misak H, Asmatulu R (2013) Galvanic corrosion of Al/Cu meshes with carbon fibers and graphene and ITO-based nanocomposite coatings as alternative approaches for lightning strikes. Int J Adv Manuf Technol 67(5):1317–1323. https://doi.org/10.1007/s00170-012-4568-3

Zhu L, Li N, Childs PRN (2018) Light-weighting in aerospace component and system design. Propul Power Res 7(2):103–119. https://doi.org/10.1016/j.jppr.2018.04.001

Zipay JJ, Lyndon NB, Thomas Modlin Jr, C, Lyndon Johnson NB, Larsen CE (2016) The ultimate factor of safety for aircraft and spacecraft-its history, applications and misconceptions

The manufacturer's authorised representative in the EU is Springer Nature Customer Service Centre GmbH, Europaplatz 3, 69115 Heidelberg, Germany. If you have any concerns regarding our products, please contact ProductSafety@springernature.com

Printed and bound by CPI Group (UK) Ltd, Croydon, CR0 4YY
30/09/2026
02225572-0010